Methods for Analysis of Musts and Wines

SECOND EDITION

Methods for Analysis of Musts and Wines

SECOND EDITION

C. S. Ough and M. A. Amerine
University of California

WILEY

A Wiley-Interscience Publication
JOHN WILEY & SONS
New York · Chichester · Brisbane · Toronto · Singapore

> A NOTE TO THE READER
> This book has been electronically reproduced from digital information stored at John Wiley & Sons, Inc. We are pleased that the use of this new technology will enable us to keep works of enduring scholarly value in print as long as there is a reasonable demand for them. The content of this book is identical to previous printings.

Copyright © 1988 by John Wiley & Sons, Inc.

All rights reserved. Published simultaneously in Canada.

No part of this publication may be reproduced, stored in a retrieval system or transmitted in any form or by any means, electronic, mechanical, photocopying, recording, scanning or otherwise, except as permitted under Sections 107 or 108 of the 1976 United States Copyright Act, without either the prior written permission of the Publisher, or authorization through payment of the appropriate per-copy fee to the Copyright Clearance Center, 222 Rosewood Drive, Danvers, MA 01923, (978) 750-8400, fax (978) 750-4470. Requests to the Publisher for permission should be addressed to the Permissions Department, John Wiley & Sons, Inc., 111 River Street, Hoboken, NJ 07030, (201) 748-6011, fax (201) 748-6008, E-Mail: PERMREQ@WILEY.COM.

To order books or for customer service please, call 1(800)-CALL-WILEY (225-5945).

Library of Congress Cataloging in Publication Data
Ough, C. S.
  Methods for analysis of musts and wines.

  Rev. ed. of: Methods for analysis of musts and wines / M.A. Amerine and C.S. Ough. c1986.
  "A Wiley-Interscience publication."
  Includes index.
  1. Wine and wine making—Analysis. I. Amerine, M. A. (Maynard Andrew), 1911-   . II. Amerine, M. A. (Maynard Andrew), 1911-   . Methods for analysis of musts and wines. III. Title.

TP548.5.A5085 1988   663'.2   87-28004
ISBN 0-471-62757-7

Printed in the United States of America
10 9 8 7 6 5 4 3 2 1

PREFACE

The first edition of *Methods for Analysis of Musts and Wines* appeared in 1980. It is a fact that since then many of the procedures needed and being used in our laboratories have changed. Three examples are increased use of HPLC and GC, sometimes coupled with mass spectroscopy; automation of dispensing, recording, and calculation of results; and greater attention to statistical analysis of the results. These and other changes will no doubt continue. For this reason we have given an outline of some new procedures, since they will surely be modified or changed in the future. However, the experienced analyst should be able, from our text, to determine whether or not the procedure is of interest to his or her laboratory. We have also deleted methods that no longer seem appropriate or necessary.

One other significant change should be noted: the increased interest of regulatory agencies (and of the public) in the composition of wines. Some of this was no doubt stimulated by the European wine frauds involving diethylene glycol and methanol, but more so by the greater awareness and knowledge due to consumer concerns. There is still a continuing interest in the detection and measurement of minute amounts of pesticide degradation products or of the pesticides themselves. Also, trace amounts of toxic compounds or carcinogens produced naturally have become a more important concern. Whatever the source of the interest of the regulatory agencies, the wine analyst must be prepared to use the most sensitive procedures, no matter how time consuming and, alas, expensive they may be. One cannot, unhappily, predict that these analysts would discover and correct inappropriate production practices that result in undesirable residues in wines *before* government agencies discover them and require their regulation.

Finally, although we have cited hundreds of research papers on wine analysis, more are appearing. We strongly recommend that wine analysts regularly read, or consult via computers, the appropriate sections of *Chemical Abstracts* and *Food Science and Technology Abstracts*.

C. S. OUGH
M. A. AMERINE

Davis, California
December 1987

CONTENTS

Introduction   1

    Sampling, 8
       Grapes. Sample preparation. Wines

1. Soluble Solids   14

    Total Soluble Solids, 14
    Polysaccharides, 21
    Pectins, 21
    Wines, 28
    Extract, 29
    Sugars, 36
       Reducing sugars. Rapid methods for sugar.
       Individual sugars

2. Acidity and Individual Acids   50

    Total (Titratable) Acidity, 51
    pH, 52
    Volatile Acidity, 53
    Fixed Acidity, 59
    Individual Nonvolatile Acids, 60
       Tartaric acid. Malic acid. Lactic acid. Citric acid. Succinic acid. Ascorbic acid (vitamin C). Other acids

3. Alcohols   80

    Ethanol, 80
       Determination of ethanol in the sample. Determination of ethanol in the distillate
    Methanol, 108
       Chemical determination. Gas chromatography (GC)

Higher Alcohols (Fusel Oil), 112
  Chemical determination. Gas chromatography
Glycerol, 119
  Chemical determination. Enzymatic analysis. Gas chromatography. Fluorometric analysis
2,3-Butanediol, 126
  Chemical determination. Gas chromatography. Enzymatic analysis
Sorbitol and Mannitol, 129
  Polarimetric determination. Enzymatic analysis. Gas chromatography. HPLC
Other Polyalcohols, 134
Terpene Compounds, 134

4. Carbonyl Compounds   140

Acetaldehyde, 141
  Chemical determination. Gas chromatography. Colorimetric procedure. HPLC. Enzymatic determination
α-Diketones and α-Hydroxyketones, 148
Acetoin and Diacetyl, 150
Hydroxymethylfurfural, 153
  Gas chromatography
Other Carbonyls, 155

5. Esters   159

Total Volatile Esters, 165
Individual Volatile Esters, 166
Methyl Anthranilate, 167
Nonvolatile Esters, 168

6. Nitrogen Compounds   172

Total Nitrogen, 176
Ammonia, 178
Amino Acids, 179
  α-Amino nitrogen. Arginine. Proline
Biogenic Amines, 186
Proteins, 187
Nitrate, 188

7. Phenolic Compounds 196

Total Phenols, 203
Grape Pigments, 206
   Anthocyanins. Diglucosides
Flavonoids and Other Phenols, 212
Hydrolyzable Tannin Phenols, 213
   Catechins. Leucoanthocyanins
Flavonols and Flavanones, 216
Nonflavonoid Phenols, 216

8. Chemical Additions 222

Sulfur Dioxide, 222
   Free sulfur dioxide. Total sulfur dioxide
Sorbic Acid and Sorbates, 235
Salicylic Acid, 241
Benzoic Acid, 241
p-Hydroxybenzoic Esters, 242
Halogenated Acids, 243
5-Nitrofurylacrylic Acid, 244
Diethyl Dicarbonate, 244
Dimethyl Dicarbonate, 244
Sodium Azide, 246
Diethylene Glycol, 246
Ferrocyanide and Cyanide, 247
Betaine, 250
β-Asarone and Other Flavors, 250
Coumarin, 251
Artificial Wines, 252
Poly(vinyl chloride), 252
Styrene, 253
Poly(vinylpyrrolidone), 254
Aflatoxins, 255
Asbestos, 256
Fungicides and Pesticide Residues, 256

9. Other Constituents 264

Ash, 264
Alkalinity of the Ash, 266
Cations, 268
   Potassium. Sodium. Calcium. Magnesium. Iron. Copper
Trace Elements, 279

Aluminum. Antimony and arsenic. Barium. Beryllium. Boron. Cadmium. Cesium and chromium. Cobalt, europium, and hafnium. Lead. Lithium. Manganese. Molybdenum. Mercury. Nickel. Rubidium, scandium, selenium, and silicon. Silver, strontium, tantalum, and tin. Titanium and vanadium. Zinc
Anions, 288
Bromide. Chloride. Fluoride. Iodide. Phosphate. Sulfate

10. Gases   302

Oxygen, 302
Carbon Dioxide, 306
Hydrogen Sulfide, 311
Nitrogen, 313
Purification of Gases, 313

11. Wine Color   316

12. General Chemical and Equipment Information and Theory   321

Measurement Equipment, 321
Extraction, 322
Evaporation and Distillation, 323
Ion Exchange, 326
Spectrophotometry, 328
Turbidimetry and Nephelometry, 331
Fluorometry, 332
Flame Photometry and Atomic Adsorption Spectrometry (AAS), 333
Chromatography, 339
   Paper and thin-layer chromatography. Gas chromatography. Supercritical chromatography. Liquid chromatography
Ion Chromatography, 352
Gel Separation, 353
Selective-Ion Electrodes, 353
   Gas-sensing electrodes. Solid-state inorganic crystal electrodes. Liquid membrane electrodes
Glass Electrodes and pH Measurement, 357
Complexometry, 358
Enzymatic Analysis, 360
   pH. Other. Sensitivity. Handling and storage of enzymes and other reagents

Index   365

Methods for Analysis of Musts and Wines

SECOND EDITION

INTRODUCTION

The physical, chemical, microbiological, and sensory analyses of musts and wines have become the cornerstones of quality control for the grape juice and wine industries. Every aspect of their production has to be controlled by microbiological, chemical, and sensory tests to reduce spoilage, determine the best and least expensive methods for handling the product, and ensure the highest quality. This extends to field tests during ripening, as well as to continuous controls during fermentation, processing, and aging.

Furthermore, legal controls on many of the constituents of wine, including ethanol, sulfur dioxide, and volatile acidity, require precise analytical results. The presence of prohibited compounds such as monochloracetic acid can be determined, and the presence of excessive amounts of a compound due to procedure or treatment (e.g., thujone in vermouth or sodium in cation-exchanged wines) may need to be established. Enological ratios have been established to prevent undue or illegal use of water, sugar, alcohol, or blending. A variety of ratios to detect sophistication based on analysis of 2000 wines is available (1). A number of others are found in the text. Often there is a need to determine rare constituents, sometimes present in only a few parts per billion. Time constraints and the small amounts to be determined may make these analyses expensive.

Modern wine making often involves blending. Here again analysis can be an important aid in achieving uniformity. The major types of commercial wine can and should be standardized by chemical analysis of critical components. Simultaneous sensory analyses are, of course, essential. As more and more of the components of wines are identified, the possibility of using analytical results to aid in predicting quality becomes difficult as the number of factors influencing "quality" increase. The whole concept of quality is so personal that precise definitions that have some meaning are difficult. This is not to say that within certain ranges and for major chemical differences some standards cannot be established. For example, if *brut* champagne is defined as a wine with less than 1.5% sugar, then that defines *brut* as far as sugar content is concerned. Recently, in order to protect certain appellations of origin, chemical proofs of their origin using statistical methods have been used. Although the differences in concentration of lithium, and so on, in the wines of one district as compared to another district may establish origin, we would be happier if some statistically significant sensory differences were established.

Chemical analysis is also a useful tool in preventing losses in quality and quantity during operations and by-product recovery. Both microbiological and sensory procedures should, of course, also be used.

The selection of the proper method for analysis is not always easy. Should one use a simple, rapid, and insensitive procedure or a slower but more accurate one?

At least two principles govern the answers to these questions. The first concerns the degree of accuracy required. For many routine control purposes a relatively imprecise procedure is an adequate guide for winery practice. For example, the amount of free sulfur dioxide in a white table wine during aging need be known only approximately. On the other hand, at the time of bottling a more accurate determination is required. In research, even more precise procedures may be needed.

Second, choice of a method depends to a large extent on the number of analyses done in a laboratory. In a large laboratory doing hundreds of iron and copper analyses each year, automated atomic absorption spectrophotometry or another, similar, large-scale procedure is preferable. On the other hand, a small laboratory making only a few metal analyses each year could not justify such an outlay and would choose some other procedure, even though it required more time.

A laboratory that must make 10 or 20 alcohol determinations every working day may choose different procedures and equipment as compared to one where only a few determinations are done once a week, once a month, or only once or twice a year. It is quite possible that in very small wineries only a few essential chemical determinations are carried out. More complicated determinations may be done more rapidly and cheaply by commercial laboratories. In research work, very small differences between constituents may be important, particularly in rate reactions. Obviously the most accurate procedures should be employed here, but such methods are often inappropriate for routine winery analyses.

The more precise procedures are often called "reference" or "standard" as contrasted with "routine" or "ordinary" methods. Collaborative analyses of California wines by winery laboratories in 1965 and 1975 (2) showed considerable technician carelessness to be a factor in poor results from some laboratories. Improved results were obtained in 1976. The problem is not unique to California wineries. In collaborative studies of the American Association of Official Analytical Chemists (AOAC), aberrant (outlier) results are sometimes obtained, despite the use of "official" procedures. This was also true in recent

European collaborative studies (3–5). The importance of a chemical analysis should not be over- or underemphasized. The results should have some useful purpose in winery operation and should comply with winery or official standards. To avoid an important and necessary analysis simply because it is difficult is irrational.

As an example of the increasingly exotic procedures being applied to foods, it is now standard to determine the concentration of ^{14}C. The normal ^{14}C content of the atmosphere decreases slowly with time. Thus fossil materials are very high in ^{14}C. It was increased by the nuclear explosions in the atmosphere in the 1960s. Some countries have set minimum limits on ^{14}C to prevent use of ethanol produced from fossil fuels in alcoholic beverages. Lower limits may be needed for alcoholic beverages produced in the Southern Hemisphere, since few nuclear explosions occurred there (6, 7). A similiar problem occurs in detecting whether spirits are made from a forbidden raw material. Grapes and grains are C_3 plants, whereas sugar cane, corn, and sorghum are C_4 plants. C_4 plants are higher in ^{13}C than are C_3 plants. Thus whiskeys, which must be made of grains and which are high in ^{13}C, are probably at least partially made from molasses or corn syrup (8).

There are a number of texts on wine analyses (9–22). Many indicate sources of error, especially Tanner (23).

German laws require considerable laboratory control not only of the finished wine but also of the musts. Practical and rapid methods were summarized (24). Jaulmes (25, 26) has reviewed the history of the official French methods for wine analysis. He emphasized the importance of accurate analyses to prevent sale of sophisticated or spoiled wines. A summary of the Receuil des Methodes International du Vin (OIV) and the European Economic Commission (EEC) methods for wine analyses now in place has recently been presented (27). The components considered and some of the types of analyses used is given in the following tabulation.

	OIV^{a,b}	EEC^a		OIV^{a,b}	EEC^a
Density at 20°C			Malic acid		
Pycnometer method	R	R	Ion exchange separation		
Hydrometer method	U	U	and colorimetric		
Hydrostatic balance	U	U	determination	U	—
Alcohol distillation and			Lactic acid		
Pycnometer method	R	Q	Ion exchange separation		
Hydrometer method	U	U	and colorimetric		
Hydrostatic balance	U	U	determination	U	U
Refractometry	U	—			
Dichromate oxidation	S				

	OIV[a,b]	EEC[a]
Total acidity		
Potentiometric titration to pH 7	R	R
Titration to pH 7 with an indicator	U	U
Volatile acidity		
Steam distillation and volumetric titration	S	S
pH		
Potentiometric measurement	S	S
Fixed acidity		
Total acidity less volatile acidity	S	S
Extract		
Vacuum distillation at 70°C	R	—
Calculation from the specific gravity of dealcoholized wine calculated with the Tabarie formula	U	S
Reducing sugars		
Luff-Schoorl method after clarification by:		
Neutral lead acetate with ion exchange	R	R
without ion exchange	U	U
Zinc ferrocyanide	U	U
Sucrose		
Qualitative detection		
Colorimetric	—	U
Thin-layer chromatography	—	R
Quantitative determination by reducing sugars before and after inversion	—	R
Citric acid		
Barium citrate precipitation, oxidation and colorimetric determination	U	S
Sorbic acid		
Steam distillation and UV spectrophotometric determination	S	S
Sulfurous acid		
Air or nitrogen entrainment, oxidation in sulfuric acid, and sulfuric acid titration	U	R
Iodometric titration	Q	U
Ash		
Extract ashing at 500–550°C	S	S
Alkalinity of ash		
Ash dissolution in a titrated acid and back-titration	S	S
Potassium		
Tetraphenylborohydride precipitation and weighing	R	—
Flame photometry	U	—
Sodium		
Flame photometry	S	S
Calcium and magnesium		
Ash dissolution and EDTA titration	U	—
Chlorides		
Potentiometric titration	R	—
Ion exchange separation and argentometric titration	U	—

	OIV[a,b]	EEC[a]		OIV[a,b]	EEC[a]
Tartaric acid			Sulfates		
Precipitation and calcium			Barium sulfate		
racemate weighing	R	R	precipitation and		
Ion exchange separation			weighing	R	—
and colorimetric			The same principle;		
determination	U	U	more useful		
Potassium monotartrate			technique	U	—
precipitation and			Glycerol		
acidimetric titration	Q	—	Oxidation in methanal		
			and colorimetric		
			titration	U	—
			2,3 Butanediol		
			Oxidation in ethanal		
			and colorimetric		
			titration	U	—

[a] R = reference method, U = usual method, Q = quick method, and S = only method.
[b] Other parameters (OIV): Ascorbic acid, cyanide, succinic acid, hydroxymethylfurfural, ammonia, carbon dioxide, preservatives, arsenic, nitrogen, boron, bromine, color, color additives, malvidine diglucoside, ethanal, iron, fluorine, manganese, mannitol, methanol, phosphorus, sorbitol, lead, zinc.

Enzymatic methods for organic acids are suggested as a definite improvement over the older methods shown above. In addition, in the case of fraudulent addition detection, the use of modern GC/MS techniques were recommended as the only plausible method available.

The paper also discusses at some length the philosophy of the OIV and EEC approaches to wine analyses. The references are almost entirely to European work. It briefly discusses gas chromatographic and HPLC methods and concludes that they may become useful in the future but the older methods are more appropriate and adequate at the present time.

There is a tendency for government agencies to adopt minimum and maximum legal limits for some constituents as quality standards. For example, for commercial grape juice, German law sets minimum limits of titratable acidity, alkalinity of the ash, and potassium and magnesium content. It also sets maximum limits on ethanol, volatile acidity, lactic acid, total sulfur dioxide, sodium, calcium, sulfate, nitrate, chloride, and free tartaric acid (28). Many countries set maximum limits for many elements. Future changes in limits are certainly expected. The present legal or suggested limits of some elements in musts and/or wines are listed in the tabulation on page 6 (29–38).

Legal or quality control limits for other constituents are suggested in the text.

Element	Range of limits (mg/L)	Reference
Aluminum	8	29, 30
Antimony	0.15–0.2	31, 32
Arsenic	0.1–1.0	14, 30, 31
Boron[a]	10–100	14, 30, 31, 33
Bromine[b]	0.5–1.0	31, 33
Cadmium	0.1–1.0	29, 31, 32, 34, 35
Chromium	0.1	29, 31
Copper	0.1–5.0	31, 34–36
Fluorine	0.5–5.0	14, 31, 33, 35
Iron	5	28
Lead	0.3–1.0	14, 29–31, 34, 36
Lithium	16.4	37
Mercury	0.02 (?)	31
Nickel	0.1–0.3	31
Selenium	0.1–2.0	31, 32
Sodium[c]	60	14
Tin	0.5–5.0	6, 31, 35, 38
Zinc	5.0–40.0	14, 30, 34, 37

[a] As boronic acid.
[b] Local limits to 2.5 allowed.
[c] Local limits for certain vineyards allowed.

Some of the procedures for the enzymatic determination of a number of compounds in beer (39, 40) are also applicable to wines. Continuous automatic analysis of wines for ethanol and sulfur dioxide has been proposed (41, 42). Not only are continuous methods of analysis practical for laboratories handling a large number of analyses, but computers can be programmed to make the necessary calculations and to record the results. Meyer (43) gives an example of how pycnometer values can be rapidly and accurately converted to percentage of ethanol. Automated and enzymatic procedures were compared with conventional ones for repeatability and reproducibility (44). In 21 laboratories the results for specific gravity, ethanol, total extract, fermentable sugar, total acid, and free and total SO_2 were similar, with reproducibility being slightly better for the automated procedures. The European Common Market has not approved automated procedures for ethanol, total sugar, and total and free SO_2 (45).

One of the advantages of high-performance liquid chromatography (HPLC) is that many constituents can be quantitatively determined in one run, thus saving much time. We cite only two examples of many: Ethanol, glycerol, glucose, fructose, tartrate, malate, lactate, succinate, acetate, and citrate were determined in one run in 25 min (46). Acetaldehyde, methanol, four higher alcohols, ethyl acetate, ethyl lactate, acetoin, 2,3-butanediol, and glycerol were measured in one run (47).

Micromethods for ethanol, reducing sugars, and SO_2 have been used to give rapid results (39, 48).

The sensory examination of wines falls outside the scope of this book. For a discussion of this subject, see References 9, 10, 49, and 50. For statistical analyses of the sensory and analytical data, see References 50 and 51.

For sensory data the chi-square distribution is used for determining the significance of differences obtained by paired, duo–trio, and triangular tests. When testing for the significance of the differences between means of scores, the t-distribution may be used. For more general testing the analyses of variance is employed. One can determine not only the significance of the difference between wines but also the relative reliability of the results of different judges. In sensory tests, care must be exercised in analyzing results of two-tailed and one-tailed tests. This problem does not arise with analyses of data from different analytical procedures.

For analytical data the procedures of Youden and Steiner (52) are standard and useful. In general, one can determine the mean, the standard deviation, the coefficient of variation, confidence limits, and the difference between population means; one can also apply analysis of variance in testing the significance of various factors (methods, analyses, etc.). The larger the standard deviation, the greater the error in determination (random error). Statistical procedures applied to analytical data obtained from two different methods provide a measure of precision in estimating differences between the two means.

Two kinds of error are present in the results of analyses: systematic errors and random errors. Systematic errors arise from differences in the skill or technique of the analysts, since one may have consistently high values and another may have consistently low values (i.e., bias). Essentially this is the difference between an observed mean and a true or target value. The method used must be specific and adequately calibrated. Tests with a standard procedure and the test procedure are called for. The t-distribution may be used to determine confidence limits for the two results. Lacking a standard method, one may compare results on normal and abnormal material between analysts or laboratories using the same method or using some reference method, analyze special samples with stated values, add pure materials and determine recovery, or add possible interfering substances. The precision of an estimator is a measure of its repeatability. Precision can be expressed in terms of the variance of an estimator, with a large variance signifying lack of precision and a sizable error of the determination (random error).

The way the individual results cluster around the population mean is usually measured by calculating the standard deviation σ, which is defined as the square root of the average of the squared differences from the mean. Since the true mean and standard deviation are not known, we estimate σ by calculating s. The square of s is called the *variance*. In calculating estimates of the variance or the standard deviation, the analytical data should not be "rounded off," since this substantially lowers the efficiency of the estimates. However, once the final analytical result has been obtained it may be rounded off, usually to two significant places. For example, if the standard deviation is calculated as 0.0154, it may be reported as 0.015. For a single laboratory and analyst the coefficient of variation (C_v) is an easily understood indication of precision or reproducibility:

$$C_v = \frac{\text{S.D.}}{\text{Mean}} \times 100 = \%$$

where S.D. stands for standard deviation.

Accuracy is the measure of the real value of a component. An analysis may be very reproducible, but the answer may not be accurate. Two usual methods of checking for accuracy include (a) use of a different method to verify values and (b) additions of known amounts to determine recovery from test samples.

The first method usually depends on a method proven by long use and extensive testing (e.g., the use of the pycnometer method for an ethanol standard analysis). The second can depend, for example, on the addition of ^{14}C-labeled compound to the test solution and comparing the recovery by the chemical method to the radioactive detection methods, which are very sensitive and absolute. Further, one should plot several levels of added test component recovered to determine if the recovery goes through the origin when the amount added is compared to the amount recovered. Many compounds are not fully extracted or recovered, but if the standard curves generated are linear and the response factors (amount recovered/amount added) are similar for the standard curve and the test material, then a good degree of accuracy can be assumed.

Internal standards are of great value in most instrumental analyses. By the addition of a compound with similar chemical attributes and functions as the component(s) in question, the variations due to handling and instrument or sampling can be accounted for. Simply speaking, the internal standard, always at a constant value, is added to the material making up the standard curve. The exact amount is added to each raw test sample prior to any extraction or sampling. The standard curve then has a plot of the ratio of the response of the test component(s) to the internal standard response on the y axis and the level of the standard material on the x axis. This decreases considerably the chance of measurement, handling, and instrument response errors as the response of the internal standard will vary in the same proportion as the component(s) in question.

SAMPLING

Grapes

Must consists of the freshly crushed grapes. It is important for the wine maker to have a thorough knowledge of the composition of the must because the composition and quality of the finished wine depend largely on the composition of the must. Furthermore, intelligent utilization of information on the composition of the must will enable the wine maker to utilize the must to its best advantage or to ameliorate it so that the best possible quality of wine can be obtained.

It is important to obtain a quantity of fruit that is as representative as possible of all the fruit being sampled, that is, a random or unbiased sample. The sample may be taken from a truckload, from a field of many acres, from a small group of vines, from a single vine, from a box of fruit, or even from a single bunch of grapes, depending on the purpose of the sampling. In each case, however, the sample of fruit that is crushed must be as representative as possible of the fruit of the load, field, or vines and is dependent on the uniformity of composition of the fruit. The size of sample required can be calculated approximately from the range in composition of the fruit. Assume, for example, that a range of 10% on either side of the mean value existed in the grapes. If one wants a sampling error of not more than ±2%, the quantity of fruit will be the ratio of the variances (the squares of the standard deviations) of these two ranges. As an approximation, the standard deviations can be taken as equal to the range. If the true mean is 0.7% (0.007 lb/lb), then the 10% standard deviation is 0.0007 and the variance is 0.00000049. The 2% standard deviation is ±0.00014 and the variance is 0.0000000196. The ratio is 0.00000049/0.0000000196 or 25 lb. This is the amount of sample required for a ±2% sampling error when the variability in composition of the fruit is ±10%. If the variability is 20%, a minimum sample of 100 lb will be required to reduce the sampling error to ±2%. For a sampling error of ±5% and a variation in composition of ±30%, a minimum sample of 36 lb is indicated.

Berg and Marsh (53) studied sampling of grapes as delivered to wineries in boxes or in gondola trucks. They recommended sampling 9–14 boxes from loads of 180–200 boxes to ensure a deviation of less than 0.05° Brix. They also found about the same error if 5-lb samples were taken from 14 boxes. For gondola trucks, five samples of 10 lb each gave a deviation of 0.0–0.7° Brix. In general, where the fruit is delivered in boxes one should sample a number of boxes equal to the square root of the total number of boxes. For example, for a load of 225 boxes, 15 should be sampled. However, this amount of sampling has been opposed by the industry because it requires too much time, hence expense. California state inspection regulation requires only two boxes from loads of up to 50 boxes, three for loads of 51–100 boxes, four for 101–250 boxes, and five for 251–400 or more boxes. From each of these boxes, 5 lb of fruit is to be crushed together. For gondola trucks the state inspection requirement is three samples, from prescribed positions, of 10 lb each, for a gondola load of one tank (4–10 tons). If the load is of two or more tanks, only three samples in all are taken. These sampling requirements may give deviations of greater than $0.5°$ Brix from the true value.

One study of field sampling (9) recommended random sampling of 200–500 single berries from all parts of the cluster and from clusters from all parts of the vine. Alternatively, the analysts used cluster sampling of 10% of the vines. These methods reduced the standard error of the mean to $\pm 0.25\%$.

At the winery, samples may be taken from the well-mixed tank of crushed grapes, provided the tank represents a single lot of fruit. Here the composition of duplicate samples should be nearly identical, and a smaller sample is sufficient.

Sample Preparation

To secure a sample for analysis, all the grapes in the selected sample must be crushed. This can be done by placing the fruit in a canvas or cloth bag and squeezing it tightly (impractical for large samples or for a large number of samples), by macerating the fruit in a fruit-juice squeezer (again impractical for large samples), by crushing the fruit in a continuous screw-type crusher by taking samples from the must line leading from the crusher, or, as indicated above, by taking a sample from a well-mixed tank of crushed fruit. The free-run juice from the bottom of a gondola truck is not a desirable sample for analysis, since it represents only the juice of the more easily crushed fruit. The most difficult problem in preparing the sample is to secure equal maceration of all the fruit. Shriveled and raisined fruit are difficult to crush, but it is particularly important that this type of fruit be crushed because its sugar content is much higher than that of normal fruit. Hand and fruit-juice squeezer crushing is inadequate. The screw-type crusher is usually satisfactory, but in exaggerated cases a Waring blender may be needed. If a Waring blender is used, the juice must be centrifuged and should be used for determination of the total soluble solids content only. This is because the Waring blender also grinds up the skins and seeds, and there is a change in the total acidity, pH, and buffer capacity of the juice (54).

The juice from these crushing procedures is more or less contaminated with skins, seeds, and particles of flesh. When a screw press is used for crushing, considerable air is also entrained. This may cause an error unless the contaminants are removed by centrifuging, settling, or filtering through cheesecloth. Early in the season the juice will settle clear in a short time, but later in the season, particularly with certain varieties, the juice will be thick and will not settle clear within a reasonable period. The gross impurities can be removed by straining the juice through a double thickness of cheesecloth. To obtain a completely clear juice, centrifuging is usually necessary. The clarified juice must be analyzed immediately, since fermentation may start very soon during warm weather. Storage in a frozen condition will keep the juice, but the juice must be heated before analysis to get the precipitated tartrates back into the solution.

Wines

Wines in small containers, such as barrels and puncheons, are normally quite homogeneous in composition, and a sample taken from any place in the container will prove satisfactory. A hand glass thief may be used to secure the sample through the bunghole. However, if there is a large air space above the wine in small containers, or even in very large tanks, the contents may not be homogeneous. The liquid near the surface may be considerably higher in volatile acidity, and the sulfur dioxide content may be different at various levels within the container. In such cases a glass thief should be lowered on a string into the container to obtain the sample, or better, the contents should be stirred. If a composite sample is taken from a number of containers, the amount taken from each container should be proportional to its size. Only clean glass thiefs should be employed. Samples should not be taken with metal thiefs or through metal spigots or faucets.

Often, very large tanks are filled from several sources, and the problems of sampling them are formidable. Mixing the contents by pumping over and taking several samples will help reduce the sampling error. Because of the variability of the base material, particularly of musts, duplicate and triplicate samples should be taken. For the most accurate results, each sample should be analyzed in duplicate. For further data on sampling, see Joslyn (16).

Only dry, clean glass or plastic bottles should be used to hold the sample, and the bottle should be filled completely, particularly if the sample is to be transferred for some distance and a delay in analysis is necessary.

When samples are being taken for collaborative analyses by two or more laboratories the total quantity of wine required should be placed in a glass container and mixed well. The samples for the individual analyses should not be taken as bottles from the cases. The necessary number of bottles should be selected, emptied into a single container, and rebottled after mixing. This is particularly true of wines from small wineries, where a given lot of wine may have been bottled from several barrels over a period of time, then placed in bins. From the bins, the wine is placed in cases; thus all the wine in a single case may not have come from a particular container. If the bottle is full and is not subjected to extreme heat or cold it will remain essentially unchanged for several weeks, except that the free sulfur dioxide, the ratio of ferrous to ferric iron, and the oxidation–reduction potential may change within a few days. With longer periods, even if the wine is free of microorganisms, chemical reactions may occur that affect the interpretation of the results.

REFERENCES

1. A. Patschky and H.-J. Schöne, *Mitt., Rebe Wein, Obstbau Fruechteverwert.* **20**, 432–435 (1970).
2. H. A. Wildenradt and A. Caputi, Jr., *Am. J. Enol. Vitic.* **28**, 145–148 (1977).
3. S. Brun, *Ann. Falsif. Expert. Chim.* **71**, 399–409 (1978).
4. C. Junge, *Bull. O.I.V.* **621**, 775–800 (1982).
5. C. Junge, *J. Assoc. Off. Anal. Chem.* **68**, 141–145 (1985).
6. A. N. Hanekom, C. S. du Plessis, J. F. de Villiers, and A. C. Houtman, *Vitis* **17**, 170–172 (1978).
7. D. J. McWeeny and M. L. Bates, *J. Food Technol.* **15**, 407–412 (1982).
8. W. A. Simpkins and D. Rigby, *J. Sci. Food Agric.* **33**, 898–903 (1982).
9. M. A. Amerine, H. W. Berg, R. E. Kunkee, C. S. Ough, V. L. Singleton, and A. D. Webb, *The Technology of Wine Making*, 4th ed. Avi Publishing Co., Westport, CT, 1979.
10. E. Vogt, revised by L. Jakob, E. Lemperle, and E. Weiss, *Der Wein; Bereitung, Behandlung, Untersuchungen*. Ulmer, Stuttgart, 1984.
11. R. Franck and C. Junge, *Weinanalytick. Untersuchung von Wein and ähnlichen alkoholischen Erzeugnissen sowie von Fruchtsäften*. Carl Heymanns Verlag, Cologne, 1983.
12. L. Mori, *Metodi Razionali di Analisi nella Moderna Tecnica Enologica*, 2nd ed. Luigi Scialpi, Rome, 1967.
13. V. I. Nilov and I. M. Skurikhin, *Khimiia Vinodeliia Kon'-iachnogo Proizvodstva*, 2nd ed. Pishchepromizdat, Moscow, 1967.
14. *Recueil des Méthodes Internationales d'Analyse des Vins*. Office International de la Vigne et du Vin, Paris, 1978.
15. M. A. Amerine and C. S. Ough, *CRC Crit. Rev. Food Technol.* **2**, 407–515 (1972).
16. M. A. Joslyn, *Methods in Food Analysis, Physical, Chemical and Instrumental Methods of Analyses* 2nd ed. Academic Press, New York, 1970.
17. *Official Methods of Analysis*, 14th ed. Association of Official Analytical Chemists, Arlington, VA, 1984, pp. 220–230.
18. *Uniform Methods of Analyses for Wines and Spirits*. American Society of Enologists, Davis, CA, 1972.
19. France, Ministère de l'Agriculture, Arrêté du 24 juin 1963 rélatif aux méthodes officielles d'analyses des vins et des moûts, *J. Offic.* **63-154**, 1037, 4551–4587 (1963).
20. J. Ribéreau-Gayon, E. Peynaud, P. Sudraud, and P. Ribéreau-Gayon, *Traité d'Oenologie*, Vol. 1. Dunod, Paris, 1976.
21. J. Blouin, *Manual Pratique d'Analyses des Moûts et des Vins*, 3rd ed. Federation des Centres d'Etudes et Information Oenologiques, Paris, 1977.
22. K. Hennig and L. Jakob, *Untersuchungen für Wein und ähnliche Getränke*, 6th ed. Ulmer, Stuttgart, 1973.

23. H. Tanner, *Schweiz. Z. Obst- Weinbau* **103**, 300–304 (1967).
24. L. Jakob, *Rebe Wein* **25**, 44–46, 78, 80, 82–83 (1971).
25. P. Jaulmes, *Mises Point Chim. Anal. Org., Pharm. Bromatol.* **14**, 182–215 (1965).
26. P. Jaulmes, *Bull. Tech. Inf. Minist. Agric. Fr.* **196**, 213–225 (1965).
27. S. Brun, J. C. Cabanis, and J. P. Mestres. *Anal. Chem. Experentia* **42**, 893–904 (1986).
28. H. J. Bielig, W. Faethe, J. Koch, S. Wallrauch, and K. Wucherpfennig, *Fluess. Obst* **44**, 215–221, 224–226 (1977).
29. W. Zipfel, *Lebensmittelrecht*. Beck-Verlag, Munich, 1985.
30. A. I. Ionescu, *Ind. Aliment. (Bucharest)* **23**, 557–560 (1972).
31. J. Schneyder, *Mitt., Rebe Wein, Obstbau Fruechteverwert.* **24**, 129–134 (1974); see also *Bull. O.I.V.* **46**(514), 1120–1125 (1973).
32. J. M. Robertson, *Rep.—N.Z.*, Dep. Sc. Ind. Res., *Chem. Div.* **CD-2222**, 1–34 (1976).
33. H. E. Haller and C. Junge, *Aktuel. Ernaehrungsmed. Klin. Prax.* **2**, 51–55 (1977).
34. K. Harju and P. Ronkainen, *Z. Lebensm.-Unters. -Forsch.* **170**, 445–448 (1980).
35. S. Wallrauch, *Fluess. Obst* **41**, 134–135 (1974).
36. S. Martina, R. Caravella, and R. M. Barbagallo, *Riv. Vitic. Enol.* **26**, 509–512 (1973).
37. A. Amati and R. Rostelli, *Ind. Agrar.* **5**, 223 (1967).
38. B. Wereszczynska-Cislo, *Przem. Spozyw.* **31**, 412–414 (1977).
39. F. Drawert and W. Hagen, *Brauwissenschaft* **23**, 300–303 (1970).
40. F. Drawert and W. Hagen, and H. Barton, *Brauwissenschaft* **23**, 432–438 (1970).
41. J. Sarris, J. N. Morfaux, P. Dupuy, and D. Hertzog, *Ind. Aliment. Agric.* **86**, 1241–1246 (1969).
42. J. Sarris, J. N. Morfaux, and L. Dervin, *Connaiss. Vigne Vin* **4**, 431–433 (1970).
43. R. Meyer, *Weinberg Keller* **18**, 323–336 (1971).
44. R. Ristow, G. Brauner-Glaesner, and K. Breitbach, *Wein-Wiss.* **40**, 271–283 (1985).
45. Ch. Junge, *Weinwirtsch., Tech.* (9), 225–227 (1984).
46. P. Pfeiffer and F. Radler, *Z. Lebensm.-Unters. -Forsch.* **181**, 24–27 (1985).
47. M. Bertuccioli, *Vini Ital.* **24**, 149–156 (1982).
48. H. Rebelein, *Allg. Dtsch. Weinfachztg.* **107**, 590–594 (1971).
49. M. A. Amerine, E. B. Roessler, and F. Filipello, *Hilgardia* **28**, 477–567 (1959).
50. M. A. Amerine, R. M. Pangborn, and E. B. Roessler, *Principles of Sensory Evaluation of Food*. Academic Press, New York, 1965.
51. M. A. Amerine and E. B. Roessler, *Wines: Their Sensory Evaluation*, revised and enlarged ed. Freeman, New York, 1983.
52. W. J. Youden and E. H. Steiner, *Statistical Manual of the AOAC*. Association of Official Analytical Chemists, Washington, DC, 1975.
53. H. W. Berg and G. L. Marsh, *Food Technol.* **8**, 104–108 (1954).
54. G. H. Carter, C. W. Nagel, and W. J. Clore, *Am. J. Enol. Vitic.* **23**, 10–13 (1972).

One
SOLUBLE SOLIDS

The soluble solids of musts and sweet wines are composed mostly of sugars. Thus besides the determination of the total soluble solids, one often determines separately the sugar content—mainly the reducing sugars—of the sample.

TOTAL SOLUBLE SOLIDS

A knowledge of the soluble solids content of the must is useful as a measure of the maturity of the grapes and, hence, as an indication of the proper time of harvesting. It is a partial guide to the rational utilization of the grapes for producing the most appropriate type of wine. Furthermore, it is an indication of the amount of amelioration needed by over- and underripe grapes. In the European Common Market countries the precise definition and determination of the soluble solids concentration of musts is especially important because it distinguishes between wines made from unripe grapes (and thus requiring additional sugar, i.e., chaptalization) and those which are not. This then changes the name on the label. (Unsugared wines are more expensive.) It also indicates how much sugar may be added. Outside of California this is also true in the United States. By appropriate formula it is possible to calculate, from the alcohol and extract of the wine, the original soluble solids of the must, at least in Austria (1). Finally, it is an approximate basis for calculating the alcohol yield. This is possible because more than 90% of the soluble solids of a wine is composed of fermentable sugars. The other materials are organic acids, sulfate, potassium, polysaccharides, pectins, protein, and other substances. The usual term for this measurement is °Brix (or °Balling).

Berg and Ough (2) summarized the °Brix for a number of musts of different varieties. They used these and the resultant sensory scores of wines made from the musts to determine optimum maturity ranges for the various varieties from different regions. A part of the basis for payment for grapes is the °Brix determination made on the grapes. The condition of the grapes, as well as the total acidity and pH of the must, may also be used in determining the price paid for the grapes. Vines overcropped or otherwise not treated by sound viticultural practices can yield substandard grapes having total soluble solids that are insufficient for making satisfactory wine. For the viticultural aspects see Winkler et al. (3). For a more detailed discussion of maturity and crop level effects on wine quality, see Amerine et al. (4).

The two most common procedures for determining the total soluble solids content of musts are hydrometry and refractometry. The Brix (or Balling) hydrometers commonly used by enologists are calibrated in degrees indicating grams of sucrose per 100 g of liquid. They usually cover a 10° range (e.g., −5° to +5° Brix) and are divided into 0.1° units. Results obtained on hydrometers calibrated to specific gravity can be converted to Brix degrees; Table 1 gives a conversion table (5).

The Mettler–Paar density meter (6) can be set up with a sampling system to measure juice densities rapidly and accurately without pretreatment. The theoretical background of this system of measurement is described later (p. 97).

The refractive index of musts also can be used as a measure of the total soluble solids, since the most important ingredients influencing the refractive index are the sugars. Temperature corrections also need be applied to the values if a temperature other than 20°C is used (Table 2). Some newer Brix refractometers are internally temperature compensated over a wide temperature range. Where a check on these two methods is desired, a sample of the grape juice must be weighed in a pycnometer, or the specific gravity can be determined with a Westphal balance. Comparison of the hydrometer to the refractometer shows that the refractometer gives slightly lower readings than the hydrometer. There is no need to filter or clarify samples before refractometer measurement (7, 8).

The hydrometers are calibrated at a specified temperature, usually 20°C (68°F). Since the temperature of the measurement is not always exactly the same, a correction must usually be applied. Table 3 gives corrections for the Brix (Balling) hydrometers; correction tables for hydrometers calibrated at 60°F (15.56°C) have been published by the federal government (9).

The method of procuring the sample is extremely important. Grapes with many raisined berries will not give a true reading from the juice if only lightly pressed. Grinding up a sample will get all the sugar into solution, but the solution will be very difficult to clarify for a hydrometer determination (10). If the grapes are to be crushed and the juice drawn off in a very short time (up to 8–12 hr) before fermentation, the juice obtained from the must line is adequate. However, if the grapes are to be fermented on the skins, the best estimate of total soluble solids is obtained about 24 hr after crushing. This gives time for the sugar in poorly crushed berries to be extracted into the juice. However, such

Table 1. Specific Gravity at 20°C (68°F) Corresponding to Readings of the Brix or Balling Hydrometers[a]

°Brix[b] (Balling)	Specific gravity	°Brix[b] (Balling)	Specific gravity	°Brix[b] (Balling)	Specific gravity	°Brix[b] (Balling)	Specific gravity
0.0	1.00000	7.0	1.02770	14.0	1.05667	21.0	1.08733
0.2	1.00078	7.2	1.02851	14.2	1.05762	21.2	1.08823
0.4	1.00155	7.4	1.02932	14.4	1.05847	21.4	1.08913
0.6	1.00233	7.6	1.03013	14.6	1.05933	21.6	1.09003
0.8	1.00311	7.8	1.03095	14.8	1.06018	21.8	1.09093
1.0	1.00389	8.0	1.03176	15.0	1.06104	22.0	1.09183
1.2	1.00467	8.2	1.03258	15.2	1.06190	22.2	1.09273
1.4	1.00545	8.4	1.03340	15.4	1.06276	22.4	1.09364
1.6	1.00623	8.6	1.03422	15.6	1.06362	22.6	1.09454
1.8	1.00701	8.8	1.03504	15.8	1.06448	22.8	1.09545
2.0	1.00779	9.0	1.03586	16.0	1.06534	23.0	1.09636
2.2	1.00858	9.2	1.03668	16.2	1.06621	23.2	1.09727
2.4	1.00936	9.4	1.03750	16.4	1.06707	23.4	1.09818
2.6	1.01015	9.6	1.03833	16.6	1.06794	23.6	1.09909
2.8	1.01093	9.8	1.03915	16.8	1.06881	23.8	1.10000
3.0	1.01172	10.0	1.03998	17.0	1.06968	24.0	1.10092
3.2	1.01251	10.2	1.04081	17.2	1.07055	24.2	1.10183

Sucrose (g/100g)[b]	Sp. gr.[a]	Sucrose (g/100g)[b]	Sp. gr.[a]	Sucrose (g/100g)[b]	Sp. gr.[a]	Sucrose (g/100g)[b]	Sp. gr.[a]
3.4	1.01330	10.4	1.04164	17.4	1.07142	24.4	1.10275
3.6	1.01409	10.6	1.04247	17.6	1.07229	24.6	1.10367
3.8	1.01488	10.8	1.04330	17.8	1.07317	24.8	1.10459
4.0	1.01567	11.0	1.04413	18.0	1.07404	25.0	1.10551
4.2	1.01647	11.2	1.04497	18.2	1.07492	25.2	1.10643
4.4	1.01726	11.4	1.04580	18.4	1.07580	25.4	1.10736
4.6	1.01806	11.6	1.04664	18.6	1.07668	25.6	1.10828
4.8	1.01886	11.8	1.04747	18.8	1.07756	25.8	1.10921
5.0	1.01965	12.0	1.04831	19.0	1.07844	26.0	1.11014
5.2	1.02045	12.2	1.04915	19.2	1.07932	26.2	1.11106
5.4	1.02125	12.4	1.04999	19.4	1.08021	26.4	1.11200
5.6	1.02206	12.6	1.05084	19.6	1.08110	26.6	1.11293
5.8	1.02286	12.8	1.05168	19.8	1.08198	26.8	1.11386
6.0	1.02366	13.0	1.05252	20.0	1.08287	27.0	1.11480
6.2	1.02447	13.2	1.05337	20.2	1.08376	27.2	1.11573
6.4	1.02527	13.4	1.05422	20.4	1.08465	27.4	1.11667
6.6	1.02608	13.6	1.05506	20.6	1.08554	27.6	1.11761
6.8	1.02689	13.8	1.05591	20.8	1.08644	27.8	1.11855

From Reference 5.
[a]Assuming specific gravity of water at 20°C as unity.
[b]Grams of sucrose per 100 g of liquid.

Table 2. Corrections for Determining Sucrose by Means of Either Abbé or Immersion Refractometer for Temperatures Other Than 20°C

| Temperature | | \multicolumn{9}{c}{Sucrose (g/100 g solution)} | | | | | | | | |
°F	°C	0	5	10	15	20	25	30	40	50	60	70
						Subtract from reading						
50.0	10	0.50	0.54	0.58	0.61	0.64	0.66	0.68	0.72	0.74	0.76	0.79
51.8	11	0.46	0.49	0.53	0.55	0.58	0.60	0.62	0.65	0.67	0.69	0.71
53.6	12	0.42	0.45	0.48	0.50	0.52	0.54	0.56	0.58	0.60	0.61	0.63
55.4	13	0.37	0.40	0.42	0.44	0.46	0.48	0.49	0.51	0.53	0.54	0.55
57.2	14	0.33	0.35	0.37	0.39	0.40	0.41	0.42	0.44	0.45	0.46	0.48
59.0	15	0.27	0.29	0.31	0.33	0.34	0.34	0.35	0.37	0.38	0.39	0.40
60.8	16	0.22	0.24	0.25	0.26	0.27	0.28	0.28	0.30	0.30	0.31	0.32
62.6	17	0.17	0.18	0.19	0.20	0.21	0.21	0.21	0.22	0.23	0.23	0.24
64.4	18	0.12	0.13	0.13	0.14	0.14	0.14	0.14	0.15	0.15	0.16	0.16
66.2	19	0.06	0.06	0.06	0.07	0.07	0.07	0.07	0.08	0.08	0.08	0.08
						Add to reading						
69.8	21	0.06	0.07	0.07	0.07	0.07	0.08	0.08	0.08	0.08	0.08	0.08
71.6	22	0.13	0.13	0.14	0.14	0.15	0.15	0.15	0.15	0.16	0.16	0.16
73.4	23	0.19	0.20	0.21	0.22	0.22	0.23	0.23	0.23	0.24	0.24	0.24
75.2	24	0.26	0.27	0.28	0.29	0.30	0.30	0.31	0.31	0.31	0.32	0.32
77.0	25	0.33	0.35	0.36	0.37	0.38	0.38	0.39	0.40	0.40	0.40	0.40
78.8	26	0.40	0.42	0.43	0.44	0.45	0.46	0.47	0.48	0.48	0.48	0.40
80.6	27	0.48	0.50	0.52	0.53	0.54	0.55	0.55	0.56	0.56	0.56	0.56
82.4	28	0.56	0.57	0.60	0.61	0.62	0.63	0.63	0.64	0.64	0.64	0.64
84.2	29	0.64	0.66	0.68	0.69	0.71	0.72	0.72	0.73	0.73	0.73	0.73
86.0	30	0.72	0.74	0.77	0.78	0.79	0.80	0.80	0.81	0.81	0.81	0.81

Table 3. Corrections for Brix (Balling) Hydrometers Calibrated at 20°C (68°F)

Temperature		Observed sugar (%)						
°C	°F	0	5	10	15	20	25	30
		Subtract						
15	59.0	0.20	0.22	0.24	0.26	0.28	0.30	0.32
16	60.8	0.17	0.18	0.20	0.22	0.23	0.25	0.26
17	62.6	0.13	0.14	0.15	0.16	0.18	0.19	0.20
18	64.4	0.09	0.10	0.10	0.11	0.12	0.13	0.13
19	66.2	0.05	0.05	0.05	0.06	0.06	0.06	0.07
		Add						
21	69.8	0.04	0.05	0.06	0.06	0.06	0.07	0.07
22	71.6	0.10	0.10	0.11	0.12	0.12	0.13	0.14
23	73.4	0.16	0.16	0.17	0.17	0.19	0.20	0.21
24	75.2	0.21	0.22	0.23	0.24	0.26	0.27	0.28
25	77.0	0.27	0.28	0.30	0.31	0.32	0.34	0.35
26	78.8	0.33	0.34	0.36	0.37	0.40	0.40	0.42
27	80.6	0.40	0.41	0.42	0.44	0.46	0.48	0.50
28	82.4	0.46	0.47	0.49	0.51	0.54	0.56	0.58
29	84.2	0.54	0.55	0.56	0.59	0.61	0.63	0.66
30	86.0	0.61	0.62	0.63	0.66	0.68	0.71	0.73
35	95.0	0.99	1.01	1.02	1.06	1.10	1.13	1.16

From Reference 5.

a long delay is not practical under the usual winery practices, and it runs the risk that some sugar will already have been fermented.

Procedure

Obtain a good juice sample either from grapes that are thoroughly crushed or from a must line or tank. Be sure the sample is representative of the whole lot by taking several samples from the lot of grapes being crushed, combining the samples, and using an aliquot for the determination. If the determination is to be made by hydrometer, either settle or centrifuge a sample of adequate size to float a Brix hydrometer in a standard hydrometer cylinder with an overflow compartment (Figure 1). Be sure that (a) the hydrometer and cylinder are clean and dry, (b) the juice is clear to the eye, (c) no fermentation has started in the juice, and (d) the hydrometers are calibrated in sucrose–water solutions. Fill the hydrometer cylinder with the clear juice. Gently, by the stem, lower the hydrometer into the

20 Soluble Solids

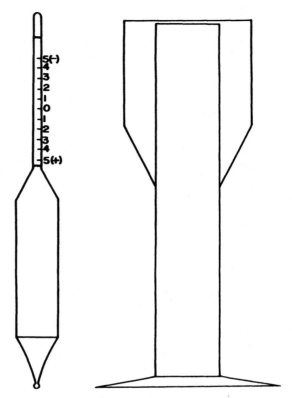

Figure 1. A −5 to +5 Brix hydrometer and hydrometer cylinder with overflow compartment. Other ranges of hydrometers are available.

liquid and give it a spin. Wait for the hydrometer to come to rest. (Be sure that it is not resting on the bottom.) Read the hydrometer to the nearest 0.1° Brix. Take the temperature and from Table 3 add or subtract the proper value.

If the Brix is to be determined by refractometer, the sample need not be clarified. Check the calibration of the refractometer at 0° Brix with water and at 20° Brix with a sucrose solution prepared by weighing 200 g of sucrose and measuring out 800 mL of water and dissolving the sugar in the water to give a 20 g/100 g sucrose solution. With a soft applicator, place a few drops of the juice onto the prism (care must be taken not to scratch the surface), close the cover, adjust light and read the °Brix. Record the temperature and make the correction (if the instrument does not have internal temperature compensation).

The refractometer cannot be used as the refractive index in the special case of wine during fermentation because the alcohol interferes. Measure the °Brix by drawing a sample from the fermenter, shake or otherwise disperse most of the carbon dioxide, place the hydrometer into the fermenting must, spin the hydrometer to clear it of bubbles on the surface, wait about 10 sec to allow for carbon dioxide to clear the wine, then read.

Other countries use scales other than °Brix, often simply percentage of soluble solids. The Australians use Baumé scale, which indicates an approximation of the alcohol yield in grams per 100 mL. Germany uses Oechsle scale, which is given as Oechsle = (sp. gravity − 1.000) × 1000. Table 4 relates these three most commonly used scales.

POLYSACCHARIDES

The polysaccharides present in grapes can be classified as homo- or heteropolysaccharides. They are mainly high-molecular-weight macromolecules. More are found in moldy grapes than in sound ones. They influence the stability of a wine as well as its filtrability, and they possibly impart some sensory properties. The pectin materials are discussed in the next section. A procedure for separation of the various pectins from hemicelluloses and celluloses is available (11).

The soluble polysaccharides of musts may be separated and quantitatively determined (12). This is done colorimetrically with phenol–sulfuric acid or carbazole–sulfuric acid. Data on the total polysaccharides (and neutral sugars and uronic acids), gums (and neutral sugars and uronic acids), and pectins (also neutral sugars and uronic acids) were measured from veraison to maturity. The pectins decrease dramatically during the later stages of ripening.

PECTINS

The pectin and gum contents of musts often indicate how easily and completely the juice can be separated from the skins and how quickly and completely the juice can be clarified.

If musts or wines are acidified and sufficient alcohol is added to bring the ethanol to about 80%, a precipitate will form. This precipitate consists mainly of colloidal material known as *pectin* and *pectic acids* (Figure 2) but also contains gum, proteins, and other alcohol-insoluble material. Two water-soluble polyuronides are found in grapes, one with the usual molecular weight for pectinic acid and a low percentage of esterification and the other of high molecular weight (11). Both contain arabinose, mannose, galactose, rhamnose, and galacturonic acid, but the latter cannot easily be precipitated by calcium. Gums

Table 4. Conversion Table for Various Hydrometer Scales

Specific gravity at 20°/20°C	Oechsle[a]	°Brix (Balling)	Baumé[b]	Specific gravity at 20°/20°C	Oechsle[a]	°Brix (Balling)	Baumé[b]
1.00000	0.0	0.0	0.00	1.08823	88	21.2	11.8
1.00078	0	0.2	0.1	1.08913	89	21.4	11.9
1.00155	1	0.4	0.2	1.09003	90	21.6	12.0
1.00233	2	0.6	0.3	1.09093	91	21.8	12.1
1.00311	3	0.8	0.45	1.09183	92	22.0	12.2
1.00389	4	1.0	0.55	1.09273	93	22.2	12.3
1.00779	8	2.0	1.1	1.09364	94	22.4	12.45
1.01172	12	3.0	1.7	1.09454	95	22.6	12.55
1.01567	15	4.0	2.2	1.09545	95	22.8	12.7
1.01965	20	5.0	2.8	1.09636	96	23.0	12.8
1.02366	24	6.0	3.3	1.09727	97	23.2	12.9
1.02770	28	7.0	3.9	1.09818	98	23.4	13.0
1.03176	32	8.0	4.4	1.09909	99	23.6	13.1
1.03586	36	9.0	5.0	1.10000	100	23.8	13.2
1.03998	40	10.0	5.6	1.10092	101	24.0	13.3
1.04413	44	11.0	6.1	1.10193	102	24.2	13.45
1.04831	48	12.0	6.7	1.10275	103	24.4	13.55
1.05252	53	13.0	7.2	1.10367	104	24.6	13.7

1.05667	57	14.0	7.8	1.10459	105	24.8	13.8
1.06104	61	15.0	8.3	1.10551	106	25.0	13.9
1.06534	65	16.0	8.9	1.10643	106	25.2	14.0
1.06968	70	17.0	9.4	1.10736	107	25.4	14.1
1.07142	71	17.4	9.7	1.10828	108	25.6	14.2
1.07404	74	18.0	10.0	1.10921	109	25.8	14.3
1.07580	76	18.4	10.2	1.11014	110	26.0	14.45
1.07844	78	19.0	10.55	1.11106	111	26.2	14.55
1.07932	79	19.2	10.65	1.11200	112	26.4	14.65
1.08021	80	19.4	10.8	1.11293	113	26.6	14.8
1.08110	81	19.6	10.9	1.11386	114	26.8	14.9
1.08198	82	19.8	11.0	1.11480	115	27.0	15.0
1.08287	83	20.0	11.1	1.11573	116	27.2	15.1
1.08376	84	20.2	11.2	1.11667	117	27.4	15.2
1.08465	85	20.4	11.35	1.11761	118	27.6	15.3
1.08554	86	20.6	11.45	1.11855	119	27.8	15.45
1.08644	86	20.8	11.55	1.12898	129	30.0	16.57
1.08733	87	21.0	11.7				

[a]The approximate sugar content on the Oechsle scale is given by dividing the degree Oechsle by 4 and subtracting 2.5 from the result. Thus for a must of 80 Oechsle: 80/4 − 2.5 = 17.5. The approximate prospective percentage alcohol by volume is given by multiplying the degree Oechsle by 0.125.

[b]Approximated by the use of this equation: °Brix = 1.8 × Baumé. These values check against those of Table 109 of Circular C440 of the National Bureau of Standards, *Polarimetry Saccharimetry and the Sugars*. Reducing sugar is approximately 2.0 less than the °Brix. The °Brix (Balling) times 0.52 will give the approximate prospective alcohol.

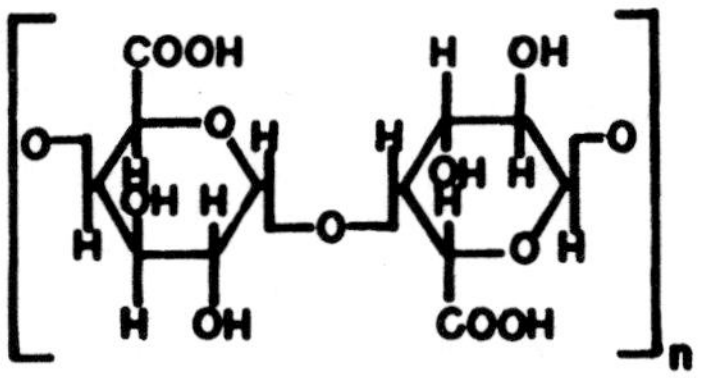

PECTIC ACID

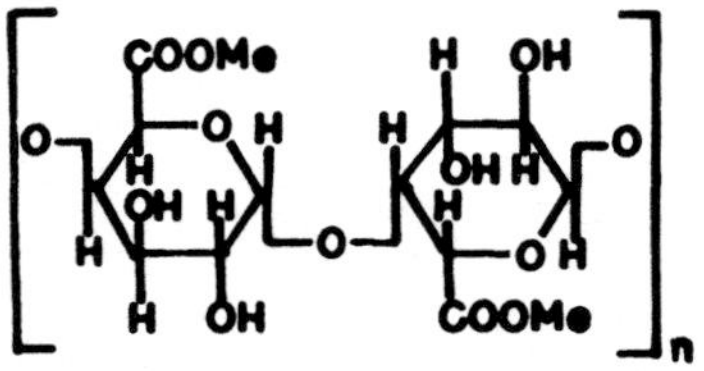

PECTIN Figure 2. Configurations of pectic acid and pectin.

and mucilages are less well defined but generally consist of complex mixtures of galactose, arabinose, and various uronic acids. Since pectins and gums exist in musts and in wines in a colloidal condition (as negatively charged particles), they tend to reduce the rate of clarification. In some cases they act as protective colloids to prevent the precipitation of suspended material—in the case of ferric phosphate casse, for example. Some French enologists believe that the pectins contribute to the softness or mellowness of the wine and that they constitute a real quality factor, but there are few sensory data to support this. Hydrolysis of pectin yields methanol, which is undesirable, particularly in fruit brandies.

Pectic substances were defined (13) as complex carbohydrates containing a large portion of anhydrogalacturonic acid units. The carboxyl groups of the acid may be partly esterified by methyl groups or partly or totally neutralized by one or more bases. Pectinic acids are used to designate colloidal polygalacturonic acids containing more than a small portion of methyl ester groups. Pectins are water-soluble pectinic acids of varying methyl ester content and degree of neutralization. Pectic acids are pectic substances composed of colloidal poly-D-galacturonic acid and are essentially free of methyl ester groups. The salts of pectic acids (pectates) are either normal or acid pectates. Pectins consist of galacturonic acid and methyl ester of galacturonic acid chains of indeterminant lengths (14). The galacturonic acid and ester chains are cross-linked with various sugars (rhamnose, arabinose, mannose, and galactose).

The connection with the sugars probably would proceed in a similar manner,

Glucose uridyl diphosphate

(structure) —2 NAD⁺ / 2 NADH→ **Glucoronic acid uridyl diphosphate** ←Epimerase→

Galacturonic acid uridyl diphosphate —Methyl esterase→ **Methyl galacturonate uridyl diphosphate** + (structure) →

Pectin-polygalacturonic segment

that is, with the uridyl diphosphate of the sugars being formed prior to reaction. Table 5 gives the molecular weight distribution of the soluble pectins.

There are natural polygalacturonases present in must, but the level of activity is relatively low. Wine yeasts commonly used for fermentation do not produce significant amounts of pectin-splitting enzymes.

Adding the pectin enzymes to crushed white grapes to allow for better juice drainage and clarity is a common practice. The majority of the commercial enzymes exhibit esterase and polygalacturonase activity. There are several types of polygalacturonase activity. One is an "endo" type, which randomly attacks the galactosidic linkages. There are two other (polymethylgalacturonase) "endo" types. One splits units that are both esterified, the other splits units in which only one component is esterified. "Exo" polygalacturonases attack only end units. Again there are two types, depending on the esterification. Figure 3 indicates the points of activity of these pectin–enzyme complexes.

There are also transeliminase enzymes, which degrade pectins by removal

Table 5. Molecular Weight Distribution of Soluble Pectins in a Grape Juice

Molecular weight	% of total
6,500– 10,000	7.0
10,000– 20,000	22.0
20,000– 30,000	15.0
30,000– 50,000	20.0
50,000–100,000	16.5
100,000–200,000	9.5
200,000–500,000	6.5
>500,000	3.5

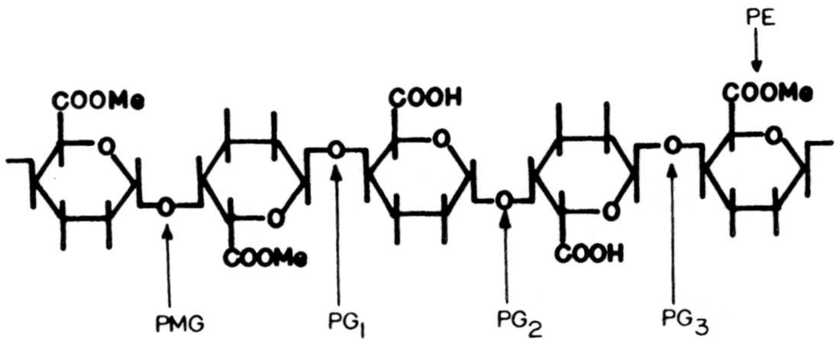

PE = Pectin esterase
PMG = Polymethylgalacturonase
Polygalacturonases (endo and exo)
  PG_1 and PG_3 when acid and ester adjacent
  PG_2 when two acids adjacent

Figure 3. A pectin chain with points of activity of the complex of pectin enzymes.

of a hydrogen from 5-position carbon and form a double bond; in the process, a molecule of water is lost. This breaks the galactosidic bond between the adjacent units. Several types exist, exhibiting preference for esterified or unesterified units.

The procedure of Solms et al. (15) is specific because it differentiates the pure pectin and the gums and indicates the percentage esterification of the pectins.

Procedure

To determine the pectic acid, evaporate 100 mL of must or wine to 25 mL (adding 8–12 g of sucrose to dry wines); cool and add 200 mL of 95% ethanol.

[Chemical structure diagram showing pectin fragment being cleaved by transeliminase into two products plus H₂O]

When the precipitate settles, filter on 15 cm qualitative filter paper and wash with 85% ethanol. Transfer the precipitate back to the original beaker using hot water and evaporate to about 40 mL. Cool to 25°C. If some water-insoluble material separates during evaporation, stir; if necessary, add a few drops of hydrochloric acid (1 + 2.5) and warm. Then cool and add 50 mL of a specially prepared sodium hydroxide solution (made by diluting 2–5 mL of 10 g/100 mL NaOH to 50 mL—the smaller the volume of precipitate, the smaller the quantity of base required). Allow to set 15 min, add 40 mL of water and 10 mL of hydrochloric acid (1 + 2.5) and boil 5 min. Filter and wash the pectic acid precipitate with hot water. This filtration should be rapid, and the filtrate must be clear. If the filtrate is cloudy, the saponification has been incomplete or done at too high a temperature, or both; then the test must be repeated.

In such cases try larger quantities of base and lower temperatures. Wash the precipitate of pectic acid back into the beaker and bring to 40 mL. Cool to below 25°C and repeat the saponification, addition of acid, and boiling as above. Filter and wash the precipitate with hot water until only a negligible amount of acid remains (not more than 500 mL of water should be required). Wash the pectic acid into a platinum dish, and dry on a steam bath and in a water oven to constant weight. Weigh, ignite, and reweigh. The loss in weight is pectic acid but is reported to be approximately the same as the tetragalacturonic acid (galacturonic acid × 0.90718). Care should be taken in all the transfers not to lose precipitate. Since the pectin material is sometimes nearly colorless, this requires careful manipulation. The original sample must, of course, be centrifuged and close filtered so that it is quite brilliant.

To determine the pure pectin by the Solms' procedure, the must is boiled to destroy any pectin-splitting enzymes. Filter 100 mL of centrifuged, boiled must through a Büchner funnel. The addition to the filter of 0.5 mm of filter aid, such as Super-Cel, before beginning the filtration often increases the filtration rate. Add 500 mL of 95% ethanol to the filtrate and let stand until the precipitate settles.

Decant off the supernatant liquid and filter on a Büchner funnel, washing with acid–alcohol (100 mL of 85% ethanol and 5 mL of concentrated HCl), then with 85% ethanol, until the wash is chloride free. Wash the precipitate off the filter with 50 mL of carbon dioxide-free water into the original beaker and titrate the free carboxyl groups to pH 7, using a pH meter with 0.1 N sodium hydroxide. Then add twice as much 0.1 N sodium hydroxide, and after 1 hr back-titrate with 0.1 N sulfuric acid.

$$\text{pectin (g/100 mL)} = 0.0176X + 0.01904Y$$

where $X = 0.1\ N$ NaOH (mL)
$Y = $ Net $0.1\ N$ NaOH reacted (mL)

$$\text{Esterified pectin (\%)} = \frac{100Y}{X + Y}$$

where X and Y are as above. The pure pectin content of musts varies from 0.05 to 0.35 g/100 mL. The percent esterification may vary from 20 to 50% but is usually about 30%.

The pectic substances can be fractionated and determined by another procedure (16). The high-molecular-weight methoxyl pectins are separated as alcohol insoluble solids. The low-molecular-weight methoxyl pectins are separated by extraction with ammonium oxalate, and the protopectin is solubilized with cold alkali. The anhydrogalacturonic content of the three fractions is then determined quantitatively by heating each extract with concentrated sulfuric acid and sodium tetraborate, thus forming a chromogen with m-hydroxydiphenyl. The absorbance of this solution is compared with that obtained from standard solutions of galacturonic acid subjected to the same procedure.

WINES

Table wines, that is, dry unfortified wines, have a specific gravity very near that of water. Wines containing sugar and musts have a specific gravity greater than 1.

Very dry sherry, because of its high alcohol content, may be somewhat lighter than water. The specific gravity of a sample reflects the net influence of the dissolved materials, such as sugars and acids, which are heavier than water, and of alcohol, which is lighter than water. Commercially, little attention is paid to the specific gravity of table wines except as a rough indication of the completion of the fermentation. However, many dessert wines, particularly those of the port, muscatel, and angelica types, are sold on the basis of their specific gravity and alcohol contents. Furthermore, the time of fortification of these wines is largely determined by their specific gravity, usually expressed in terms of °Brix.

The total soluble solids content of wines can also be determined by hydrometry, calibrated either to specific gravity or °Brix, and the precautions and procedure are the same as for determining the °Brix of musts described above. The correction values listed in Table 3 can be applied only for dry wines of relatively low alcohol content; theoretically these corrections apply accurately only to water-sucrose mixtures. Tables 6 and 7 should be used for temperature corrections with wines having alcohol contents of 17 or 19%, respectively (17).

EXTRACT

The higher the initial sugar content of the must, the higher the nonalcohol residue of the resulting wine, that is, the extract composed of the nonvolatile components of the wine (sugars, fixed acids, organic salts, and other substances). Thus the extract content of the wine is an indication of the sugar content of the original must. Red wines have higher residual (sugar-free) extracts than do white wines. Rosés are intermediate. The values depend on variety (18) and on handling conditions (19). A comparison of spontaneous yeast fermentations to pure

Table 6. Corrections for Hydrometers in Dessert Wines Containing 17% v/v Ethanol

Temperature		Observed °Brix					
°C	°F	0	2.5	5.0	10.0	12.5	15.0
		Subtract					
15	59.0	0.35	0.43	0.45	0.45	0.49	0.50
16	60.8	0.31	0.34	0.37	0.37	0.38	0.41
17	62.8	0.20	0.22	0.24	0.26	0.27	0.27
18	64.4	0.10	0.10	0.16	0.17	0.20	0.23
19	66.2	0.06	0.06	0.06	0.06	0.07	0.07
		Add					
21	69.8	0.06	0.06	0.06	0.06	0.07	0.07
22	71.6	0.10	0.10	0.14	0.17	0.20	0.23
23	73.4	0.23	0.25	0.26	0.26	0.27	0.27
24	75.2	0.31	0.35	0.36	0.37	0.38	0.41
25	77.0	0.27	0.44	0.44	0.46	0.48	0.51

From Reference 17.

Table 7. Corrections for Hydrometers in Dessert Wines Containing 19% v/v Ethanol[a]

Temperature		Observed °Brix					
°C	°F	0	2.5	5.0	10.0	12.5	15.0
		Subtract					
15	59.0	0.40	0.45	0.48	0.50	0.53	0.55
16	60.8	0.33	0.35	0.38	0.40	0.41	0.43
17	62.8	0.24	0.25	0.26	0.28	0.29	0.30
18	64.4	0.15	0.15	0.18	0.20	0.22	0.24
19	66.2	0.08	0.08	0.08	0.08	0.09	0.10
		Add					
21	69.8	0.08	0.08	0.08	0.08	0.09	0.10
22	71.6	0.15	0.15	0.17	0.20	0.22	0.24
23	73.4	0.24	0.26	0.28	0.28	0.29	0.30
24	75.2	0.33	0.37	0.38	0.40	0.41	0.43
25	77.0	0.40	0.45	0.46	0.48	0.50	0.53

From Reference 17.
[a]For very accurate work the specific gravity is best determined in a pycnometer. A 50-mL pycnometer with attached thermometer is convenient. A Westphal balance may also be used for the determination by specific gravity.

culture fermentations showed no significant difference in residual extract (20). Values for (sugar-free) extracts range from 0.7 g/100 mL for low-alcohol German wines to more than 3 g/100 mL for late-harvested red wines, with an average value of about 2.0 g/100 mL. European countries have traditionally used the alcohol/extract ratios to detect addition of water, alcohol, or sugar to wine before bottling.

The Office International de la Vigne et du Vin (OIV) (21) defines *extract* as the nonvolatile materials. However, the physical conditions for definition of the nonvolatile constituents must be precise. A number of enological formulas used to detect falsification include the extract content or the extract less the reducing sugar. Gilbert (22) reviewed a number of these formulas and gave a short formula, with all components expressed in grams per liter:

$$E = R - 0.9T - 0.8A - 0.05ET$$

where E = (sugar-free) extract
R = indicated extract − reducing sugar

T = titratable acidity
A = ash
ET = ethanol

A formula he recommended for official use, again with all components in grams per liter, is:

$$E = R + 0.125L + 0.75V - 0.9T - 0.8A - 0.05ET - 0.2TA$$

where E = (sugar-free) extract
R = indicated (sugar-free) extract
L = lactic acid
V = volatile acid
T = titratable acid
A = ash
ET = ethanol
TA = tartaric acid

These formulas are of importance in Europe, where wine sophistication is a more serious problem than in other areas. In the United States, the extract content is usually determined by calculation, with the help of the Taberié formula (5), according to the following procedure.

Procedure

Determine the alcohol content of the wine and, using Table 8, establish the specific gravity of an aqueous ethanol solution with the same ethanol concentration; this is d_a.

Determine the specific gravity of the wine using a hydrometer; this is d_w. If the hydrometer is calibrated to °Brix (Balling), convert to specific gravity with the help of Table 1. Calculate the density of the residue d_r using the following formula:

$$d_r = d_w - d_a + 1.0000$$

Convert the values of d_r to the amount of the extract, expressed as grams per 100 mL, with the help of Table 9.

The procedure of the OIV (21) is the same as that given above except that the value of d_w is corrected for acetic acid and sulfur dioxide:

$$d_{w'} = d_w - \frac{(A)(0.0086)}{1000} - \frac{(B)(0.6)}{1000}$$

where A = volatile acidity (meq/L)
B = sulfur dioxide content (g/L)

Table 8. Proof, Percent Ethanol by Volume and Weight, and Specific Gravity at 20°C[a,b]

Proof	Alcohol (vol %)	Alcohol (wt %)	Alcohol (g/100 mL)	Specific gravity
0.0	0.00	0.00	0.00	1.00000
1.0	0.50	0.40	0.40	0.99925
2.0	1.00	0.795	0.79	0.99851
3.0	1.50	1.19	1.19	0.99777
4.0	2.00	1.593	1.59	0.99704
5.0	2.50	1.99	1.98	0.99633
6.0	3.00	2.392	2.38	0.99560
7.0	3.50	2.80	2.78	0.99490
8.0	4.00	3.194	3.18	0.99419
9.0	4.50	3.60	3.58	0.99350
10.0	5.00	3.998	3.97	0.99281
11.0	5.50	4.40	4.37	0.99214
12.0	6.00	4.804	4.76	0.99149
13.0	6.50	5.21	5.16	0.99084
14.0	7.00	5.612	5.56	0.99020
15.0	7.50	6.02	5.96	0.98956
16.0	8.00	6.422	6.35	0.98894
17.0	8.50	6.83	6.75	0.98832
18.0	9.00	7.234	7.14	0.98771
19.0	9.50	7.64	7.54	0.98711
20.0	10.00	8.047	7.93	0.98650
21.0	10.50	8.45	8.33	0.98590
22.0	11.00	8.862	8.73	0.98530
23.0	11.50	9.27	9.13	0.98471
24.0	12.00	9.679	9.52	0.98412
25.0	12.50	10.80	9.92	0.98354
26.0	13.00	10.497	10.31	0.98297
27.0	13.50	10.90	10.71	0.98239
28.0	14.00	11.317	11.11	0.98182
29.0	14.50	11.72	11.51	0.98127
30.0	15.00	12.138	11.90	0.98071
31.0	15.50	12.54	12.30	0.98015
32.0	16.00	12.961	12.69	0.97960
33.0	16.50	13.37	13.09	0.97904
34.0	17.00	13.786	13.49	0.97850
35.0	17.50	14.19	13.89	0.97797
36.0	18.00	14.612	14.28	0.97743
37.0	18.50	15.02	14.68	0.97690
38.0	19.00	15.440	15.08	0.97638

Table 8. (*Continued*)

Proof	Alcohol (vol %)	Alcohol (wt %)	Alcohol (g/100 mL)	Specific gravity
39.0	19.50	15.84	15.47	0.97585
40.0	20.00	16.269	15.87	0.97532
41.0	20.50	16.67	16.26	0.97479
42.0	21.00	17.100	16.66	0.97425
43.0	21.50	17.51	17.06	0.97372
44.0	22.00	17.933	17.46	0.97318
45.0	22.50	18.34	17.86	0.97262

[a] Referred to water at the same temperature. Multiply by 0.99908 to convert to specific gravity referred to water at 4°C.
[b] This table was taken from the *AOAC Official Methods* (5) and was subsequently taken from the *U.S. Bureau of Standards Circular No. 19*, dated 1924. There are small differences between these values and those of Jaulmes given in Reference 21.

The volume of water needed to bring the volume back to 100 mL with an alcoholic solution containing sugar is not quite the same as that needed for an aqueous solution containing sucrose (23). The error is small for wines with 5–10% sugar and not over 11% alcohol volume.

Ruppert (24) recommends the Mettler–Paar DMA 55D density meter, interfaced to the Hewlett-Packard 97S programmable calculator, a Mettler–Paar SP automatic sample changer, an Anton–Paar DT 100-20 digital thermometer, a Neslab EX-100 UHP circulating bath, and a Neslab EN 350 flow-through refrigerated bath cooler. This was connected to a T/S 1000 Timex Sinclair 16K microcomputer based on the 280 8-bit microprocessor. The density meter was used on the dealcoholized wine extract (brought to volume) and the alcoholic content of the distillate (also brought to volume). Ruppert (24) programmed an equation relating to the data in Tables 8 and 9 (25) and the Taberié equation. The values of the percent alcohol and extract were automatically reported in the computer printout.

When the most accurate results are required, the pycnometer method of the OIV (21) should be used. For details on the use of the pycnometer see p. 94. The residue from an alcohol distillation may be used. Pour the residue into a 100-mL volumetric flask, rinse the distillation flask two or three times, and add to the volumetric flask. Bring to volume and temperature and mix well. Fill the pycnometer with as much of the extract solution as is required. Pycnometer flasks are available in varying volumes. The pycnometric procedure has been tested in a collaborative study and was found (26) to give excellent repeatability

Table 9. Specific Gravity and Extract Concentration at 20°C[a]

Specific gravity	Extract (g/100 mL)	Specific gravity	Extract (g/100 mL)	Specific gravity	Extract (g/100 mL)
1.000	0.00	1.040	10.37	1.080	20.84
1.001	0.26	1.041	10.63	1.081	21.10
1.002	0.51	1.042	10.90	10.82	21.36
1.003	0.77	1.043	11.16	1.083	21.62
1.004	1.03	1.044	11.42	1.084	21.89
1.005	1.29	1.045	11.68	1.085	22.15
1.006	1.54	1.046	11.94	1,086	22.41
1.007	1.80	1.047	12.20	1.087	22.68
1.008	2.06	1.048	12.46	1.088	22.94
1.009	2.32	1.049	12.72	1.089	23.20
1.010	2.58	1.050	12.98	1.090	23.47
1.011	2.84	1.051	13.24	1.091	23.73
1.012	3.10	1.052	13.50	1.092	23.99
1.013	3.36	1.053	13.76	1.093	24.25
1.014	3.62	1.054	14.03	1.094	24.52
1.015	3.88	1.055	14.29	1.095	24.78
1.016	4.13	1.056	14.55	1.096	25.04
1.017	4.39	1.057	14.81	1.097	25.31
1.018	4.65	1.058	15.07	1.098	25.57
1.019	4.91	1.059	15.33	1.099	25.84
1.020	5.17	1.060	15.59	1.100	26.10
1.021	5.43	1.061	15.89	1.101	26.36
1.022	5.69	1.062	16.12	1.102	26.63
1.023	5.95	1.063	16.38	1.103	26.89
1.024	6.21	1.064	16.64	1.104	27.15
1.025	6.47	1.065	16.90	1.105	27.42
1.026	6.73	1.066	17.16	1.106	27.68
1.027	6.99	1.067	17.43	1.107	27.95
1.028	7.25	1.068	17.69	1.108	28.21
1.029	7.51	1.069	17.95	1.109	28.48
1.030	7.77	1.070	18.21	1.110	28.74
1.031	8.03	1.071	18.48	1.111	29.00
1.032	8.29	1.072	18.74	1.112	29.27
1.033	8.55	1.073	19.00	1.113	29.53
1.034	8.81	1.074	19.26	1.114	29.80
1.035	9.07	1.075	19.52	1.115	30.06
1.036	9.33	1.076	19.78	1.116	30.33
1.037	9.59	1.077	20.05	1.117	30.59
1.038	9.85	1.078	20.31	1.118	30.86
1.039	10.11	1.079	20.58	1.119	31.12

[a]The values for extract in this table are slightly less than those in Table 1 when the °Brix (Balling) in that table is multiplied by the specific gravity to give grams per 100 mL. See Franck and Junge (25) for specific gravity and extract values from 1.120 to 1.200.

and reproducibility. The results for wines of 1% or less sugar content are better than for sweeter wines.

When the percent ethanol and °Brix (or density) of a wine are known it is possible, from Taberié's formula (p. 31), to prepare a nomogram to ascertain the extract (in grams per liter); see Figure 4 (27). Simply place a ruler across the nomogram connecting percent ethanol and °Brix (or specific gravity). Then read the extract from where the ruler crosses the extract curve. Of course, if

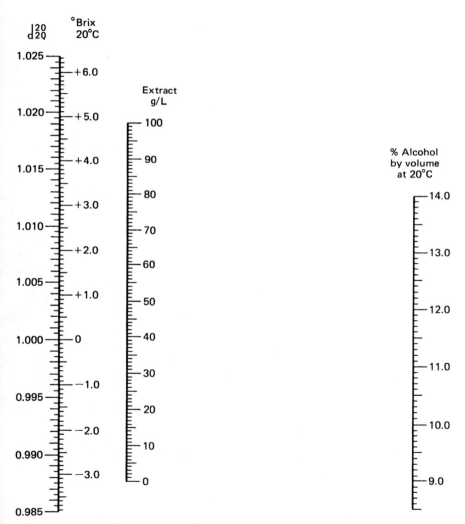

Figure 4. Nomograph relating °Brix, extract, and ethanol for dessert and appetizer wines (27).

°Brix (or specific gravity) and the extract are known, the same procedure may be used to estimate the percent ethanol of the wine. The values determined are an approximation of the actual values, although in some cases they are remarkably close.

SUGARS

In varieties of *Vitis vinifera* the predominant sugars are glucose and fructose. Small amounts of sucrose and other sugars are present; in some varieties of *V. labruscana* and its hybrids, sucrose may constitute as much as 25% of the total sugar. In some countries, sucrose may be added when the must is deficient in sugar. After the grapes are crushed and during alcoholic fermentation, the sucrose is hydrolyzed and fermented. In either case, very little sucrose remains in the finished wine.

The sugar content of ripening grapes is an important factor in determining the time of harvest. However, since most (>90%) of the dissolved solids are sugar, the total soluble solids determination is often taken as an approximate measure of the sugar content (p. 14).

Truly "dry" wines contain less than 0.1% reducing sugar, and much of this is probably due to nonfermentable reducing sugars such as pentoses. From the sensory point of view, "dryness" is a relative term, varying with the wine type and the individual. With some low-alcohol white table wines the threshold for sweetness is about the same as in water (i.e., about 0.4%). However, in red wines it may be as high as 1–1.5%. The most sensitive people have thresholds less than these values, and the least sensitive have higher thresholds.

Sparkling wines labeled "brut" usually have a sugar content of less than 1.5%. Some rosés and many "dry" sherries have a sugar content of as much as 1.5–2.5%. Dessert wines, such as ports, sweet sherries, and muscatels, may have 10–15% sugar. The Malaga wines of Spain may contain up to 20% sugar. Some Russian table wines stabilized by the Dellé procedure have 15–20% sugar.

In winery operation the sugar determination is frequently made to ascertain the completeness of the fermentation, to conform to legal or commercial requirements on the sugar content of a wine type (in quality control) and on the "cuvée" (base wine) for sparkling wines.

Because of recently developed techniques for sterile filtration and bottling, many "dry" wines are now sold with reducing sugar levels exceeding 0.5 g/100 mL. These levels of sugar yield wines of better consumer acceptance in many cases, particularly in certain areas. The amounts of sugar are critical, and the sugar content must be determined on all such wines before bottling.

Reducing Sugars

Reducing sugars are those with the following configurations:

(a) Terminal aldehyde: $-C\overset{\displaystyle O}{\underset{\displaystyle H}{\lessgtr}}$

(b) α-Hydroxy ketone: $-C\overset{\displaystyle O}{\underset{\displaystyle CH_2OH}{\lessgtr}}$

(c) Hemiacetal: $-\underset{|}{\overset{|}{C}}-O-\underset{|}{\overset{OH}{\underset{|}{C}}}-$

A number of chemicals can cause these sugars to be oxidized. The degree of oxidation depends on the oxidizing agent and the conditions of oxidation. The reactions are not stoichiometric. Therefore careful attention to detail and consistency is imperative if reproducibility is to be expected. The sugar is generally oxidized to acid.

The Lane–Eynon procedure is an official procedure of the AOAC (5). It depends on first determining how much of a standard sugar solution is required to react under specified conditions with a measured volume of alkaline copper sulfate solution, the so-called Soxhlet's reagent. In the second part of the determination an aliquot of dealcoholized and clarified wine is added to the same volume of Soxhlet's reagent, and the volume of the standard sugar solution which is required to react with the residual Soxhlet's reagent is determined. The difference represents the sugar content of the sample aliquot. Methylene blue is used as the indicator because it changes from blue to colorless at an oxidation–reduction potential that corresponds closely to that for the completion of the reaction between sugar and copper. The tartrate acts to complex the Cu^{2+} and hold more in solution. The sodium hydroxide causes oxidation as well as allowing isomerization for aldehyde formation.

Procedure

Prepare Fehling's A solution by dissolving 34.639 g of cupric sulfate pentahydrate ($CuSO_4 \cdot 5H_2O$) in water in a 500-mL volumetric flask and diluting to volume. Allow to settle clear and filter through cellulose. Prepare Fehling's B solution by dissolving 173 g of sodium potassium tartrate (Rochelle salt) and 50 g of sodium hydroxide in water in a 500-mL volumetric flask and diluting to volume. Allow to settle clear and filter through cellulose. Store in a Pyrex or alkali-resistant bottle. Prepare a 0.50% aqueous sugar solution.

For standardization, pipet 25 mL of the Soxhlet's reagent (consisting of equal parts of Fehling's A and B) into a narrow-mouth 250-mL Erlenmeyer flask. Place a few glass beads in the flask to reduce bumping. Pipet in 20 mL of 0.5% sugar solution. Fill a 25-mL buret, with an offset delivery, with the same solution. Place the Erlenmeyer flask on a wire gauze over a Bunsen burner and bring the contents to boil in 3 min. While swirling the contents of the flask, add the 0.5% sugar solution from the buret until only a faint blue color remains. Now add 5 drops of 1% methylene blue solution and continue the titration. The end point is the disappearance of all the blue color. Detection of the correct end point requires practice. It is usually best seen by the disappearance of blue color in the foam. Less than 3 min should elapse between the start of boiling and the completion of the titration. About 24 mL of 0.5% glucose solution is required for 25 mL of properly prepared Soxhlet's reagent. The determination should be repeated until the results do not differ by more than ±0.2 mL.

Prepare a wine sample free of non-sugar-reducing substances and with a sugar content of not over 1% in the following way: 50 mL of dealcoholized dry wine, or the equivalent, is either pipetted or washed into a 100-mL volumetric flask; the solution of the residue from the ethanol determination (p. 90) may be used. Add 5 mL of saturated neutral lead acetate solution, enough decolorizing charcoal to decolorize the wine, and 2 drops of glacial acetic acid. The amount of charcoal needed varies with the wine; as little as 0.1 g will clear most musts and white wines, but 0.5 g may be necessary with deep-colored red wines. Do not add a large excess because some sugar may be absorbed. Mix the contents thoroughly and let stand 10 min. Bring to volume with water. For sweet wines use only 5 mL of dealcoholized wine and for musts only 2 mL; in these cases, 2 mL of the lead acetate solution and a pinch of charcoal are sufficient.

Place 0.4 g of disodium hydrogen phosphate or sodium oxalate per milliliter of lead acetate used in a 400-mL beaker and filter the contents of the 100-mL volumetric flask into the beaker. Refilter if charcoal passes through the paper. Stir while filtering to prevent caking of the phosphate or oxalate. This solution will settle fairly clear in a few minutes. Add a pinch more of phosphate or oxalate to determine whether the precipitation of lead is complete. Refilter this solution into another beaker if necessary.

Now pipet 20 mL of the clarified wine solution into an Erlenmeyer flask containing 25 mL of the Soxhlet's reagent. Bring to a boil and titrate with 0.5% sugar to a faint blue. Add 5 drops of 1% methylene blue solution and complete the titration to a full brick-red end point. The same 3 min time to bring to a boil and 3 min for completion of the titration apply here as in determining the sugar titer.

$$\text{Reducing sugar (g/100 mL)} = \frac{(A - B)(0.005)(100)}{v}$$

where A = volume of 0.5% sugar used for Soxhlet's reagent (mL)
B = volume of 0.5% sugar used for the wine sample (mL)
v = volume of the wine in the final aliquot (mL)

If the 20-mL aliquot of clarified wine is found to contain too much sugar for the Soxhlet's reagent, use a lesser volume (e.g., 5 or 10 mL) and dilute with water to 20 mL.

The AOAC (5) has a special table to facilitate the calculation of reducing sugars, but the directions for preparation of reagents and the titration procedure must be adhered to very closely.

Lay (28) tested the Fehling's reagent, Luff–School, and Rebelein "5-min" methods. The Luff–School method was the most accurate, but the Rebelein "5-min" method was of comparable reliability. A modified Luff–School method is given below. Junge (25) reported very good repeatability and reproducibility down to 1%. Below 1% this report recommends that the analyst should indicate how much uncertainty exists. This has some importance where wines are legally defined by sugar content, namely, the "soft" wines.

Procedure

Prepare an alkaline citrate solution. Dissolve 120 g of sodium carbonate, 70 g of sodium citrate ($Na_3C_6H_5O_7 \cdot 2H_2O$), and 25 g of cupric sulfate pentahydrate ($CuSO_4 \cdot 5H_2O$) in water and bring to 1 L. Allow the solution to settle; then decant off the clear supernatant. The solution should be 10.2 N when titrated against acid and should have a pH of 10.3–10.4. Add 420 g of potassium iodide to a 1-L flask. Add 3 mL of 1 N sodium hydroxide and sufficient water to bring to volume. Solution should remain clear. The iodide is not easily oxidized in alkaline solution. Add 25 mL of concentrated hydrochloric acid to a 100-mL flask and bring to volume with water. Prepare a 0.1000 N sodium thiosulfate solution and standardize against a 0.1000 N potassium dichromate solution (4.9035 g $K_2Cr_2O_7$/L). Prepare a 1 g/100 mL starch solution and a 1.0 g/100 mL glucose standard solution. Add 50 g of potassium thiocyanate to a 100-mL flask and bring to volume with water.

To a 500-mL reflux flask add 2 glass beads, 25 mL of distilled water (blank) or 1, 2, 3, 4, or 5 mL of standard dextrose solution and sufficient water to make 25 mL, and 25 mL of copper solution. Reflux (with condenser attached) for 20 min (timed), cool in a water bath, and add 10 ml of potassium iodide solution and 25 mL of hydrochloric acid solution, and mix well to dissolve precipitate and expel carbon dioxide. Titrate liberated I_3^- with 0.1000 N thiosulfate to a pale yellow, add 5 mL of starch indicator, and carefully titrate to creamy end point. Add 4 mL of potassium thiocyanate solution, and titrate to a white end point.

Subtract the titration values for the standard solutions from the blank. Draw a curve (milligrams of reducing sugar vs. milliliters of thiosulfate). The curve is linear up to 45 mg.

For wine do similarly. If the wine is "dry," decolorize with 2 g of decolorizing charcoal (Darco KB) per 25 mL. The chart on p. 40 gives recommended dilutions and factors for different sugar levels.

The rapid "5-min" method (29, 30), given below, compares favorably to the more lengthy methods. The author suggests that this method is stoichiometric as set up; hence additional heating does cause variations in the results.

Sample	Reducing sugar range (g/100 mL)	Dilution	To use milligrams	Corrections for nonfermentable substances		Factor[a]
				White	Red	
Juice or wine	15–22	5	4	−0.3	−0.5	0.5
Wine	10–15	10	3	−0.45	−0.75	0.333
Wine	5–10	20	2	−0.6	−1.0	0.25
Wine	1–5	—	1	−1.5	−2.5	0.1
Wine	1.0	—	4	−6.0	−10.0	0.025
Dry wine	<0.15[b] or 0.25[c]	—	10	+2.5	+10.0	0.01

[a]Read milligrams from standard curve, subtract proper correction value for nonfermentable substances, then multiply by factor to get reducing sugar in grams per 100 mL.
[b]White wine.
[c]Red wine.

Procedure

Place 41.92 g of cupric sulfate pentahydrate ($CuSO_4 \cdot 5H_2O$) in a 1-L flask and dissolve in distilled water; add 10 mL of 1 N sulfuric acid and bring to volume. Dissolve 250 g of sodium potassium tartrate (Rochelle salt) in 400 mL of water. Dissolve 80 g of sodium hydroxide in 400 mL of water. Put both these solutions into a 1-L flask and bring to volume with water. Put 300 g of potassium iodide and 100 mL of 1 N sodium hydroxide into a 1-L flask. Add water and bring to volume. Make up a 16% sulfuric acid solution (175 mL of 95% H_2SO_4 brought to 1 L with water). Dissolve 10 g of soluble starch in 500 mL of boiling water and cool. Dissolve 20 g of potassium iodide and 10 mL of 1 N sodium hydroxide in 500 mL and mix with the dissolved starch solution. Add 13.777 g of $Na_2S_2O_3 \cdot 5H_2O$ to a 1-L flask with 50 mL of 1 N sodium hydroxide and bring to volume with distilled water.

Into a 200-mL Erlenmeyer flask add 10 mL of cupric sulfate solution, 5.0 mL of Rochelle salt solution, a boiling stone, and 2.0 mL of wine (containing no more than 2.8 g of reducing sugar/100 mL). Boil for $1\frac{1}{2}$ min and cool under a water tap. Add 10 mL of potassium iodide solution, 10 mL of 16% sulfuric acid, and 10 mL of starch solution. Titrate with thiosulfate to a creamy yellow. The sugar concentration is read directly from the buret. If the wine was diluted, multiply by the dilution factor. For a red wine, decolorize with a minimum amount of charcoal. Run a blank and subtract the titration values obtained from the blank. The answers are in grams per liter of reducing sugar.

An automatic method for sugar determination in foodstuffs was suggested (31). Hexacyanoferrate(III) was the oxidizing agent. The results compared favorably to those of the Lane–Eynon method. The procedure was tested in wine, and the

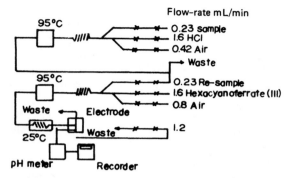

Figure 5. Diagram for an automatic setup for sugar determination (31).

system was found to be satisfactory (32). The change in the oxidation–reduction potential of trivalent hexacyanoferrate to the bivalent form is recorded. This potential is proportional to the amount of reducing sugar.

Figure 5 shows an automated system. If total sugar is required, the system is used as illustrated. If only reducing sugar is desired, the upper segment is disconnected and the sampler is connected to the resample position.

Procedure

Prepare standard solutions of 0.01–0.1 g/100 mL of glucose. Make up 0.1 N hydrochloric acid and 1 N sodium hydroxide solutions. Weigh out 30 g of reagent grade potassium hexacyanoferrate(III) and 15 g of the bivalent form and dissolve in 5 L of deionized water. When ready to use, mix equal volumes of this with 1 N sodium hydroxide solution (make fresh daily).

Set up for reducing sugar configuration and sample standards at rate of 30 per hour. Draw a standard curve of concentration versus millivolts.

Wine samples should be diluted to fall into desired range and run directly. The value determined is multiplied by the dilution factor.

Mauer (33) gives a rapid and simple colorimetric procedure using Fehling's solution. The OIV reference procedure (21) is to pass the wine through a column filled with Dowex 3 ion-exchange resin, 20/50 mesh. The column is prepared by two complete cycles of 1 N hydrochloric acid and 1 N sodium hydroxide solution. The resin is then transferred to a beaker and treated with 4 N acetic acid. It is replaced in the column and more 4 N acetic acid is passed through; subsequently, the column is washed with water until neutral. To regenerate the column, 2 N sodium hydroxide solution is passed through and finally the column is washed with water until neutral. Mercuric acetate is used to further clarify, and hydrogen sulfide must be used to remove excess mercury.

Neutral lead acetate, basic lead acetate, and zinc ferrocyanide may also be used as clarifying agents.

With the availability of high-performance liquid chromotography (HPLC), sugars are now commonly determined by this method. In most cases, various sugars, alcohol, glycerol, 2,3-butanediol, and several organic acids are determined at the same time. For recent procedures see References 34–38.

Among the advantages of HPLC is that it can separate sucrose from glucose and fructose, and so on, and it is rapid. Pfeiffer and Radler (38), for example, used a stationary phase of a cation-exchange resin Trennsäule Aminex HPX 87H and a mobile phase of dilute sulfuric acid plus an injection system, a refractive index detector, a high-temperature oven, an ultraviolet (UV) detector with an integrator, a 300 × 7.8-mm column, and a microguard to prevent contamination. For ethanol, glycerol, glucose, fructose, tartrate, malate, lactate, succinate, acetate, and citrate, each sample required 25 min.

A simple industry-tested and effective HPLC method of determining sugars is given below.

Procedure

Weigh out 2.000 g of sucrose, 12.000 g of D-fructose, and 12.000 g of D-glucose into a volumetric flask, dissolve, and bring to volume. Divide into 1-mL serum ampules and keep frozen until ready to use. Shake the ampules well after thawing out the standards. Inject 10 μL of the standard into the HPLC with water as the mobile phase. The analytical column suggested is a 300 × 10-mm Bio-Rad Aminex Q15S with two guard columns, an Ion Exclusion Bio-Rad 40 × 4.6-mm Aminex HPX-85H, and an Anion/OH 40 × 4.6-mm Aminex A-25. Flow rate is 0.7 mL/min for the water which has been triple distilled and filtered through a 0.45-μm filter. It should be degassed before use. The other pertinent conditions are: pressure, 120 bar; temperature of columns, 85.0°C; refractive index detector attenuated ×64; length of run, 16 min. The approximate retention times for the three sugars are: glucose, 9.2 min; sucrose, 7.2 min; and fructose, 11.5 min. Run replications. If reproducibility is adequate determine the area per grams of sugar for each.

Prepare test samples by degassing and then filtering through a 0.45-μm filter. For samples greater than 15 g/100 mL for any one sugar component, dilute accurately so the amount is not over about 12 g/100 mL.

From the peak area for each sugar calculate the sugar concentrations:

$$\frac{\text{Peak area}_{\text{sugar}} \times \text{Area}_{\text{sugar}}}{g_{\text{sugar}} \times F} = \frac{g \text{ sugar}}{100 \text{ mL}}$$

where F = dilution factor.

Rapid Methods for Sugar

Often it is necessary to determine only whether a wine has less than 1% sugar. Sugar test pills are useful for wines containing up to 2% sugar. The procedure requires only a few minutes and is especially useful when one needs to establish the approximate sugar content in the range of 0.1–2.0%. The tests should be carried out in duplicate. In using the sugar test pills or solutions, the following precautions should be noted: The tablets should be kept dry; cloudy wines should be filtered, while red wines should be decolorized by pretreating a 50-mL aliquot with about 1 g of activated carbon and filtering. The test has been used to determine the completeness of fermentation, to detect residual sugar in distilling material, and to determine whether Swiss wines have more than 0.4% sugar, the level at which Swiss wines have to be labeled "legérement doux" or "avec sucre résiduel" (39).

Individual Sugars

There is a wealth of material on the relative portions of glucose, fructose, and sucrose in grapes. In 26 samples, the following values were reported (40): 5.0–8.0% (average 5.4%) fructose, 5.2–8.2% (average 6.7%) glucose, 0.0 to 2.24% (average 0.6%) sucrose, and 0% to traces of sorbitol. There exist a number of methods for determining the individual sugars. These included enzymatic, colorimetric, fluorometric, radioisotopic, and electrochemical approaches. Similar procedures can be used for most of the sugars. The most specific methods include paper and thin-layer chromatography and gas and liquid chromatography.

Enzymatic Methods

The enzymatic procedure (41) permits determination of glucose and fructose. Glucose reacts with adenosine triphosphate (ATP) in the presence of hexokinase (HK) to produce glucose-6-phosphate:

$$\text{Glucose} + \text{ATP} \xrightleftharpoons{\text{HK}} \text{Glucose-6-phosphate} + \text{ADP}$$

Addition of glucose-6-phosphate dehydrogenase (G-6-PDH) in the presence of oxidized nicotinamide adenine dinucleotide (NADP) produces gluconic acid-6-phosphate and an equivalent amount of the reduced form (NADPH):

$$\text{Glucose-6-phosphate} + \text{NADP}^+ \xrightleftharpoons{\text{G-6-PDH}} \text{Gluconic acid-6-phosphate} + \text{NADPH} + \text{H}^+$$

The amount of NADPH can be measured from the absorbance at 340 nm and is equivalent to the glucose originally present. In the case of fructose the reaction is:

$$\text{Fructose} + \text{ATP} \underset{}{\overset{\text{HK}}{\rightleftharpoons}} \text{Fructose-6-phosphate} + \text{ADP}$$

This is converted to glucose-6-phosphate in the presence of phosphoglucose isomerase (PGI):

$$\text{Fructose-6-phosphate} \underset{}{\overset{\text{PGI}}{\rightleftharpoons}} \text{Glucose-6-phosphate}$$

Finally, sucrose must be hydrolyzed to glucose and fructose using invertase. Generally no preliminary treatment of musts or wines is needed, except that the sample should be diluted so that it contains no more than 3–100 μg of glucose plus fructose per cuvette (42). This method was found satisfactory for wines (43). The procedure given here is now the official first action of the AOAC (44).

Procedure

Make reaction solution I by mixing 8 mL of triethanol–HCl buffer (0.94 mol/L, pH 7.6, containing $MgSO_4$, 12.5 mmol/L), 1 mL of Na_2NADP (10 mg/mL), and 1 mL of $Na_2ATP \cdot 3H_2O$ (50 mg/mL containing $NaHCO_3$, 50 mg/mL). This is stable for a week at 4°C. Make solution II from an enzyme suspension containing 2 mg/mL of hexokinase (about 140 U/mg) and 1 mg/mL of glucose-6-phosphatase dehydrogenase [~140 U/mg in $(NH_4)_2SO_4$ of 3.2 mol/L and pH 6]. Solution III is an enzyme suspension containing 2 mg/mL of phosphoglucose isomerase (~350 U/mg) in $(NH_4)_2SO_4$ (of 3.2 mol/L and pH 6). Kits for this determination can be purchased from Boehringer Mannheim Biochemicals, 7941 Castleway Drive, P.O. Box 50826, Indianapolis, IN 46250.

Dilute the untreated wine sample so that it contains less than 0.5 g/L of glucose and fructose. The precision of the fructose determination is less if the glucose : fructose ratio is more than 5 : 1. Label one cuvette (either glass or disposable with a 1-cm light path) "sample" and the other "blank." Pipet 1.0 mL of solution I into each cuvette. Pipet 0.1 mL of wine into the "sample" cuvet. Rinse pipet three times with solution before transfer. The tip of the pipet should be 1 mm above the solution in the cuvette. Add 2.0 mL of water to the "blank" cuvette and 1.9 mL of water to the "sample" cuvette. Mix each cuvette and wait 5 min at 20–25°C. Determine A_1 of each solution at 340 nm (or Hg 365 nm) versus an air "blank." Start reaction by adding 0.02 mL of solution II to each cuvette. Mix and hold at 20–25°C until reaction has stopped (~12 min). Determine A_2 of both cuvets. Add 0.02 mL of solution III to each cuvette, mix, and hold at 25°C until reaction has stopped (~12 min). Determine A_3 of both cuvettes. Determine the glucose and fructose concentrations from the formulas

$$\Delta A \text{ (glucose)} = (A_2 - A_1)(\text{sample}) - (A_2 - A_1)(\text{blank}) \quad (1)$$

$$\Delta A \text{ (fructose)} = (A_3 - A_2)(\text{sample}) - (A_3 - A_2)(\text{blank}) \quad (2)$$

$$C(\text{g/L}) = \frac{V \times \text{MW}}{\epsilon \times d \times v \times 1000} \quad (3)$$

where V is the final volume (in milliliters), v is the sample volume (in milliliters), MW is the molecular weight (glucose + fructose = 180.16), d is the light path (in centimeters), and ϵ is the absorption coefficient of NADPH at 340 nm = 6.3 (L $\times$ mol^{-1} $\times$ cm^{-1}). At 365 nm (Hg), ϵ = 3.5 (1 $\times$ mmol^{-1} $\times$ cm^{-1}).

Therefore grams of glucose per liter of sample solution equals 5.441 $\times$ ΔA (glucose)/ϵ, and grams of fructose per liter of sample solution equals 5.477 $\times$ ΔA (fructose)/ϵ. If the sample was diluted, multiply by the dilution factor.

In a collaborative test with 18 laboratories the results of three were disregarded as outliers [by Dixon's (45) method]. The repeatability and reproducibility were larger than desired (even without the outlier data). The authors suggest this is due to lack of experience with pipeting small volumes. They recommend glass Mohr and glass serological pipets ≥ 1 mL and microcapillary pipets for volume < 0.100 mL. Mechanical pipets with either disposable plastic tips (i.e., Eppendorf) or with positive displacement capillaries may also be used if care is taken to calibrate them according to the manufacturers' recommendations.

As Hurst et al. (46) state, "Sugar analysis by HPLC is a technique whose time has come." They used a Waters Associates' Model ALC/6PC 201 which contained a Model R401 differential refractometer and an M6000A solvent delivery system. The injection system is a Waters Intelligent Sample Processor (WISP) Model 71 with a capacity of 43 sample vials. Each vial can be individually programmed if desired. The recorder has a variable-input, variable-speed Sargent Welch Model XKR. The Shimadzu E1A data processor is coupled with the HPLC and autoinjector. The carbohydrate separation is on a μ Bondapack/carbohydrate column (4 mm $\times$ 30 cm, from Waters Associates). Acetonitrile and water (80/20 v/v) is the mobile phase. A guard column is used. Recoveries for glucose, fructose, sucrose, maltose, and lactose were 93.3–102.5%. Each sample took 15 min.

To determine sucrose, the analysate must first be hydrolyzed to glucose and fructose, then determined as above. To do this, follow the procedure given below.

Procedure

Dissolve 6.9 g of citric acid and 9.1 g of trisodium citrate · 2H$_2$O in 150 mL of double-distilled water; adjust pH to 4.6 with 0.2 N sodium hydroxide and bring to 200-mL volume with water. Dissolve 10 mg β-fructosidase in 2 mL of double-distilled water.

Set up three cuvets; one for the blank, one for glucose, and one for glucose plus sucrose. To the blank and to the glucose plus sucrose cuvets, add 0.20 mL of citrate buffer and 0.02 mL of β-fructosidase. To the glucose and glucose plus sucrose cuvettes, add 0.10 mL of sample. Keep solutions at 20–25°C for 15 min; then proceed as with previous procedure, but do not add any further samples. Add 1.70, 1.82, and 1.60 mL of water to the blank, glucose, and glucose plus sucrose cuvets, respectively, to bring each to 1.92 mL. Then add the same amounts of other buffers and reagents as given for the glucose and fructose determination with the same time and temperature restrictions.

Calculate the sucrose by determining $\Delta E_{sucrose} = \Delta E_{glucose + sucrose} - \Delta E_{glucose}$.

$$C \text{ (g/100 mL)} = 1.075 \left(\frac{\Delta E_{sucrose}}{63} \right) (F)$$

The total amount of glucose and sucrose in the cuvette should be 5–150 mg.

Glucose can be determined enzymatically by another method (47). It involves the oxidation of β-D-glucose to gluconolactone by β-D-NAD-oxidoreductase. The NADH formed is measured. Standard clinical test sets are available with full directions.

Pocock et al. (48) compared the Lane–Eynon, Rebelein, and enzymatic procedures for reducing sugars in wines. The Lane–Eynon and Rebelein approaches were satisfactory, but the enzymatic approach gave low results. (No explanation is offered for the failure of the enzymatic procedure in their tests.) However, in a comparison of Rebelein's titrimetric procedure with a continuous colorimetric enzymatic method, comparable results were obtained with both red and white wines (49).

Other Methods

Thin-layer or paper chromatography offers a quick and easy method of detection of trace amounts of sugars (50). Besides the usual monosaccharides, a mannoheptulose, a sedoheptulose, and four other heptuloses were found. The identification of the first two agreed with the previous findings (51) that also identified altroheptulose. Thin-layer chromatography was used to separate sucrose from glucose and fructose (52). This has some application in detection of sucrose additions to wine. Chebregzabher et al. (53) review thin layer chromatography of sugars. High-pressure liquid chromatography also can be used to separate sucrose, glucose, and fructose, as well as other sugars (54). High-pressure liquid chromatography has been used to rapidly (25 min) determine the quantity of sucrose, fructose, and glucose, as well as glycerol and ethanol, in wine samples with no pretreatment (55).

Bertrand et al. (56) found sucrose, lactose, trehalose, isomaltose, and questionable gentiobiose and melibiose in wines. Trehalose was found in wines from in amounts ranging from 0 to 611 mg/L. At medium (25°C) fermentation temperatures, less trehalose was formed than at low (15°C) or high (35°C) fermentation temperatures. Average values for seven white wines were 143 mg/L; for 29 red wines, 78 mg/L; and for eight wines from grapes with *Botrytis cinerea*, 240 mg/L. The other disaccharides rarely exceed 50 mg/L and usually are less than 5 mg/L. Lactose, maltose, and sucrose from about 10-50 mg/L were identified in two wines using paper chromatography (57).

Using similar isolation techniques and detection by gas-liquid chromatography of trimethylsilyl derivatives, the presence of a large number of sugars was determined in a Hungarian dry wine and in three sweet table wines (58). The latter wines should be rich in sugar. Table 10 lists sugars for the four wines and the ranges found. It is noteworthy that no trehalose was found. Also no sucrose was detected.

For 63 wines of nine different varieties and eight vintages for five residual sugars, Table 11 (59) summarizes the results. The red wines were consistently higher in these residual sugars than were the white wines. The author used isolation by ion exchange, then thin-layer separation and densitometer quantification. The results are in good agreement with earlier results (57).

Table 10. Sugars Found in a Hungarian Dry Table Wine and Three Sweet Table Wines[a]

Sugar	Range (mg/L)
Fructose	930–265,000
Glucose	560–250,000
Galactose	80–2,490
Mannose	20–370
Cellobiose	20–70
Melibiose	Trace–10
Raffinose	0–10
Sucrose	0–0
Arabinose	210–2,420
Rhamnose	110–1,210
Xylose	110–1,460
Ribose	120–620
Fucose	20–90

[a] The dry wine has the lowest values. An essence type was the sweetest and in general had the greatest amount of sugars.

Table 11. Residual Sugars Measured in Italian Wines

Statistic	Sugar (mg/L)				
	Arabinose	Xylose	Ribose	Galactose	Rhamnose
Number of samples	63	63	63	63	22[a]
Range	10–134	6–24	Trace–31	Trace–119	2–15
Average	62.3	15.6	18.2	39.8	8.5

From Reference 59.
[a] Does not include 41 trace amounts.

REFERENCES

1. F. Bandion, *Mitt. Klosterneuberg* **29**, 74–80 (1979).
2. H. W. Berg and C. S. Ough, *Am. J. Enol. Vitic.* **28**, 235–238 (1977).
3. A. J. Winkler, J. A. Cook, M. Kleiwer, and L. A. Lider, *General Viticulture*, 2nd ed. Univ. of California Press, Berkeley, 1974.
4. M. A. Amerine, H. W. Berg, R. E. Kunkee, C. S. Ough, V. L. Singleton, and A. D. Webb, *The Technology of Wine Making*, 4th ed. Avi Publishing Co., Westport, CT, 1979.
5. *Official Methods of Analysis*, 14th ed. Association of Official Analytical Chemists, Arlington, VA, 1984, pp. 220–230.
6. H. W. van Gend and P. P. van Leyenhorst, *Z. Lebensm.-Unters. -Forsch.* **163**, 8–10 (1977).
7. L. Jakob and W. Schrodt, *Wein-Wiss.* **28**, 169–180 (1973).
8. G. M. Cooke, *Am. J. Enol. Vitic.* **15**, 11–16 (1964).
9. U.S. Internal Revenue Service, *Wine*, Par 240 of Title 26, IRS Publ. No. 146. Code of Federal Regulations, U.S. Govt. Printing Office, Washington, DC, 1970.
10. M. A. Amerine and E. B. Roessler, *Hilgardia* **28**, 93–114 (1958).
11. J. Mourgues, *Ann. Technol. Agric.* **28**, 121–149 (1979).
12. D. Dubourdireu, D. Hadjinicoliaou, and P. Ribéreau-Gayon, *Connaiss. Vigne Vin* **15**, 29–40 (1981).
13. W. H. Forgarty and O. P. Ward, *Prog. Ind. Microbiol.* **13**, 59–119 (1974).
14. L. Usseglio-Tomasset, *Ann. Accad. Agric. Torino* **118**, 259–284 (1976).
15. J. Solms, W. Buchi, and H. Deuel, *Mitt. Geb. Lebensmittelunters. Hyg.* **43**, 303–307 (1952).
16. G. L. Robertson, *Am. J. Enol. Vitic.* **30**, 182–186 (1979).
17. P. Jaulmes, *Analyse des Vins*, 2nd ed. Coulet, Dubois et Poulain, Montpellier, 1951.
18. Z. Lipka and J. F. Schopfer, *Rev. Suisse Vitic., Arboric., Hortic.* **4**, 96–100 (1972).
19. K. Wucherpfennig, *Weinwirtschaft* **112**, 687–693 (1976).
20. K. Wagner and P. Kreutzer, *Weinwirtschaft* **112**, 842–844 (1976).
21. *Recueil des Methods Internationales d'Analyse des Vins*, 4th ed. Office International de la Vigne et du Vin, Paris, 1978.
22. E. Gilbert, *Weinwirtschaft* **112**, 118–127 (1976).

23. F. Alvarez-Nazario, *J. Assoc. Off. Anal. Chem.* **65**, 765–767 (1982).
24. J. R. Ruppert, *J. Assoc. Off. Anal. Chem.* **69**, 709–722 (1986).
25. R. Franck and C. Junge, *Weinanalytik. Untersuchungen von Wein und ähnlichen alkoholisches Erzeugnissen sowie von Fruchtsäften.* Carl Heymanns Verlag, Cologne, 1983.
26. C. Junge, *J. Assoc. Off. Anal. Chem.* **68**, 141–143 (1985).
27. J. M. Vahl, *Am. J. Enol. Vitic.* **20**, 262–263 (1979).
28. A. Lay, *Weinberg Keller* **20**, 29–36 (1973).
29. H. Rebelein, *Allg. Dtsch. Weinfachztg.* **110**, 590–594 (1971).
30. H. Rebelein, *Chem., Mikrobiol., Technol. Lebensm.* **2**, 112–121 (1973).
31. D. G. Porter and R. Sawyer, *Analyst (London)* **97**, 569–575 (1972).
32. M. Kallay, *Elelmiszervizsgalati Kozl.* **22**, 43–47 (1976).
33. R. Mauer, *Weinberg Keller* **18**, 39–47 (1971).
34. A. Rapp and A. Ziedler, *Dtsch. Lebensm.-Rundsch.* **79**, 393–398 (1979).
35. W. Flak, *Mitt. Klosterneuburg* **31**, 204–208 (1981).
36. B. W. Li and P. J. Schuhmann, *J. Food Sci.* **48**, 633–635, 653 (1983).
37. K. Yokotsuka, N. Nozaki, and T. Matsudo, *J. Inst. Enol. Vitic. Yamanashi Univ.* **19**, 39–56 (1984).
38. P. Pfeiffer and F. Radler, *Z. Lebensm.-Unters. -Forsch.* **181**, 24–27 (1985).
39. J. F. Schopfer and R. Regamey, *Rev. Suisse Vitic. Arboric.* **3**, 107–110 (1971).
40. R. E. Wrolstad and R. S. Schallenberger, *J. Assoc. Off. Anal. Chem.* **181**, 24–27 (1981).
41. F. Drawert and G. Kupfer, *Z. Anal. Chem.* **211**, 89–94 (1965).
42. U. Bergmeyer, J. Bergmeyer, and M. Grass, *Methods of Enzymatic Analysis*, 10 vols. Verlag Chemie, Weinheim, 1983.
43. A. Lay, *Weinberg Keller* **24**, 369–374 (1977).
44. G. Henniger and L. Mascaro, *J. Assoc. Off. Anal. Chem.* **68**, 1020–1024 (1985).
45. W. J. Dixon and F. J. Massey, *Introduction to Statistical Analysis*, 2nd ed. McGraw-Hill, New York, 1957.
46. W. J. Hurst, R. A. Martin, Jr., and B. L. Zoumans, *J. Food Sci.* **44**, 892–895 (1979).
47. A. Temperli, H. Schärer, and U. Kunsch, *Fluess. Obst* **43**, 257–258 (1976).
48. K. R. Pocock, A. V. Hood, B. Wilson, and B. C. Rankine, *Winemaker* **172**, 52, 54 (1978).
49. H. R. Bruner and H. Tanner, *Schweiz. Z. Obst- Weinbau* **121**, 9–13 (1985).
50. M. Castino, *Vini Ital.* **14**(76), 69–73, 76–80 (1972).
51. P. Esau and M. A. Amerine, *Am. J. Enol. Vitic.* **15**, 187–189 (1964).
52. M. Matta, G. Gaetano, and M. R. Simone, *Riv. Vitic. Enol.* **29**, 148–158 (1976).
53. M. Chebregzabher, R. Rufini, B. Monaldi, and M. Lato, *J. Chromatogr.* **127**, 133–162 (1976).
54. J. D. Palmer and W. B. Brandes, *J. Agric. Food Chem.* **22**, 709–712 (1974).
55. A. Rapp, O. Bachmann, and A. Ziegler, *Dtsch. Lebensm.-Rundsch.* **71**, 345–348 (1975).
56. A. Bertrand, M. O. Dubernet, and P. Ribéreau-Gayon, *C. R. Hebd. Seances Acad. Sci., Ser. D* **280**, 1907–1910 (1975).
57. P. Esau and M. A. Amerine, *Am. J. Enol. Vitic.* **17**, 265–267 (1967).
58. F. Drawert, G. Leupold, and V. Lessing, *Z. Lebensm.-Unters. -Forsch.* **162**, 407–414 (1976).
59. L. Usseglio-Tomasset and G. Amerio, *Vini Ital.* **20**, 27–33 (1978).

Two
ACIDITY AND INDIVIDUAL ACIDS

Grapes contain appreciable amounts of various organic acids. During ripening there is a marked decrease in the concentration of several of the acids; the titratable acidity decreases and the pH increases. Grape juice and musts are thus dilute acid solutions, mainly of tartaric, malic, and citric acids. Without the acids, commercial grape juices and wines would be impossible: They would taste flat, the colors would be abnormal, and spoilage would occur easily.

Wines contain the acids of musts and a number of acids produced during and after alcoholic fermentation: acetic, propionic, pyruvic, lactic, succinic, glycolic, galacturonic, glucuronic, gluconic, mucic, oxalic, fumaric, and so on. Without the acids, fermentation would produce undesirable by-products, and the resulting wines would spoil during and after fermentation. The wines would also have abnormal color and flavors and would taste flat and unpalatable. The sour taste of the acids in wines is modified by ethanol, sugar, and various cations. The degree of sourness is also related to the total titratable acidity, pH, relative amount of dissociated and undissociated acids, buffer capacity, and the relative amount of each of the different acids. The acids are all more or less sour, and some have characteristic flavors. The fatty acids, especially, have pronounced odors that are usually undesirable (1). For a general discussion of the organic acids of musts and wines and their microbial metabolism, see Radler (2).

It is essential for the rational harvesting of grapes to follow the changes in acidity and pH; this allows the viticulturist to harvest at the optimum level. After fermentation and during aging it is no less important to follow the changes in certain of the fixed acids and in the level of volatile acids. During the malolactic fermentation, the malic and lactic acid contents must be followed. During stabilization of both commercial grape juice and wines, the changes in tartrates and pH are of critical importance.

TOTAL (TITRATABLE) ACIDITY

Commercial standards dictate a grape juice acidity of about 0.6–0.9% (as grams of tartaric acid per 100 mL of juice). Dry table wines have titratable acidities in the same range. Sweet and dessert wines usually are in the range of 0.4–0.65%. However, under cool climatic conditions, table wines may be higher. Also botrytized sweet wines frequently exceed 1.0%.

Wine producers need to know the titratable acidity of musts to determine the proper amount of sulfur dioxide to add and also to decide on whether correction of the acidity is necessary. In the eastern United States and elsewhere except in California, musts may be diluted with water to reduce excess acidity, provided it is not reduced below 0.5%. The titratable acidity is used during processing and finishing operations to standardize the wines and to follow undesirable changes due to bacteria, yeasts, and so on.

The acids present in musts and wines (tartaric, malic, lactic, acetic, etc.) are relatively weak organic acids. Thus when musts or wines are titrated with a strong base, the true end point will be greater than pH 7.0, usually between 7.8 and 8.3. Therefore, the definition of the OIV (3) based on titration to pH 7.0 is theoretically incorrect. In the following procedure (4), which is the accepted procedure of the American Society of Enologists (5) and the AOAC (6), the titration end point is at pH 8.2. Results obtained by titrating to pH 7.0 will, of course, be lower than those obtained by titrating to pH 8.2.

Procedure

Place 200 mL of boiling water in a 500-mL wide-mouth Erlenmeyer flask; add 1 mL of a 1% phenolphthalein indicator solution (in 70% ethanol) and titrate with 0.1 N sodium hydroxide solution to a faint but definite pink color. Pipet 5 mL of must into the flask, and titrate to the same color. If the end point is determined electrometrically, titrate to a pH of 8.2. In this case use a 250-mL beaker, de-gas the wine, add 100 mL of water, and cool the solution to room temperature before titration.

If the titratable acidity is expressed as tartaric acid, calculate as follows:

$$\text{Tartaric acid (g/100 mL)} = \frac{(V)(N)(75)(100)}{(1000)(v)}$$

where V = volume of sodium hydroxide solution used for titration (mL)
N = normality of sodium hydroxide solution
v = sample volume (mL)

Automatic titrators are available to add the base, sense the end point, and print out the calculated acidity values.

The titratable acidity is sometimes expressed as other acids; for example, in France it is expressed as sulfuric acid. Values can be interconverted from one acid to another by multiplying one value with the following factors:

Expressed as	Tartaric	Malic	Citric	Lactic	Sulfuric	Acetic
Tartaric	1.000	0.893	0.853	1.200	0.653	0.800
Malic	1.119	1.000	0.955	1.343	0.731	0.896
Citric	1.172	1.047	1.000	1.406	0.766	0.938
Lactic	0.833	0.744	0.711	1.000	0.544	0.667
Sulfuric	1.531	1.367	1.306	1.837	1.000	1.225
Acetic	1.250	1.117	1.067	1.500	0.817	1.000

In Germany the titratable acidity is sometimes expressed as the volume of 0.1 N base needed to titrate 100 mL of the sample. This value can be calculated from the acidity expressed as grams of acid per 100 mL by dividing the latter by 0.1 (meq weight). The meq wt values of the acids are: tartaric, 0.075; malic, 0.067; citric, 0.064; lactic, 0.090; sulfuric, 0.049; and acetic, 0.060.

This determination is subject to the interference of dissolved carbon dioxide, sulfur dioxide, sorbic acid, or nonacid compounds such as esters and sugars. The carbon dioxide error is minimized by adding the must or wine to boiling water as in the procedure above, or by agitation and application of a vacuum in other procedures. The interference of amphoteric compounds between pH 8 and 8.4 is small. The error due to low levels of sulfur dioxide or sorbic acid is also small, and normally it is not critical.

pH

In biological systems the pH is often of greater significance than the total acidity. The determination of the pH of wines is far more important to the wine maker than is commonly realized. It is particularly important in its effect on microorganisms, on the tint (or hue) of color, on the taste, on the oxidation-reduction potential, on the ratio of free to bound sulfur dioxide, on susceptibility to iron phosphate cloudiness, and so on. Of these, the resistance to disease is perhaps the most important. Table wines usually should have a pH not exceeding 3.6, and dessert wines more than 3.8. The pH of musts for table wines should fall in the range 3.1–3.6, and the permissible range for dessert wines is from 3.4 to about 3.8. Late in the season some California musts of low-acid table grapes have a pH higher than 4.0. Grape harvesting should take into account these limits. The pH of the new wine, free of carbon dioxide, is usually higher than that of the must from which it was produced.

There is no apparent direct or predictable relationship between the pH and the total titratable acidity. There is an empirical relationship between the pH and the ratio of potassium bitartrate to total tartaric acid (7). This indicates that the pH is primarily dependent on the degree of neutralization of the tartaric acid. Insufficient information is available on the factors influencing the migration of potassium into the fruit during maturation. This varies from variety to variety and, to a lesser extent, between seasons (8). There is no comparable correlation of pH and malic acid content (9).

The pH of a must or wine can be measured by a pH meter, that is, potentiometrically, using an ion-selective electrode. For most winery needs, a determination to ±0.03 pH unit is satisfactory. Ordinary pH meters are adequate for this degree of accuracy, and the directions for the particular instrument used should be followed carefully. The pH meter should be calibrated with a saturated pure potassium acid tartrate solution; its pH is 3.55 at 30°C and is 3.56 at 25°C and at 20°C.

VOLATILE ACIDITY

The volatile acidity consists of the fatty acids found in wine (formic, acetic, butyric, etc.). It should not include the steam-distillable lactic, succinic, or sorbic acids, nor carbon dioxide or sulfurous acid. Accurate determination of the volatile acidity of wines is one of the important determinations normally conducted by wine makers. This is because both federal and California laws, as well as those of several other states and most foreign countries, have specific regulations concerning the maximum amount of volatile acidity permitted in wines that are offered for sale or that are imported. Table 12 summarizes present maximum limits.

Table 12. Maximum Permitted Volatile Acidity[a] in Wines

Wine type	United States		France[c]	Germany	Italy[d]
	Federal[b]	California[b]			
Red table wines	0.140	0.120	0.110	0.160	
White table wines	0.120	0.110	0.110	0.120	
All other wines	0.120	0.110	0.110	0.250	

[a] Expressed as grams of acetic acid per 100 mL of wine.
[b] Exclusive of sulfur dioxide.
[c] This is the limit for wine merchants; for certain wines a limit as high as 0.120 may be established.
[d] For export the limit is 0.090. (New regulations for internal sale are 0.01 g of acetic acid per 100 mL of wine per % alcohol v/v.)

High acetic acid content seems to be less objectionable in old wines than in young ones. These limits have been made because wines of high volatile acidity indicate that spoilage organisms, particularly *Acetobacter*, are present and that the wine may eventually turn to vinegar. Small, but measurable, quantities of acetic acid are formed during normal, bacteria-free, alcoholic fermentation, usually not exceeding 0.030 g/100 mL. The bacteria involved in the malolactic fermentation may also produce small amounts of acetic acid, presumably by decomposition of citric acid.

Actually, spoilage becomes sensorily noticeable below the legal limits, usually at about 0.060–0.090 g/100 mL. In fact, it is not the acetic acid alone that is primarily responsible for the spoiled odor but rather the high ethyl acetate and acetic acid content and small amounts of other spoilage products. Since the two major compounds are normally produced simultaneously and proportionately, the amount of acetic acid usually is a good measure of the amount of ethyl acetate present, hence of the degree of spoilage of the wine. Exceptions do occur when primarily acetic acid or ethyl acetate is produced. The esterification equilibrium is only slowly achieved in wines.

Wines made from high-sugar musts are naturally higher in volatile acidity than those of normal sugar content. They may also undergo excessive malolactic and acetic fermentation during fermentation and storage. Volatile acidities as high as 0.123 g of acetic acid per 100 mL were considered tolerable, but not desirable, in sweet Bordeaux white table wines (10).

The wine maker needs to know the amount of volatile acidity present to meet the legal limits, to follow the development of volatile acidity during storage, and as a measure of spoilage.

The main spoilage acid present is acetic, although small quantities of propionic, butyric, and valeric acids have been reported. These acids can be distilled in a current of steam, provided sufficient steam is used. Lactic acid is not, properly, considered to be a spoilage acid; however, it is slightly volatile with steam, depending on the type of apparatus used, the quantity of steam passed through the wine, and the percentage of lactic acid in the wine, as well as other factors. The definition of the volatile acidity is somewhat difficult because varying proportions of lactic acid will be present in the distillate. Fortunately, in the usual Sellier tube the amount of lactic acid distilled over is quite small, about 5%. Concordant results can be obtained only if such variables as rate of distillation, length of distillation, and pH of the wine, as well as volume of the wine are kept constant from one wine to the next. A rectifying column is employed (11) to reduce this error. The procedures of the OIV (3) also use rectifying columns.

Another difficulty is the presence of sulfurous acid, which partially distills over with the acetic acid. In wines high in sulfur dioxide, this correction is important. As indicated in Table 12 the federal and California limits are "exclusive of sulfur dioxide."

A final source of error is the carbon dioxide, either that present in the original wine or that which dissolves in the distillate during or after distillation. Carbon dioxide present in the wine can be partially removed by shaking the wine in a flask while under a vacuum or by bringing it to a boil under an air condenser and cooling immediately. Another possibility is to boil the distillate of the determination for 30 sec.

The determination of volatile acidity is carried out in the apparatus shown in Figure 6, known as the Cash electric volatile acid assembly.

Procedure

Carefully pipet 10 mL of juice or wine into the funnel of the Cash still. Set the two-way stopcock to allow the sample to flow into the inner steam distillation tube. Rinse sample in with a few milliliters of water. Turn the stopcock 180° to allow access to the water reservoir in the still. Adjust water level in the still, turn on water to condenser, and turn on electric immersion heater. When water boils, turn stopcock 90° to seal off still and allow steam distillation to begin. Collect 100 mL of the distillate in a 250-mL Erlenmeyer wide-mouth flask. Bring the distillate to 100°C on a hot plate (do not boil), swirl the flask several times to remove any carbon dioxide, add a few drops of 1% phenolphthalein indicator (1 g dissolved in 100 mL of 80% v/v ethanol), and titrate to a faint pink end point

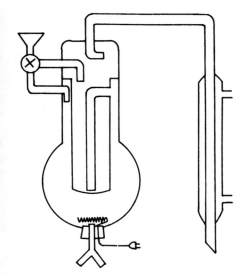

Figure 6. Cash steam distillation apparatus.

using 0.1 N NaOH. Express the volatile acidity as acetic acid:

$$\text{Acetic acid (g/100 mL)} = A = \frac{(V_1)(N_1)(60)(100)}{(1000)(v)}$$

where V_1 = volume of sodium hydroxide solution used for titration (mL)
N_1 = normality of sodium hydroxide solution
v = sample volume (mL)

In France both the volatile and the total acidity are expressed as grams of sulfuric acid per liter. To convert this value to grams of acetic acid per 100 mL, multiply by 0.1225.

Normally no correction is made in the volatile acidity for the amount of sulfur dioxide distilled over. In the case, however, where the volatile acidity approaches the legal limit, such a correction may be important.

Procedure

Immediately after the titration above is completed, cool the distillate, add 1 mL of dilute sulfuric acid, 2-3 mL of freshly prepared starch solution, and a crystal of potassium iodide. Rapidly titrate to a faint blue end point using a 0.002 N iodine solution. The amount of iodine used is equivalent to the *free* sulfurous acid present in the distillate and is to be expressed as acetic acid (g/100 mL):

$$B = \frac{(V_2)(N_2)(32)(100)(60)(2)}{(1000)(v)(64)}$$

where V_2 = volume of iodine solution used for titration (mL)
N_2 = normality of iodine solution
v = sample volume (mL)

Add 10 mL of N sodium hydroxide solution to the same flask, boil 2-3 min, cool, acidify with 5 mL of N sulfuric acid, add 2 or 3 mL of starch, and titrate to a faint blue end point with the 0.002 N iodine solution. The amount of iodine used corresponds to the bound sulfurous acid present. Express the result as acetic acid (g/100 mL):

$$C = \frac{(V_3)(N_2)(32)(100)(60)(2)}{(1000)(v)(64)}$$

where V_3 = volume of iodine solution used for titration (mL)
N_2 = normality of the iodine solution
v = sample volume (mL)

Calculate volatile acidity free of sulfurous acid, expressed as acetic acid:

$$\text{Volatile acidity (g/100 mL)} = A - (B + C)$$

The sulfur dioxide error can be avoided by treating 10 mL of wine with 1 mL of hydrogen peroxide and 1 g of sodium arsenite before distillation.

The back-titration can be avoided by complexing sulfur dioxide with 0.5–1.0 mL of a 1% solution of red mercuric oxide directly into the Sellier or Cash distillation tube before distillation. Dissolve 1 g of red mercuric oxide in 100 mL of 10% sulfuric acid (12). The 1 mL of mercuric oxide solution will bind about 250 mg/L of sulfur dioxide. Unfortunately, elemental mercury accumulates during the distillation and may be a health hazard.

The useful suggestion has been made (13) to treat the wine with 0.5 mL of 2 N sodium hydroxide and, after 5 min, with 0.5 mL of hydrochloric acid–ferric chloride solution [prepared by adding 20 g of ferric chloride ($FeCl_3 \cdot H_2O$) to 400 mL of hydrochloric acid ($d = 1.19$) and making to 1 L with distilled water]. This results in the oxidation of the sulfur dioxide before distillation.

It has been demonstrated (14) that shaking under a vacuum or heating is insufficient to remove carbon dioxide from bottle-fermented sparkling wines. A double-titration procedure of the distillate is also recommended (14): first to pH 6.33 using bromocresol purple and then to pH 8.4 with phenolphthalein. At pH 6.33, only half the first hydrogen of carbonic acid is titrated, but already 97% of the volatile organic (and sulfurous) acids are neutralized.

Procedure

Cool the distillate, add bromocresol purple indicator solution, and titrate to the end point. Now add phenolphthalein indicator solution and titrate to a faint pink end point. Use 0.1 N sodium hydroxide solution as the titrant. Express the acidity as acetic acid:

$$\text{Acetic acid (g/mL)} = \frac{(V_1 - V_2)(N)(60)(100)}{(1000)(v)(0.94)}$$

where V_1 = volume of sodium hydroxide solution required to titrate to the bromocresol purple end point (mL)
V_2 = total volume of sodium hydroxide required to titrate to the phenolphthalein end point (mL)
N = normality of the sodium hydroxide solution
v = sample volume (mL)

Appreciably lower results for the volatile acidity of sparkling wines are obtained using this additional correction.

An enzymatic procedure for this determination that has been developed (15) indicates that acetate constitutes only 24–83% of the volatile acidity. A proposed rapid enzymatic procedure for acetic acid uses four coupled enzyme-mediated reactions. The NADH produced is measured at 340 nm. It is rapid and specific and is recommended when many samples are to be analyzed, since 200–300 samples can be handled per day.

The final version of this enzyme system (16) requires about 30 min; however, it is acetate specific. The reactions are:

$$\text{Acetate} + \text{ATP} \xrightarrow{\text{Acetate kinase}} \text{Acetylphosphate} + \text{ADP}$$

$$\text{Acetylphosphate} + \text{CoA} \xrightarrow{\text{Phosphotranseacetylase}} \text{Acetyl CoA} + P_i$$

$$\text{Phosphoenolpyruvate} + \text{ADP} \xrightarrow{\text{Pyruvate kinase}} \text{Pyruvate} + \text{ATP}$$

$$\text{Pyruvate} + \text{NADH} + H^+ \xrightarrow{\text{Lactate dehydrogenase}} \text{Lactate} + \text{NAD}^+ + H_2O$$

Polyvinylpyrrolidone is added to protect the enzymes from the phenolic substances in the wine.

A continuous flow technique has been offered (17). It is enzymatic but develops a color in the visible range, 570 nm.

Recently, sorbic acid has been used as an antiseptic agent by a number of wineries. It is almost completely volatile with steam. If the level of sorbic acid in the wine is 0.02 g/100 mL, as much as 0.011% may appear in the distillate (calculated as acetic). The official French (18) and OIV procedure calls for direct colorimetric determination of the sorbic acid on a small aliquot of the distillate. Since 1 g of sorbic acid equals 8.92 meq or 0.536 g as acetic acid, one can easily calculate the amount of sorbic acid in the distillate.

The Technicon Autoanalyzer-2R has been used to determine volatile acidity (19). The wine is automatically sampled and mixed with 5% tartaric acid; 0.1 volume of dilute hydrogen peroxide is added to oxidize sulfur dioxide, and the sample is distilled at reduced pressure at 95°C with nitrogen gas, using a microrectifying column. The color of the distillate was determined with bromophenol blue at 450 nm with continuous automatic measurement of the optical density. About 30 samples were analyzed per hour.

Acetic, butyric, and propionic acids may be quantitatively determined by gas–liquid chromatography (20). The limit of detection is 0.5 mg/L using the neutralized distillate from the volatile acidity determination (21). In New Zealand, wines having more than 5 mg/L of butyric acid were considered to be objectionable (21).

Acetic acid was determined by direct injection of wines onto packed columns (22, 23). Carbopack C, modified by the addition of 0.3% Carbowax 20M and

0.1% phosphoric acid as a liquid phase, has been reported to give excellent results (24). Propionic acid was the preferred standard. The carrier gas was saturated with formic acid to keep the acid in the nonionized form.

Shinohara (25) measured hexanoic, octanoic, and decanoic acids by extraction, concentration, and then GC analysis using internal standard. The levels found for the three acids were 0.9–6.1, 0.7–7.9, and 0.0–3.0 mg/L, respectively. Totals exceeding 20 mg/L gave an undesirable odor to the wine. Briefly the wine has heptanoic and pelargonic acids added as internal standards. The wine is then neutralized and concentrated on a rotary evaporator. It is acidified with sulfuric acid, and the fatty acids are extracted with an ethyl acetate-n-pentane (2:1 v/v) mixture. The extract is further concentrated and injected onto a gas chromatographic column for quantification.

FIXED ACIDITY

A sound definition of fixed acidity depends on clearly specified and compatible definitions for the total and volatile acidities. The fixed acidity is simply the total (titratable) acidity less the volatile acidity or the nonvolatile acids of the wine. These include malic, tartaric, citric, lactic, succinic, and inorganic acids.

Normally the fixed acidity is calculated directly from the total (titratable) and volatile acidities as determined separately. This is not strictly correct because in the usual method for determining volatile acidity, 100% of the volatile acidity is not determined. Furthermore, the percentage of volatile acidity not determined varies from one wine to another.

The determination of fixed acidity is particularly important in California wines, since there are minimum limits of 0.25 g/100 mL calculated as tartaric acid for dessert wines and 0.3 and 0.4 g/100 mL for white and red table wines.

If the fixed acidity is to be calculated, the volatile acidity cannot be subtracted from the titratable acidity directly, because these qualities are ordinarily reported on a different basis: the volatile acidity as acetic acid and the titratable acidity as tartaric acid. Therefore, the volatile acidity must be converted first to the latter, by multiplying its value, originally expressed as acetic acid, by 1.25, the ratio of the two equivalent weights (75/60).

The direct determination of the fixed acidity can be utilized when the total and volatile acidities have not been, or are not to be, determined. This is done by evaporating off the volatile acidity and titrating the residual acids. The primary problem is that only by steam distillation can a high percentage of the volatile acidity be removed. To increase this percentage, several evaporations may be made. Evaporating 25 mL of wine down to 5 mL, adding 25 mL of hot water, reevaporating, and repeating two or three times will remove most of the volatile acidity. The final 5 mL is diluted to 100 mL with water, and an aliquot is used for determining the nonvolatile or fixed acidity. Procedures in which the wine is evaporated to dryness in the presence of sodium chloride are not recommended because of the decomposition of acid salts and sugars and other changes that occur at temperatures above 100°C.

Fixed acidity is required in certain enological formulas used in France and other European countries to detect addition of water in wines. In normal wines the fixed acidity is closely related to the original total acidity of the must, although certain changes occur during and after fermentation, including precipitation of potassium acid tartrate, utilization of malic and citric acids by yeasts or bacteria, and formation of acetic, lactic, and succinic acids. However, after fermentation the changes in fixed acidity are useful in distinguishing between the activity of acid-destroying microorganisms and volatile acid-forming organisms. In a few cases, acid formation may occur at the expense of glycerol. The new methods of analyses (3, 18, 26) may result in a lower fixed acidity. Minimum levels of free tartaric acid have been proposed: 0.15 g/100 mL for white wines and 0.10 for red ones (26).

INDIVIDUAL NONVOLATILE ACIDS

The individual nonvolatile acids can be separated by paper or thin-layer chromatography. The paper chromatographic procedure (27) used n-butanol as the moving phase while the paper is saturated with water vapor. Formic acid is added to suppress the ionization of the acids, since ionization would prevent the separation of the acids. The spots are made visible by spraying with bromocresol green solution.

The following simplified procedure (28) was designed primarily for detection of the absence or presence of malic acid (i.e., to follow the malolactic fermentation). The procedure utilizes the same solvents and indicator as used in the previous technique.

Procedure

Mix, in a separatory funnel, 100 mL of reagent grade n-butanol, 100 mL of water, 10.7 mL of reagent grade formic acid, and 15 mL of a 1% aqueous bromocresol green solution. Shake thoroughly and allow it to separate into two layers, then discard the lower, aqueous layer.

Use 20 × 30-cm rectangles of Whatman® No. 1 or Schleicher and Schüll® No. 2043-B chromatographic paper, and spot on the same place five times about 10-μL aliquots of the wine sample, waiting between spottings until each spot is dry. The individual spots can be at 2.5-cm intervals.

Place about 70 mL of the freshly prepared solvent mixture in the bottom of a wide-mouth chromatographic jar, form a cylinder of the paper, and place it in the jar with the spotted edge at the bottom. Close and wait until the solvent reaches nearly the upper edge of the paper (about 6 hr). Remove the paper and dry in a well-ventilated area. Identify the yellow spots by their R_f values:

Tartaric acid	0.28
Citric acid	0.45
Malic acid	0.51
Ethyl acid tartrate	0.59
Lactic acid	0.78
Succinic acid	0.78
Ethyl acid malate	0.80
Bromocresol green front	0.87

Use 0.3% standard solutions of the acid as control solutions.

The thin-layer chromatographic procedures are more sensitive and they require less sample, about 1 μL (29, 30). A 4:2:5 butanol-formic acid-water mixture is used as the moving phase, and the spots are fast developed by spraying with a 0.02% solution of Acridin in 1:1 ethanol, then observed under UV light at 254 nm. The R_f values reported (30) are: tartaric acid, 0.13; citric acid, 0.23; malic acid, 0.35; lactic acid, 0.76; and succinic acid, 0.86. A variation (31) similar to the paper method, but using thin layer, takes 45 min to 1 hr. A faster thin-layer method (32) that takes only 10 min is available.

Gas chromatographic determination of nonvolatile organic acids requires that the acids be derivatized in some manner to make them volatile. This has been accomplished usually by esterfication or silylation. Separation from other components has been done prior to derivatization. This is usually done by lead precipitation or by collection on anion resins (33-36).

The method of De Smedt et al. (37), as well as a similar one (38), seems to have the best features of the other methods but does not require the separation step. It can also be used to quantitate sugars.

Procedure

Dilute sample as necessary (depending on sugar concentrations) and add an equal volume of internal standard (1.0 g/L vanillic acid, 0.5 g/L α-methyl-D-mannoside in 20% v/v ethanol). Transfer 50 μL to a 2-mL vial, make basic with one drop of aqueous ammonia, and evaporate to dryness under a nitrogen gas flow at 20-25°C. Treat residue with 50 μL of anhydrous pyridine, 100 μL of N,O-bis(trimethylsilyl)trifluoroacetamide (BSTFA) containing 1% trimethylchlorosilane (TMCS). Close with Teflon® seal and heat 1 hr at 80°C. Inject 1 μL into gas chromatograph. Use a 30-m, 0.25 mm i.d. methyl silicone-bonded capillary column. Use a 1:10 column split and program from 60°C to 280°C at 4°C/min. Keep the injector port and flame ionization detector at 290°C. Maintain carrier gas flow at 3 mL/min.

Make a series of standards covering the anticipated normal ranges, and treat as above. Calculate the response factors and dilution factors:

$$SC \ (mg/L) = \frac{SCPA}{ISPA} \times CRF \times DF$$

where SC = sample component concentration
SCPA = sample component peak area
ISPA = internal standard peak area (vanillic acid)
CRF = component response factor (mg/L)
DF = dilution factor

Clean the glass insert in the injector port if the ratio of the two internal standards vary more than 1.55. If this doesn't correct, remove a coil or two from the front of the column.

High-performance liquid chromatography (HPLC) has been used with C_{18} reverse-phase columns to separate and quantify the organic acids (39–42). Generally, problems result in interferences between the acids and the sugars. Precolumn derivatization has been suggested (43, 44). Several workers (45, 46) used HPLC with direct injection of juice or wine onto cation columns. The most effective HPLC separation and measurement has been achieved by (a) collection of the acids on an anion column, elution, and then separation and (b) by detection using a cation column (47, 48). The method (49) suggested is accurate.

Procedure

After treatment of the sample with NH_4OH to adjust pH to 8–9, place 5 mL onto anion-exchange column (7-mm i.d. × 8 cm packed with 2 g of Bio-Rex 5 200–400 mesh or equivalent). Wash with 20 mL of distilled water. Elute with 2 mL of 10% H_2SO_4 then 10 mL of water. Filter 10 mL of the eluate through 0.45 μm membrane and C_{18} Sep Pack® (Waters Associates). Inject 5–25 μL onto Aminex HPX-87H® (Bio-Rad Laboratories) cation column. Keep column temperature 65°C and use 0.05 N H_2SO_4 as the mobile phase. Detect with UV at 210 nm. Prepare acid standard curves for comparison.

Tartaric Acid

Since tartrate is relatively resistant to respiratory oxidation in grapes or to bacterial action in wines, it is usually the most significant part of the acid fraction of grapes and wines. Normally half or more of the total acidity of musts and wines is due to tartaric acid and its acid salts. The tartrate content, as tartaric acid, varies from 0.2 to 0.8%, the percentage depending on variety, season, and region. Tartaric acid is the strongest of the organic acids present in grapes; thus it buffers the pH to somewhat lower values than do the other organic acids.

This property is of importance to the tint of color of the wine, in the resistance of wines to disease, and in the acid taste of the wine. The determination of tartrates is also of interest in maturity and varietal studies in certain types of spoilage and in making an acid–base balance.

Metatartaric acid is sometimes used for stabilizing wines. It decomposes slowly to tartaric acid in storage. To include it in the tartrate determination, the OIV (3) recommends adding 0.4 mL of acetic acid to 10 mL of wine and heating at 100°C for 30 min.

The reference procedure of the OIV (3) precipitates the tartrate as calcium racemate. Since L-tartaric acid is difficult to obtain and a double precipitation is required, the procedure has not been popular in the United States, although it is probably the most accurate method (50, 51).

The official procedures of the AOAC (6) are based on precipitation as bitartrate. This method was first used by Pasteur and has been modified since by many researchers (52, 53). For routine purposes, the following procedure is adequate.

Procedure

Determine the titratable acidity (p. 51) and the ethanol content (Chapter 3) of the must or wine sample, and calculate from these values the volumes of 2.67 N potassium acetate solution, ethanol, and water to be added:

$$2.67\ N \text{ Potassium acetate solution (mL)} = A = \frac{(\text{Ac} - 0.6)}{0.4}$$

$$95\% \text{ Ethanol (mL)} = B = 26.5 - [(0.52)(E\%)]$$

$$\text{Water (mL)} = C = 30 - (A + B)$$

where Ac = titratable acidity (g/100 mL)
$E\%$ = ethanol content (vol %)

For musts $E\% = 0$, $B = 26.5$ mL. If Ac ≤ 0.6 g/100 mL, omit the potassium acetate addition.

Pipet 50 mL of must or wine into a 100-mL beaker and add 8 g of finely powdered potassium chloride. Then add the C mL of water, 1.3 mL of glacial acetic acid, A mL (if any) of 2.67 N potassium acetate solution, and B mL of 95% ethanol. Mix and place in the refrigerator for several hours. Meanwhile, place 30 g of filter paper in 1 L of water, add 50 mL of concentrated hydrochloric acid, and allow this to disintegrate. Wash with hot water until neutral. Suspend in 2 L of water. Coat a fritted-glass filter with this suspension, and filter the stored sample solution. Wash six times with about 3-mL portions of a washing solution not exceeding a total volume of 20 mL. Prepare the washing solution in a 500-

mL volumetric flask by dissolving 50 g of potassium chloride in 325 mL of water and diluting to volume with 95% ethanol.

Dissolve the tartrate crystals from the filter with hot water into the original beaker. Heat the contents of the beaker to boiling, add 1-2 drops of a phenolphthalein indicator solution, and titrate with 0.1 N sodium hydroxide solution to a distinct pink color. Express the tartrate content as tartaric acid.

$$\text{Tartrate as tartaric acid (g/100 mL)} = \frac{(V)(N)(150)(100)}{(1000)(v)}$$

where V = volume of sodium hydroxide solution used for titration (mL)
N = normality of sodium hydroxide solution
v = sample volume (mL)

The usual procedure of the OIV (3) is essentially the colorimetric metavanadate method (54), which has been modified (55) to improve its accuracy.

Procedure

Wash 20 mL of the strong anion-exchange resin Duolite A101D into a 500 × 12-mm Pyrex column fitted with a sintered-glass disk and a needle control valve. Charge the resin with 20 bed-volumes of 30% acetic acid, then wash the resin with 30 mL of 0.5% acetic acid and 50 mL of water. Use a flow rate of 2.5-5.0 mL/min. Add a 10-mL wine sample to the top of the resin column and carry it through the column with 25 mL of 0.5% acetic acid; then wash with 50 mL of water. Elute the adsorbed acids with 0.5 M sodium sulfate solution: Lactic acid will elute in the first 30 mL; tartaric acid will elute in the 46-75 mL fraction. Collect the samples in graduated test tubes.

Pipet a 25-mL aliquot of the fraction containing tartrates into a 50-mL volumetric flask. Add 0.5 mL of glacial acetic acid and 2 mL of a 3% sodium metavanadate solution, then dilute to volume with water. Mix and let stand for 80 min at room temperature, then measure the absorption in 1 cm matched cuvettes at 480 nm. Use a blank consisting of a mixture of 25 mL of 0.5 M sodium sulfate solution, 0.5 mL of glacial acetic acid, and 2 mL of 3% sodium metavanadate solution, diluted to 50 mL with water. It is essential that the blank be prepared at the same time as the sample. Compare the absorbance with a standard curve prepared using tartaric acid solutions carried through the procedure.

A simple, rapid colorimetric procedure for tartaric acid in musts of wines has been developed (56).

Procedure

To a sample (10 mL of wine or 3 mL of must) in a 200-mL Erlenmeyer flask, add 15 mL of 0.1 N silver nitrate (in 30% acetic acid) and 0.5 g of decolorizing charcoal. Shake 10-15 sec and while swirling the flask add 15 mL of 1% ammonium vanadate solution (10 g of NH_4VO_3 dissolved in 150 mL of N NaOH, 200 mL of 27% sodium acetate added, and diluted to 1 L with distilled water). Filter through a folded filter, discarding the first 5 mL; then collect 10 mL. Allow this to stand 30 min. In the spectrophotometer, determine the optical density at 530 nm in a 1-cm cuvette. Use distilled water treated as above for the blank. A standard curve is prepared with diluted tartaric acid solution using the same procedure.

The procedure above is covered by a German patent (57). It has been recommended for control use because of its practicality, accuracy, and rapidity (58, 59). Sugars, tannins, galacturonic acid, gluconic acid, α-ketoglutarate, and citrate do not interfere. However, low and variable results were found with carbon treatment (60).

Tartaric acid may be determined in a continuous flow system (61). However, a correction factor for malic acid interference is necessary:

$$y = 0.08x + b$$

where y = concentration of tartaric acid (g/L)
x = concentration of malic acid (g/L), determined enzymatically
b = concentration of tartaric acid (g/L) as determined by the continuous flow method

The Rebelein procedure has been modified for red wines to prevent interference of anthocyanins (62). Essentially a correction is made for the color of the wine by running a blank rather than using charcoal to remove color. The results compared favorably with the longer and more expensive racemate procedure.

Procedure

Prepare 30% acetic acid solution (solution 1). For the metavanadate solution, dissolve 10 g of ammonium metavanadate in 150 mL of N sodium hydroxide and add 200 mL of 27% sodium acetate. Make to 1000 mL with distilled water (solution 2). Prepare an ammonium chloride solution by dissolving 4.5 g of ammonium chloride in 150 mL of 1 N sodium hydroxide and add 200 mL of 27% sodium acetate. Make to 1000 mL with distilled water. Add 1 vol of this to 1 vol of solution 1 to form solution 3. Adjust the pH to the same pH as a mixture of equal volumes of solutions 1 and 2 using 1 N sodium hydroxide.

Prepare four test tubes as follows:

	Amount in test tube (mL)			
	A	B	C	D
Wine or must or standard tartrate	1	0	1	0
Water	0	1	0	1
Solution 1	10	10	0	0
Solution 2	10	10	0	0
Solution 3	0	0	20	20

Mix, then after 15 min determine the absorption at 500 nm at 1 cm. Prepare a standard curve with 0–5 g/L of tartaric acid. Solution B and D need not be prepared for white wines or the standard tartaric acid solutions. Net absorption is $A - C$. For red wines, net absorption is $(A - B) - (C - D)$.

Passage through a resin adsorptive column before development of the color has been recommended (49).

Procedure

Pack 20 mL of dry FSP-4022 resin (Bio-Rad Laboratories) into a Bio-Rex chromatography column (15-mm i.d. × 250 mm long, equipped with a type-MD metering drip tip; Bio-Rad Laboratories). Place a styrofoam plug (about 7 mm thick × 18 mm diameter) on the top of the resin to prevent flotation. Attach a Pasteur capillary pipet (cut to a length of about 30 mm) to the end of the metering drip tip to accommodate the neck of a 50-mL volumetric flask. Condition the resin by several alternate washings with 2 bed-volumes of deionized water and 70% denatured alcohol (ethanol) with final washing with deionized water.

Pipet 2 mL of wine or tartaric acid standard solution onto the top of the column and allow it to enter the resin slowly (1–2 mL/min). Follow by two to three portions of 2–4 mL of deionized water. Once the wine is adsorbed onto the resin, increase the flow rate to maximum and pass two portions (about 15–18 mL) of deionized water through the column to give 40–45 mL of decolorized diluted wine in the 50-mL volumetric flask. The sample is now ready for color development as described below. Using two columns, 10–12 samples per hour can be prepared by this treatment.

Remove the adsorbed color from the resin by washing with 10–20 mL of 70% denatured ethanol alcohol followed by two 20-mL portions of deionized water at maximum flow rate. The column now is ready for the next sample, after the metering valve is reduced to the original flow rate. Reduce the column flow rate upon application of the alcohol, but return it to normal upon washing with water. Cracks and bubbles formed in the resin column as a result of local volatilization of the alcohol can be partially removed by applying air pressure to the column during washing with water. They do not interfere with the decolorization of wine, since excess resin is used. When not in use for long periods (1 week or more), keep the resin in 70% denatured alcohol.

Add 3 mL of buffer solution (500 mL of saturated sodium acetate plus 500 mL of glacial acetic acid) and 2 mL of 5% sodium metavanadate solution to the resin sample already collected in a 50-mL volumetric flask. Bring the contents to volume with deionized water and mix, then allow the color to develop for 90 min in a covered water bath at 25°C. Measure color absorbance at 520 nm (10-mm light path, deionized water as reference). Prepare a standard curve with 0.1 to 0.6% tartaric acid solutions treated as above.

Malic Acid

Of the two main acids present in grapes, malic acid is perhaps the more interesting from a physiological point of view. At high (40.5°C) temperature it is respired from grapes, whereas tartaric acid is not respired until a temperature of about 57.2°C is reached. Consequently, during July and August, malic acid steadily decreases in both relative and absolute amounts in the fruit; thus at maturity it only amounts to 10–40% of the acid fraction of grapes, whereas early in the season it may constitute as much as 60%.

Malic acid is also significant in wine in that it is more easily attacked by microorganisms than are tartaric and succinic acids. In European wine making one type of bacterial attack is commonly utilized to reduce excessive acidity. This is the malolactic fermentation. However, its continued activity may reduce the total acidity unduly, resulting in flat-tasting wines, sometimes with undesirable odors.

The malic acid content of wines is much influenced by the absence or presence of the malolactic fermentation. In cool regions or after a period of years in high malic acid varieties, if no malolactic fermentation occurs, the malic acid content of the wine may be 0.5 g/100 mL or higher. In red wines that have undergone a complete malolactic fermentation, malic acid may be absent altogether.

Two enzymatic procedures have also been developed; both appear to be more rapid and less subject to manipulative errors than oxidation procedures. In the first procedure (63–65), malic acid is quantitatively fermented by *Schizosaccharomyces pombe* to ethanol and carbon dioxide:

$$COOHCH_2CHOHCOOH \longrightarrow CH_2CH_2OH + 2CO_2$$

The amount of malic acid is obtained by determining the total acidity before and after fermentation with *S. pombe*. A trial fermentation of $L(-)$-malic acid should be made to be sure the yeast culture is active.

The second procedure (66) utilizes the enzyme L-malic dehydrogenase to catalyze the reaction between malic and nicotinamide adenine dinucleotide (NAD^+), forming oxaloacetate and the reduced form of the dinucleotide (NADH):

$$L(-)\text{-Malic acid} + NAD^+ \rightleftharpoons \text{Oxaloacetate} + NADH + H^+$$

The oxaloacetic acid is fixed by hydrazine in alkaline solution. The extent of the reaction is measured spectrophotometrically by the change in the extinction coefficient at 340 nm resulting from the reduction of NAD^+ to NADH.

Procedure

To a small test tube add 0.9 mL of hydrazine–glycine buffer, pH 9.5 [suspend 7.5 g of glycine, 5.2 g of hydrazine sulfate, and 0.2 g of EDTA in a small volume of water, add 51 mL of 2 N NaOH, and dilute to 100 mL with water (the solution is stable for about 1 week if kept at 0–4°C)]. Add 1.0 mL of water and 1.0 mL of wine or grape juice containing about 10–40 µg of malic acid (dilute juice or wine 100–200 ×). Add 0.1 mL of NAD$^+$ (40 mg/mL) and mix well. Read optical density at 340 nm (E_1). Add 4 µL of commercial malic dehydrogenase solution, then mix. Incubate at 37°C for 30 min, cool at 25°C for 6 min, and read optical density at 340 nm (E_2). Prepare a standard curve using 0.2, 0.4, 0.6, 0.8, and 1.0 mL vols of a 40 mg/mL standard solution of malic acid.

$$\Delta A = E_2 - E_1$$

ΔA = absorbance due to effect of malic acid

Read, from the standard curve, the malic acid concentration; then multiply by the dilution factor.

The results should be interpreted with caution because the enzyme has been shown (67) to be somewhat nonspecific. The enzymatic results are usually lower than those by oxidation procedures (68). A modified enzymatic version has been suggested (69) using glutamate oxaloacetate transaminase (GOT) to remove the oxaloacetic acid rather than using hydrazine. A further modification (70) was reported using organic dyes to allow readings in the visible range for use with certain automated equipment. Gas–liquid chromatographic procedures for malic acid have been used (71, 72).

A fluorometric procedure for malic acid in musts or wines has been developed (73). The sample is passed through an ion exchanger "Ionac CGA 541" in the acetate form. The malic acid is eluted with 0.5 M sodium sulfate. The color is developed with orcinol in sulfuric acid. The colored product, 7-hydroxy-5-methylcoumarin, is strongly fluorescent and can be determined by fluorometry. A standard curve is prepared with pure malic acid solutions. This avoids interference from tartaric and lactic acids, which are present in the procedure of the OIV (3).

Lactic Acid

D(−)-Lactic acid is a constant, but minor, by-product of alcoholic fermentation and is usually present in concentrations of 0.01–0.1 g/100 mL. Much larger amounts of L(+)-lactic acid are produced by the decarboxylation of malic acid in the malolactic fermentation, particularly in red wines, where it can amount to 0.3–0.6 g/100 mL. Generally, the higher the total lactate the greater the portion of L(+).

One of the methods of following the course of the malolactic fermentation involves the appearance of a large lactic acid spot and the gradual reduction or disappearance of the malic acid spot in paper chromatography.

The usual procedure of the OIV is to separate the lactate by passing through an anion-exchange resin in the acetate form. Then the lactate is oxidized to acetaldehyde with ceric sulfate, and the acetaldehyde formed is determined colorimetrically at 570 nm with sodium nitroprusside and piperidine. If the wine contains more than 250 mg/L of sulfur dioxide the sulfurous acid may interfere. The OIV suggests correcting the result as follows: Place 15 mL of the eluate in a glass-stoppered flask, then add 5 mL of 27% sodium acetate and 2 mL of 1.55 N sulfuric acid. Add nitroprusside and piperidine, mix, and determine the color extinction at 570 nm for apparent lactic acid. True lactic acid (g/L) = (lactic determined) − (apparent lactic acid) × 0.4.

The colorimetric procedure (74) also oxidizes the lactic acid to acetaldehyde and forms the color with p-hydroxyphenyl. It is sensitive to lactic acid in the concentration range of 1–10 μg/mL. D-Tartaric acid, L-malic acid, citric acid, acetic acid, ethanol, glucose, fructose, and sulfur dioxide do not interfere.

Procedure

Dilute the wine with water to contain lactic acid in the range of 10–100 μg/mL. To 1.0 mL of the diluted sample in a 20 × 180-mm test tube, add 1 mL of 20% copper sulfate pentahydrate solution, dilute to 10 mL with water, add about 1 g of calcium hydroxide powder, shake vigorously, let stand for 30 min, then centrifuge. Pipet 1 mL of the supernatant into a 20 × 180-mm test tube and add 0.05 mL of a 4% copper sulfate pentahydrate solution. While swirling the test tube, add 6 mL of concentrated sulfuric acid. Place the test tube in a boiling water bath for 7 min and cool to below 20°C in an ice-water bath. Add 0.1 mL of a 1.5% p-hydroxydiphenyl solution in 0.5% sodium hydroxide solution stored in a brown bottle, then shake. Incubate for 15 min in a 30°C water bath, shake, and continue to incubate for 15 min more; then place in a boiling water bath for 1.5 min to dissolve the precipitate, and cool under running water. Read the absorbance at 560 nm. Prepare a standard curve using lithium lactate solutions with concentrations of 0, 10, 20, 30, 50, and 70 μg/mL.

Enzymatic procedures for total lactic acid and L(+)- and D(−)-lactic acids are recommended (75). Alkaline hydrolysis cannot be used because of the formation of lactic acid from residual sugars.

Kanbe et al. (76) have developed a rapid automated enzymatic procedure for determining D(−)- and L(+)-lactate in wines.

Procedure

Make a pH 9.5 buffer from 7.5 g of glycine, 5.2 g of hydrazine sulfate, 0.2 g of EDTA, and 4 g of NaOH in 150 mL of H_2O. Dissolve 177 mg of NAD^+ in 100 mL of H_2O. Prepare a lactate standard (0.1, 0.2, 0.3, 0.4, and 0.5 mg/mL) from L(+)-lactate(Li). Prepare an L(+)-lactate dehydrogenase (LDH) solution (0.25 units/μL) diluted with buffer solution made without hydrazine sulfate. Place 0.1 mL of sample or standard, 1.5 mL of buffer solution, and 50 μL of LDH solution into a 10-mL cuvette and mix. Add 1.5 mL of NAD$^+$ solution after a 1.5-min delay, then heat at 37°C for 30 min. For the blank to zero instrument, exclude the 50 μL of LDH solution. Record absorbance at 340 nm. Calculate concentrations from the standard curves. To measure D(−)-lactic acid, use D(−)-lactic acid for the standards and D(−)-lactic dehydrogenase in place of L(+)-dehydrogenase.

Citric Acid

Citric acid rarely is determined by enologists because it is present in very small amounts (0.01–0.07%). Citric acid is added to wines to increase acidity and to complex iron. Its determination is needed in countries that limit addition of citric acid, usually to 0.05 g/100 mL. The presence of more than about 0.1% would indicate addition of citric acid in amounts of more than 0.05%. In a study of 127 California table wines, Fong et al. (77) found that most wines had less than 0.05%. Untreated laboratory samples contained less than 0.02%. Patschky (78) reported less than 0.035% in 912 wines. Red table wines contain less than white. Of 1170 Hungarian wines, about 95% contained less than 0.04% (79).

The classic oxidation procedures for the determination of small amounts of citric acid in wine require $1\frac{1}{2}$–2 days. Citramalic acid formed during fermentation may interfere (80).

The newer procedures (81–86) involve enzymatic or colorimetric measurements. Rebelein's colorimetric procedure (87), the OIV normal procedure, can be finished in an hour and is given here. The reference procedure of the OIV (3) uses ion exchange with fractional elution to separate citramalic acid. The citric acid is then oxidized to produce acetone. The acetaldehyde is oxidized to acetic acid, and the acetone is determined by titration with iodine.

Procedure

Prepare Grier's reagent by adding 1.5 g of sulfanilic acid to 50 mL of glacial acetic acid and diluting it to 250 mL with water. To prepare the diazo reagent, mix a 5-mL aliquot of Grier's reagent with 1.0 mL of a 2% sodium nitrite solution. Keep at 0°C during preparation. Use immediately.

To 5 mL of wine or must, in a 40–50-mL centrifuge tube, add 1 mL of concentrated ammonium hydroxide and 1 mL of 20% barium chloride solution, then mix. After 2 min add 15 mL of 96% ethanol, mix, and centrifuge for 3–4 min. Discard the supernatant liquid. Wash the precipitate three times with 2 mL of a mixture of 140 mL of water and 300 mL of 96% ethanol, mixing thoroughly each time with a rubber-coated stirring rod; centrifuge each time, discarding the supernatant liquid.

To the washed precipitate, add 10 mL of 7.1% sodium sulfate solution and heat on a boiling water bath for 10 min. Stir thoroughly. Transfer to a 50-mL volumetric flask and, after cooling, dilute to volume with 7.1% sodium sulfate solution. Transfer to an Erlenmeyer flask containing 0.2 g of charcoal, leave for 5 min, then filter.

To each of two Erlenmeyer flasks add 10 mL of 27% sodium acetate solution, 2.0 mL of the filtrate, and 2 mL of the diazo reagent. To the first flask add 5 mL of glacial acetic acid; to the second flask add 5 mL of a 1% lead tetraacetate solution. Start counting time following the addition of the lead tetraacetate solution. Exactly 5 min later, filter both flasks; exactly 13 min later, measure the absorbance of the content of the second flask at 420 nm, using the contents of the first flask as the blank. Establish the amount of citric acid present from a calibration curve prepared by analyzing standard citric acid solutions. Citramalic acid does not interfere.

Bergner-Lang (88) reported high results with enzymatic procedures in red wines or in wines containing sugar and preferred Rebelein's (87) colorimetric method. The high results were due to nonspecificity of citrate lyase and interference of Maillard reaction products.

Others (89, 90) have reported excellent results.

Procedure

Measure 11 mL of glycine pH 7.8 buffer, made up with 0.6 mM $ZnCl_2$, 1 mL of NADH(Na_2), 12 mg/mL of $NaHCO_3$, and 0.24 mL of enzyme suspension (0.5 mg/mL malate dehydrogenase at 1200 units/mg and 2.5 mg/mL of lactate dehydrogenase, from rabbit muscle, at 550 units/mg in 3.2 M ammonium sulfate at pH 6-7). Prepare a citrate lyase solution (12 units per 0.3 mL redistilled water).

Dilute samples to less than 0.4 g/L of citric acid. Pipet 1.0 mL of buffer solution into each cuvette, 2.0 mL of redistilled water in the cuvette for the "blank," and 1.8 mL into the "sample" cuvette. Pipet 0.2 mL of sample into sample cuvette. Mix and hold for 5 min at 20-25°C, then read 340 nm absorbance of each and record E_1^b (for blank) and E_1^s (for sample). Add 0.02 mL of citrate lyase solution to each cuvette, mix, and hold as above; then read at 340 nm, and record absorbance for E_2^b and E_2^s, and calculate the citric acid from the extinction coefficient:

$$\text{Citric acid (mg/L)} = \frac{(V)(192.1)(\Delta E)(F)}{6.3v}$$

where $\Delta E = (E_1^s - E_2^s) - (E_1^b - E_2^b)$
V = volume in cuvette (3.02 mL)
v = sample volume (0.2 mL)
F = dilution factor

Succinic Acid

Although succinic acid is present in musts and is an important by-product of alcoholic fermentation, it is rarely determined. Musts of botrytized grapes are higher in succinic acid than those of sound grapes (91). However, *Auslese* wines are no higher in succinic acid than ordinary wines. About one part of succinic acid is produced per 100 parts by volume of ethanol. Thus in a wine containing 10 vol % of ethanol, 0.1 g/100 mL of succinic acid can be expected. The reported concentrations are in the range of 0.050–0.399 g/100 mL for musts and 0.05–0.75 g/100 mL for table wines. Fortification of dessert wines naturally reduces their succinic acid content.

The procedures for determining this acid are based on its resistance to oxidation by permanganate, on the insolubility of its barium salt in 80% ethanol, and on enzymatic analysis with succinyl-coenzyme A-synthetase. Since some of the succinic acid is bound, alkali hydrolysis is recommended. Succinic acid is also soluble in ether, and it forms an insoluble silver salt.

A semienzymatic method was suggested (92) using succinic dehydrogenase and a color reagent. A more acceptable enzymatic method is available (91, 93).

$$\text{Succinate} + \text{CoA} + \text{Guanosine-5-P} \xrightarrow{\text{Succinyl CoA synthetase}} \text{Succinyl-CoA} + \text{GDP} + P_i$$

$$\text{GDP} + \text{PEP} \xrightarrow{\text{Pyruvate kinase}} \text{Pyruvate} + \text{GTP}$$

$$\text{Pyruvate} + \text{NADH} + H^+ \xrightarrow{\text{Lactate dehydrogenase}} \text{Lactic acid} + \text{NAD}^+$$

The NADH is monitored at 340 nm. A variation using inosine-5-P (ITP) instead of GTP worked well with wine analyses (94).

The recommended procedure of the OIV (3) separates the anions from the cations by passing the wine through Dowex® 2 or Amberlite® IRA-400 anion-exchange resin. The colorimetric procedure (95) avoids interference of α-ketoglutaric acid. The anions are eluted from the column with an ammonium carbonate solution. Anions other than succinate are oxidized by permanganate, the volatile acids are removed by steam distillation, and the succinate is extracted with ether in a liquid–liquid extractor. Finally silver nitrate is added and the excess is titrated with potassium thiocyanate.

The technique of Marignan (96) as modified by Lafon (97) may be used. The procedure titrates only 70% of the total succinic acid present; thus a correction is necessary in the calculation. However, it is reported that 97% recovery can be achieved with this correction (98).

Procedure

Pipet 20 mL of the wine sample into the 100-mL beaker and evaporate to about 10 mL on a steam bath. Cool and transfer the contents into the liquid-liquid extraction chamber using 1 mL of 1:3 sulfuric acid and 9 mL of water to wash the beaker. Place 10 mL of water in the receiving flask, then extract with ether for 5 hr. The time required depends on the efficiency of the extraction and must be established by trial. Disconnect the flask and evaporate the ether. Add 1 mL of 1:3 sulfuric acid and 10 mL of saturated potassium permanganate solution, then boil for 15 min. Cool the solution and add a few drops of saturated sodium bisulfite solution to destroy the excess potassium permanganate. To remove the volatile acids, steam distill until 300–400 mL of distillate is collected. Now the solution contains only sulfuric and succinic acids. Using a pH meter, adjust the pH to 4.18 first with saturated barium hydroxide and then with 0.05 N barium hydroxide. Then titrate the succinic acid from pH 4.18 to 7.5 with 0.05 N barium hydroxide.

$$\text{Succinic acid (g/100 mL)} = \frac{(V)(N)(118.07)(100)(100)}{(1000)(70)(v)}$$

$$= \frac{(V)(N)(16.867)}{v}$$

where V = volume of barium hydroxide solution used for the titration (mL)
N = normality of the barium hydroxide solution
v = sample volume (mL)

Ascorbic Acid (Vitamin C)

Only very small quantities of ascorbic acid (0.5–15 mg per 100 g of fruit) are present in musts, and only up to 2 mg/L is found in wines, unless the substance has been added as an antioxidant. Although ascorbic acid has been employed as an antioxidant in the United States and Germany in grape juices and wines, its use during fermentation is of questionable value.

To find ascorbic acid in musts, the analyst must use grapes that are freshly harvested and do the determination quickly. The recommended procedure is based on the use of a colored dye, 2,6-dichlorophenolindophenol, to oxidize the ascorbic acid. Another procedure (99) is based on the formation of a blue complex of phosphotungstic acid and ascorbic acid in an acid solution. Sugars, ferrous ion, and sulfur dioxide do not interfere. Since formaldehyde blocks the 3,5-enolic hydroxy group of ascorbic acid, the color caused by other reducing compounds can be determined by adding formaldehyde before the Folin-Ciocalteu reagent. Since formaldehyde has a slight reducing value for phosphotungstic acid, a second blank determination must be carried out in this case.

Procedure

Prepare an approximately 0.01 N solution of ascorbic acid and standardize it against a standard iodine solution, using starch as an indicator. Using the freshly standardized ascorbic acid solution, standardize an approximately 0.002 N 2,6-dichlorophenolindophenol solution. Titrate the ascorbic acid into a measured volume of the dye. The end point is from a blue to a faint pink color. Practice is required to recognize the end point.

De-stem and weigh 25 g of freshly picked grapes. Place them with 10 g of acid-washed sand, 10 mL each of N sulfuric acid, and 2% metaphosphoric acid solution in a mortar, and grind up the grapes without delay. Decant the liquid into a 100-mL volumetric flask, regrind the residue with a mixture of equal parts of the sulfuric and metaphosphoric acid solutions, and decant the liquid into the same volumetric flask.

If necessary, carry out a third extraction. Dilute the liquid in the volumetric flask to volume using 8% acetic acid. Filter, then dilute 10 mL to 50 mL using 8% acetic acid. Transfer the solution to a buret. Pipet exactly 0.5 mL of the freshly standardized dye into a beaker and run the prepared grape juice from the buret into it until a faint pink end point is reached. Carry out the determinations in duplicate or triplicate and also run a blank containing only the acids.

$$\text{Ascorbic acid (mg/100 g of fruit)} = \frac{(W)(100)(5)(4)}{v}$$

where W = amount of ascorbic acid equivalent to 0.5 mL dye (mL)
v = volume of juice used in the final aliquot (mL)

According to the procedure of the OIV (3), total ascorbic acid is oxidized to dehydroascorbic acid, which is precipitated as the 2,4-bis-dinitrophenylhydrazone using 2,4-dinitrophenylhydrazine. The precipitate is extracted with ethyl acetate, and the derivative is separated by thin-layer chromatography. Its color is estimated in a sulfuric acid solution. The procedure is lengthy and time-consuming.

A shorter and apparently equally accurate procedure using iodine oxidation and thin-layer chromatography has been used (100). The sum of the L-ascorbic acid, dehydro-L-ascorbic acid, and 2,3-diketo-L-gulonic acid was taken as the total ascorbic acid. Interference with tannins was eliminated by adsorption on polyamide. The reaction product with ascorbic acid and 2,4-dinitrophenylhydrazine was extracted and separated by thin-layer chromatography. The red 2,4-dinitrophenylhydrazine derivative was eluted and evaluated spectrophotometrically. For rapid approximate determination of "active" ascorbic acid, the OIV (3) recommends fixing the sulfur dioxide with acetaldehyde and using iodine to titrate the "active" ascorbic acid. The method is not very specific.

Gas chromatographic procedures for ascorbic acid are available (101, 102). A new colorimetric procedure has also been proposed (103).

Other Acids

Many other organic acids are found in musts and wines. Few normally are determined in winery control laboratories, but some may prove to be important in research. For determination of formic acid in fruit juices, a photometric procedure with an accuracy of ±5% has been recommended (104). Pyruvic, α-ketoglutaric, and D-gluconic acids have been determined by enzymatic methods (105–107). In 73 European wines, 0.01–5.15 g/L of gluconic acid was found (108). Wines made from nonbotrytized grapes are low in gluconic acid (less than 0.5 g/L) (109). Wines from botrytized grapes were higher in gluconic acid (up to 5.9 g/L). The presence and amount of this acid might become the basis for a quality control procedure for wines claiming to be made from botrytized grapes. The fluorometric procedure (110) using pyridoxamine and zinc(II) ion may also be useful. Dilution would be needed. Details for the enzymatic determination of gluconic acid are available (108).

The enzymatic reactions are:

$$\text{D-Gluconate} + \text{ATP} \xrightarrow{\text{Gluconic kinase}} \text{Gluconate-6-phosphate} + \text{ADP}$$

$$\text{Gluconate-6-phosphate} + \text{NADP}^+ \xrightarrow{\text{6PGDH}} \text{Ribulose-5-phosphate} + \text{NADPH} + \text{CO}_2 + \text{H}^+$$

The quantity of NADPH formed is equivalent to the quantity of gluconate oxidized. The increase of optical density at 340 nm is used to determine the gluconate concentration.

Galaturonic acid (about 95% of the total uronic acids) and glucuronic acid are present in musts (mainly in a colloidal state) and in wines (mainly free) (111). Red wines are higher in uronic acids than are white wines. Musts contain up to 0.4 g/100 mL, and wines contain up to about 0.23 g/100 mL. Normal table wines had 0.02–0.14 g/100 mL in one study (112). Those made from botrytized grapes and fortified wines contained up to 0.75 g/100 mL. German commercial grape juices had 0/17–0.29 g/100 mL. Whether galacturonic acid is lowered during fermentation is not known.

The colorimetric procedure (111) for determination of galacturonic acid was found to be simple and rapid and to give very good results (109). The wine is passed through an anion-exchange resin in the acetate or formate form. It is eluted with 2 N acetic or formic acid. The blue color developed by heating with sulfuric acid and naphthoresorcine. The color is extracted with toluene, and the absorption is measured at 565 nm. Glucuronic acid, which is present in small amounts, interferes. In another colorimetric procedure (112) the wine is heated with 2,4-dinitrophenylhydrazine, the hydrazone converted to the osazone. Thin-layer chromatography is used to separate it from the hydrazones of ascorbic, pyruvic, and α-ketoglutaric acids.

A gas chromatographic method to determine uronic acids is available (113). Recently, five uronic acids of wines were separated by HPLC (114).

REFERENCES

1. G. C. Whiting, *J. Inst. Brew.* **82**, 84–92 (1976).
2. F. Radler, *Dtsch. Lebensm. -Rundsch.* **71**, 20–26 (1975).
3. *Receuil des Méthodes Internationales d'Analyse des Vins*, 4th ed. Office International de la Vigne et du Vin, Paris, 1978.
4. J. F. Guymon and C. S. Ough, *Am. J. Enol. Vitic.* **13**, 40–45 (1962).
5. *Uniform Methods of Analyses for Wines and Spirits*. American Society of Enologists, Davis, CA, 1972.
6. *Official Methods of Analysis*, 14th ed. Association of Official Analytical Chemists, Arlington, VA, 1984, pp. 220–230.
7. E. Brémond, *Contribution à l'Etudé Analytique et Physico-chimique de l'Acidité des Vins*. Imprimeries La Typo-Litho et Jules Carbonel Réunies, Alger, 1937.
8. M. A. Amerine and A. J. Winkler, *Proc. Am. Soc. Hortic. Sci.* **40**, 313–324 (1942).
9. Z. D. Rabinovich, *Sadovod. Vinograd. Vinodel. Mold.* **29**(1), 47–48 (1974).
10. S. Lafon-Lafourcade, P. Ribéreau-Gayon, and A. Joyeux, *C.R. Seances Acad. Agric. Fr.* **63**, 551–558 (1977).
11. P. Jaulmes, *Ann. Falsif. Fraudes* **43**, 110–117 (1950).
12. G. J. Pilone, B. C. Rankine, and C. J. Hatcher, *Aust. Wine Brew. Spirit Rev.* **91**, 62, 64, 66 (1972).
13. J. Schneyder and G. Pluhar, *Mitt. Klosterneuburg* **27**, 14–17 (1977).
14. M. Pato and A. Salvador, *Méthode pour le Titrage des Acides Volatiles des Vins, avec Deduction de l'Acide Carbonique*. Edição Vitivinícola de Beira Litoral, Anadia, Portugal, 1947.
15. W. Postel, F. Drawert, and G. Maccagnan, *Chem., Mikrobiol., Technol. Lebensm.* **1**, 11–14 (1971).
16. L. P. McCloskey, *Am. J. Enol. Vitic.* **31**, 170–173 (1980).
17. B. Doneche and P. J. Sanchez, *Connaiss. Vigne Vin* **19**, 161–169 (1985).
18. France, Ministère de l'Agriculture, Arreté du 24 juin 1963 rélatif aux méthodes officielles d'analyses des vins et des moûts, *J. Off.* **63-154**, 1037, 4551–4587 (1963).
19. M. Dubernet, *Connaiss. Vigne Vin* **10**, 297–309 (1976).
20. G. C. Cochrane, *J. Chromatogr. Sci.* **13**, 440–447 (1975).
21. J. M. Robertson, B. L. Kirk, and A. C. Crum, *Rep. N.Z. Dep. Sci. Ind. Res., Chem. Div.* **CD-2247**, 1–43 (1976).

22. C. S. Ough and M. A. Amerine, *Am. J. Enol. Vitic.* **18,** 157-164 (1967).
23. M. Bertuccioli, *Vini Ital.* **138,** 149-156 (1982).
24. B. Trombella and A. Ribeiro, *Am. J. Enol. Vitic.* **31,** 294-297 (1980).
25. T. Shinohara, *Agric. Biol. Chem.* **49,** 2211-2212 (1985).
26. P. Sudraud and S. Clermont, *Ann. Falsif. Expert. Chim.* **54,** 7-15 (1965).
27. J. W. H. Lugg and B. T. Overell, *Aust. J. Sci. Res., Ser. A* **1,** 98-111 (1948).
28. R. E. Kunkee, *Wines Vines* **49**(3), 23-24 (1968).
29. M. Bourzeix, J. Guitraud, and F. Champagnol, *J. Chromatogr.* **50,** 83-91 (1970).
30. H. Tanner and M. Sandoz, *Schweiz. Z. Obst- Weinbau* **108,** 182-186 (1972).
31. J. R. Stamer, L. D. Weirs, and L. D. Mattick, *Food Chem.* **10,** 235-238 (1983).
32. M. Salgues and J. Andre, *Vignes Vins* **(261),** 36-37 (1977).
33. G. L. Brunelle, R. L. Shoenman, and G. E. Martin, *J. Assoc. Off. Anal. Chem.* **50,** 329-333 (1967).
34. L. R. Mattick, A. C. Rice, and J. C. Moyer, *Am. J. Enol. Vitic.* **21,** 179-183 (1971).
35. J. J. Ryan and J. A. Dupont, *J. Agric. Food Chem.* **21,** 45-49 (1973).
36. W. W. D. Wagner, C. S. Ough, and M. A. Amerine, *Am. J. Enol. Vitic.* **22,** 167-171 (1971).
37. P. De Smedt, P. A. P. Liddle, B. Cresto, and A. Bossard, *J. Inst. Brew.* **87,** 349-351 (1981); see also *Ann. Falsif. Expert. Chim. Toxicol.* **72,** 633-642 (1979).
38. J. E. Marcy and D. E. Carroll, *Am. J. Enol. Vitic.* **33,** 176-177 (1982).
39. G. Shen, R. Wang, and M. Ying, *Shipin Yu Fajiao Gongye* **(6),** 14-20, 40 (1984).
40. C. Droz and H. Tanner, *Schweiz. Z. Obst- Weinbau* **118,** 434-438 (1982).
41. W. Flak and G. Pluhar, *Mitt. Klosterneuburg* **33,** 60-68 (1983).
42. J. P. Goiffon, A. Blachere, and C. Remeniac, *Analusis* **13,** 218-225 (1985).
43. E. Mentasti, M. C. Gennaro, C. Sarzanini, C. Baiocchi, and M. Savigliano, *J. Chromatogr.* **322,** 177-189 (1985).
44. W. Steiner, E. Muller, D. Frohlich, and R. Battaglia, *Mitt. Geb. Lebensmittelunters. Hyg.* **75,** 37-50 (1984).
45. A. Rapp and A. Ziegler, *Chromatographia* **9,** 148-150 (1976).
46. Y. Shimazu and M. Watanabe, *Wein-Wiss.* **31,** 45-53 (1976).
47. D. J. Woo and J. R. Benson, *Am. Lab.* **16**(1), 50, 52-54 (1984).
48. P. Pfeiffer and F. Radler, *Z. Lebensm.-Unters. -Forsch.* **181,** 24-27 (1985).
49. J. D. McCord, E. Trousdale, and D. D. Y. Ryu, *Am. J. Enol. Vitic.* **35,** 28-29 (1984).
50. P. Martinere and P. Sudraud, *Connaiss. Vigne Vin* **2,** 41-51 (1968).
51. L. Usseglio-Tomasset, *Riv. Vitic. Enol.* **26,** 375-384 (1973).
52. E. Negre, A. Dugal, and J. M. Evesque, *Ann. Technol. Agric.* **7,** 31-101 (1958).
53. C. von der Heide and K. Henning, *Ber. Ver. Forschungsanst. Wein. Obst. Gartenbau Geisenheim* pp. 25-27 (1934).
54. J. R. Matchett, R. R. Legault, C. C. Nimmo, and B. K. Notter, *Ind. Eng. Chem.* **36,** 851-857 (1944).
55. G. Hill and A. Caputi, Jr., *Am. J. Enol. Vitic.* **21,** 153-161 (1970).
56. H. Rebelein, *Chem., Mikrobiol., Technol. Lebensm.* **2,** 33-38 (1973).
57. H. Rebelein, Verfahren zur Bestimmung des Weinsäuregehaltes von Flüssegkeiten, German Pat. 2,243,337 (Cl. G. 01n), February 28, 1974, Appl. Pat. 22 43 397.2, September 4, 1972, 11 p.

58. H. Tanner and Z. Lipka, *Schweiz. Z. Obst- Weinbau* **109**, 684-692 (1973); see also *ibid.* **108**, 251-254 (1972); *Rev. Suisse Vitic., Arboric., Hortic.* **6**, 5-10 (1974).
59. S. Chauvet and P. Sudraud, *Rev. Fr. Oenol.* **66**, 58-59 (1977).
60. G. F. Pilone, *Am. J. Enol. Vitic.* **28**, 104-107 (1977).
61. J. Trossais and C. Asselin, *Connaiss. Vigne Vin* **19**, 249-259 (1985).
62. M. Vidal and J. Blouin, *Rev. Fr. Oenol.* **70**, 39-46 (1978).
63. E. Peynaud and S. Lafon-Lafourcade, *C.R. Hebd. Seances Acad. Sci.* **158**, 5542 (1964).
64. E. Peynaud and S. Lafon-Lafourcade, *Ann. Technol. Agric.* **14**, 49-59 (1965).
65. E. Peynaud, *Weinberg Keller* **12**, 71-73 (1965); also *Wines Vines* **56**(3), 28 (1965).
66. K. Mayer and J. Busch, *Mitt. Geb. Lebensmittelunters. Hyg.* **54**, 60-65 (1963).
67. C. Poux and M. Caillet, *Ann. Technol. Agric.* **18**, 359-366 (1969).
68. D. W. Olschimke, W. Niesner, and C. Junge, *Dtsch. Lebensm.-Rundsch.* **65**, 383-384 (1969).
69. L. P. McCloskey, *Am. J. Enol. Vitic.* **31**, 212-215 (1980).
70. A. Lonvaud-Funel, B. Doneche, and D. Blueze, *Connaiss. Vigne Vin* **14**, 207-217 (1980).
71. C. Russo, A. Zamorani, and S. Campisi, *Ind. Agrar.* **10**, 374-375 (1972).
72. B. H. Gump, S. Saguandeekul, G. Murray, and J. T. Villar, *Am. J. Enol. Vitic.* **36**, 248-251 (1985).
73. M. Castino, *Vini Ital.* **16**, 43-47 (1974).
74. G. J. Pilone and R. E. Kunkee, *Am. J. Enol. Vitic.* **21**, 12-18 (1970).
75. W. Postel, F. Drawert, and W. Hagan, *Z. Lebensm.-Unters. -Forsch.* **150**, 267-273 (1973).
76. C. Kanbe, Y. Ozawa, and T. Sakasai, *Agric. Biol. Chem.* **41**, 863-867 (1977).
77. D. C. Fong, M. A. Amerine, and C. S. Ough, *Am. J. Enol. Vitic.* **25**, 222-224 (1974).
78. A. Patschky, *Allg. Dtsch. Weinfachztg.* **109**, 308-309 (1973).
79. Z. Jeszensky and P. Szalka, *Borgazdasag* **24**, 136-139 (1976).
80. G. Dubus, *Connaiss. Vigne Vin* **5**, 43-73 (1971).
81. V. Dimotaki-Kourakou, *Ann. Falsif. Expert. Chim.* **55**, 148-149 (1962).
82. M. Castino, *Riv. Vitic. Enol.* **6**, 247-257 (1967).
83. K. Mayer and G. Pause, *Mitt. Geb. Lebensmittelunters. Hyg.* **56**, 454-458 (1965).
84. J. Schneyder, *Mitt., Rebe Wein, Obstbau Fruechteverwert.* **19**, 122-123 (1965).
85. F. Addeo, *Sci. Tecnol. Alimenti* **2**, 87-92 (1972).
86. K. Mayer and G. Pause, *Lebensm.-Wiss. Technol.* **2**, 143 (1969).
87. H. Rebelein, *Dtsch. Lebensm.-Rundsch.* **63**, 337-340 (1967).
88. B. Bergner-Lang, *Dtsch. Lebensm.-Rundsch.* **73**, 279-280 (1977).
89. A. Seppi and A. Sperandio, *Riv. Soc. Ital. Sci. Aliment.* **12**, 478-482 (1983).
90. G. Henniger and L. Mascaro, *J. Assoc. Off. Anal. Chem.* **68**, 1024-1027 (1985).
91. W. R. Sponholz and H. H. Dittrich, *Wein-Wiss.* **32**, 38-47 (1977).
92. R. Pires and K. Mohler, *Z. Lebensm.-Unters. -Forsch.* **143**, 96-99 (1970).
93. G. Michal, H. O. Beutler, G. Lang, and U. Guentner, *Fresenius' Z. Anal. Chem.* **279**, 137-138 (1976).
94. A. Joyeux and S. Lafon-Lafourcade, *Ann. Falsif. Expert. Chim. Toxicol.* **72**, 317-320 (1979).
95. M. Castino, *Vini Ital.* **11**, 509-511, 513, 515, 517, 519, 521 (1969).
96. R. Marignan, *Le Dosage de l'Acide Succinque dans les Vins*. Aristide Quellet, Montpellier, 1944.

97. M. Lafon, *Ann. Technol. Agric.* **4**, 169–221 (1955).
98. J. Ribéreau-Gayon, E. Peynaud, P. Sudraud, and P. Ribéreau-Gayon, *Traite d'Oenologie*, Vol. 1. Dunod, Paris, 1976.
99. F. W. Muller and H. Kretzdorn, *Dtsch. Wein-Ztg.* **91**, 314–316 (1955).
100. H. Thaler and U. Gieger, *Mitt. Geb. Lebensmittelunters. Hyg.* **58**, 473–495 (1967).
101. R. Gerstl and K. Ranfft, *Z. Lebensm.-Unters. -Forsch.* **154**, 12–17 (1974).
102. J. E. Schlack, *J. Assoc. Off. Anal. Chem.* **57**, 1346–1348 (1974).
103. A. S. Hammam, *J. Appl. Chem. Biotechnol.* **26**, 611–617 (1976).
104. H. Tanner, *Schweiz. Z. Obst- Weinbau* **112**, 38–42 (1976).
105. K. Mohler and S. Looser, *Z. Lebensm.-Unters. -Forsch.* **140**, 149–154 (1969).
106. W. R. Sponholz, B. Wunsch, and H. H. Dittrich, *Z. Lebensm.-Unters. -Forsch.* **172**, 264–268 (1981).
107. C. Delfini, *Riv. Vitic. Enol.* **36**, 307–315 (1983).
108. W. Postel and F. Drawert, *Chem., Mikrobiol., Technol. Lebensm.* **1**, 151–155 (1972).
109. S. Chauvet and P. Sudraud, *Inst. Oenol. Rapp. Act. Rech. Bordeaux, 1976–1977* pp. 32–33 (1978).
110. M. Takeda, T. Kinoshita, and A. Tsuji, *Anal. Biochem.* **72**, 184–190 (1976).
111. L. Usseglio-Tomasset and G. Gabri, *Atti Accad. Ital. Vite Vino, Siena* **27**, 151–172 (1975).
112. W. Arndt and H. Thaler, *Mitt., Rebe Wein, Obstbau Fruechteverwert.* **24**, 325–340 (1974).
113. A. Bertrand and D. Dubourdieu, *Ann. Falsif. Expert. Chim.* **71**, 303–312 (1978).
114. W. R. Sponholz and H. H. Dittrich, *Vitis* **23**, 214–224 (1984).

Three
ALCOHOLS

Ethanol is the resulting product of yeast fermentation of natural carbohydrates. Because of the amount and the simplicity of its formation, the relative lack of toxicity of the fermentation by-products, the biological stability of dry or fortified wines, and the pleasing physiological effects, ethanol, per se, has been studied in detail.

Besides ethanol, a number of other monoalcohols and polyalcohols are present in wines, as listed in Table 13. These substances come either from the grapes or are formed in the wine during fermentation. The sugar polyalcohols originate in the grape, but also may be formed by other means from the parent sugar.

The higher alcohols are responsible for some of the complex sensory attributes of wine. Knowledge of the amount of some of these components is used to assess the trueness of source material for the wine or the freedom of the wine from falsification. These assessments have legal or quasi-legal implications.

For general information on the history, legal, chemical, and sensory aspects of the alcohols in wine, References 1–5 are recommended.

The monoalcohols found in wines are all colorless liquids. They vary in mobility from very free flowing (methanol) to very viscous (2-phenyl ethanol). Their odors also vary: Methanol is nearly odorless, 1-propanol has a pleasant sweetish odor, *n*-butanol has a more penetrating and heady odor, the amyl alcohols are still more aromatic and penetrating, and 2-phenyl ethanol gives off a very clinging roselike aroma. The polyalcohols are more viscous and have little or no odor, and the six-carbon-sugar polyalcohols are solids at room temperature.

ETHANOL

Alcoholic fermentation is the anaerobic biological degradation of glucose and fructose to ethanol and carbon dioxide, with a number of side products. Figure 7 shows the metabolic intermediates between glucose and fructose to ethanol.

Table 13. The Most Important Alcohols Present in Wines Other Than Ethanol

Monoalcohols	Polyalcohols
Methanol	Glycerol
1-Propanol	2,3-Butanediol (*levo*)
1-Butanol	2,3-Butanediol (*meso*)
2-Methyl-1-propanol (isobutyl alcohol)	1,2,3,4,5,6-Hexanehexol (*levo*) (D-sorbitol)
2-Methyl-1-butanol (*levo*) (active amyl alcohol)	1,2,3,4,5,6-Hexanehexol (*levo*) (D-mannitol)
3-Methyl-1-butanol (isoamyl alcohol)	1,2,3,4,5,6-Cyclohexanehexol (*meso*) (mesoinositol)
1-Hexanol	
2-Phenyl ethanol (β-phenethyl alcohol)	

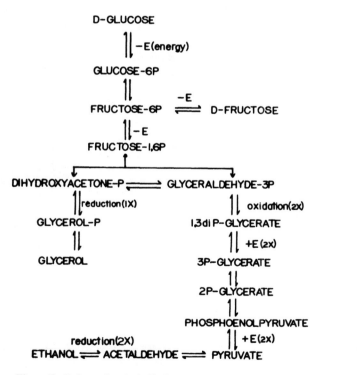

Figure 7. Pathway for alcoholic fermentation by yeast.

The process is essentially identical to glycolysis, the pathway by which glycogen is converted to lactic acid to furnish energy for animal cells. Small amounts of alcohol can also be produced from the decomposition of L-malic acid by *Schizosaccharomyces* sp. yeast or, to a lesser extent, by other microorganisms:

$$COOHCH_2HCOHCOOH \longrightarrow CH_3CH_2OH + 2CO_2$$

Under normal circumstances this degradation seldom occurs. Small amounts of ethanol can also be formed by enzymatic reduction of L-malic acid during "macération carbonique" treatments. This technique is not extensively practiced.

The amount of ethanol formed depends on a number of factors: (a) amount of fermentable sugar, (b) yeast species, (c) fermentation temperature, (d) nutrient level of the grapes, and (e) general fermenting conditions.

If, because of temperature, nutrients, or yeast species, the wine does not ferment all the sugar, the ethanol content will be less than expected. As the temperature increases, the losses in ethanol can be attributed to several sources, namely, (a) the greater amounts of sugar required for carbon skeletons for increased yeast growth and (b) losses due to evaporation and carbon dioxide entrainment as rates of fermentation increase. Figure 8 summarizes data showing these effects of temperature fermentation (6).

There are a number of formulas to predict the amount of ethanol that will be formed. The °Brix is usually the base for these calculations. The possible variations in these types of estimate are exemplified by the data (Figure 9) (7) taken for white and red grape varieties from two climatic regions. These differences have recently been confirmed (8). In addition they showed significant differences between hand and machine harvest, varieties, and maturity on yield of alcohol per °Brix. The multiplication factor for the °Brix to give alcohol percent by volume is usually taken as 0.59 (9). This is a reasonable estimate but is not valid for red grapes from warm areas. A better estimate for these types is 0.54 vol % ethanol/°Brix.

The only significant loss that can occur in the winery, besides that by evaporation, is by the aerobic action of acetic acid bacteria and aerobic yeast action:

$$CH_3CH_2OH \xrightarrow[\text{Bacteria}]{\{O\}} CH_3COOH + H_2O$$

$$CH_3CH_2OH \xrightarrow[\text{Yeast}]{\{O\}} CH_3C\begin{smallmatrix}\diagup O \\ \diagdown H\end{smallmatrix} + H_2O$$

This problem rarely occurs in the modern winery when the tanks and barrels are filled to exclude air, the storage temperature is relatively low, and the proper level of sulfur dioxide is maintained.

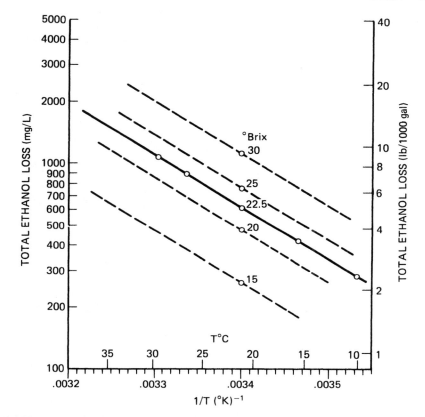

Figure 8. The linear relationship between the reciprocal absolute temperature and the logarithm of the total ethanol lost for the initial sugar contents indicated (6).

The amounts of ethanol present after a fermentation usually vary from 8–9% to as high as 18–18½% v/v. The low levels are attributed to grapes of low sugar content, and the upper levels are due to grapes of very high concentration of sugar with raisins present. The sugar in the latter slowly dissolves and acts as a "syrup addition" type of fermentation. *Saccharomyces cerevisiae* strain Montrachet is able to ferment to this level. Normal ranges for table wines in California are 10–13½% for white grapes and 11–14% for red grapes.

Grapes grown in California may not have sucrose added, but ethanol can legally be added to the wines, if necessary, to bring the ethanol levels of the wine to within the legal range. Wines of less than 10% v/v ethanol may readily become bacterial or grow spoilage yeast if not properly cared for (i.e., failure to use antiseptics and germ-proof filtration).

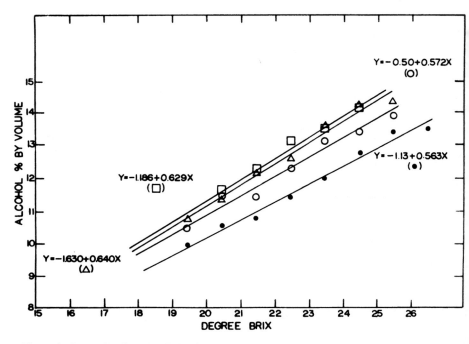

Figure 9. Regression lines for °Brix of must plotted against the resulting percentages of alcohol, v/v, formed for dry white wine region I (□), dry white wine region IV (△), dry red wine region (○), and dry red wine region IV (●) (7).

The amount of ethanol present is very important for the taste. Ethanol has little odor but does add a mouth-feel and a slight sweet taste to table wines. This is especially true for red wines. If the alcohol is less than 10–10½% v/v ethanol, dry wines tend to taste "thin" and red wines taste bitter. White wines having more than 14.0% v/v ethanol tend to have an "alcoholic" character sometimes described as "hot." Further information on the sensory attributes, problems, and methods of tasting can be obtained (10).

Tax considerations partially determine ethanol concentrations in wines. Table 14 gives the legal limits and ranges for California and the United States as of February 1987. For table wines containing up to 14.0% v/v ethanol, the tax is $0.17 per wine gallon. If more than 14.0% v/v ethanol is present, the dessert wine tax is imposed even though the wine is normally considered a table wine, such as a late-harvested Zinfandel containing 15% v/v ethanol. The taxes on sparkling and carbonated wines are $3.40 and $2.40 per wine gallon, respectively. Any deviation from the limits set causes the wine to go into a different tax category, and no tolerance for error is allowed in this determination. Cali-

Table 14. Legal Ranges and Limits for Ethanol in Wines

	Ethanol (% v/v)		
	California (min.)	United States	
		min.	max.
Table wines (including sparkling and nonsparkling wines and carbonated wine)	7.0	7.0	14.0
Dessert Wines	14	14	24
Angelica, port, muscatel	18.0	18.0^a	24.0^b
Sherry	17.0	17.0^a	24.0^b

Data courtesy of Wendell Lee of the Wine Institute, San Francisco.
^{a}Light sherry, Angelica, and so on, may be in excess of 14 and not over 17 and 18, respectively.
^{b}The tax rate for dessert wines goes from $0.67 per wine gallon to $2.25 per wine gallon at more than 21% v/v ethanol. The tax is $10.50 per proof gallon for more than 24% v/v ethanol.

fornia limits the lower ethanol level, not for tax purposes but in an effort to improve wine quality. Other states have different control laws. The Internal Revenue Service excise tax on wines is in addition to any state taxes.

Determination of Ethanol in the Sample

In securing a sample for ethanol analysis, care should be taken to avoid evaporation losses, dilution, or contamination. If blending or fortification has been done recently, adequate mixing to ensure complete homogeneity of the contents of the container is essential.

A number of techniques are available to determine the ethanol concentration. These include methods based on boiling point measurement; distillation followed by chemical oxidation, specific gravity, refractive index, or other properties of the distillate; gas chromatographic separation and quantification; enzymatic procedure; and others. For France (11), Germany (12), the OIV (13), and the United States (14), official or semiofficial methods are available.

Boiling Point Depression

The depression of the boiling point of water–alcohol mixtures has long been used as a measure of the alcohol concentration. A history of the methods of determining the alcohol content of liquids from Arnaldo di Vallanova in the thirteenth century was given by Marco and Leoci (15). The original concept of the ebullioscope is attributed to the Danish physicist Federico Groning, in 1823. Malligand's ebullioscope dates from 1875. It was modified by Salleron in 1881.

There are now a number of modifications; those specifically authorized by the federal BATF (16) for use in the United States are the latest models of the Malligand, S. B. Torino, Arnoldo-Sala, Ebulliometer Levesque, Tag-Twin (all with shields), Juerst, Lefco, Salleron-Dujardin, and Braun types.

Wines should be diluted so that their boiling point is within 96–100°C (204.8–212°F) and should differ by less than 4°C (7.2°F) from the boiling point of pure water. This corresponds to an ethanol concentration below 5 vol %. In practice, however, this is seldom done. The sample should also be diluted so that the sugar content is less than 2%. The theoretical error (in vol % alcohol read) should then be about ±0.1%. Unless extreme care is taken in diluting, the errors caused in these steps will be larger than direct measure of the undiluted sample. If no other method is available, one can reduce the error by distilling the wine and bringing it to temperature then to volume. While using this solution in the ebullioscope, Tey (17) obtained fewer errors. This, of course, takes more time, negating the chief advantage of the method.

Figure 10 illustrates the main parts of a typical ebulliometer. The wine is placed in a shielded, enclosed vessel having a side-arm where heat is applied. A thermometer extends down into the vapor space above the liquid. A condenser is mounted above, with the liquid return down the side of the vessel.

It is important that the ebulliometer be kept free of scale. It should be boiled out with a solution of sodium hydroxide (2 g/100 mL) after about 50 determinations, then rinsed free of sodium hydroxide with water. The adequacy of the rinsing should be tested by boiling several water samples. If the maximum temperature of the successive boilings is constant, the caustic soda has been completely removed.

Most of the ebulliometers come with alcohol lamps, which should be filled with denatured alcohol and occasionally cleaned. The cotton wicks should be trimmed and should be long enough to serve the purpose. A micro-Bunsen burner can also be used instead of the alcohol lamp.

The special thermometers used in the ebulliometers are fragile and should be handled with care. If there is a break in the mercury in the column, it should be removed by careful tapping, gentle cooling, and heating, or swinging. If these methods fail, the bulb should be heated very carefully to cause the break to rise into the open space at the top of the thermometer or until the two portions unite.

A procedure for the Braun-type ebulliometers is given below. The procedure for other types is similar; however, the instruction manual should be consulted for details.

Procedure

Rinse the ebulliometer thoroughly and drain. Add 25 mL of water through the thermometer opening, then place the thermometer back in its opening, being sure

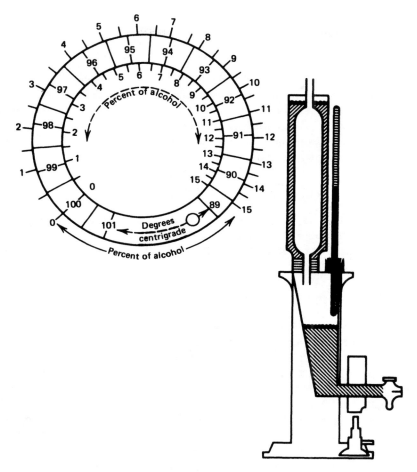

Figure 10. Typical ebulliometer with calculator.

the rubber stopper fits snugly. Fill the condenser jacket with cold water. Light the burner and adjust the flame so the tip touches the horizontal tube. When the boiling water reaches a constant temperature, record the temperature. Take the calculator and set the revolving disk so that the temperature value recorded is just opposite zero on the outer circle. Lock the calculator by tightening the screw.

Dilute the wine sample to bring its alcohol content within about 5% (bp = 96–100°C). Drain the condenser and boiling vessel and use the diluted wine to rinse the vessel thoroughly. Add 50 mL of the diluted wine sample to the vessel through the thermometer opening. Insert the thermometer and connect the condenser. Fill the condenser with cold water, then boil the sample as described above. When a constant temperature is reached, record its value. Discontinue the heating if the condenser becomes hot.

Locate the boiling point of the diluted wine sample on the inner circle of the calculator and read the alcohol value opposite to it. Multiply this value by the dilution factor to determine the alcohol concentration in volume percent.

Several procedures have been used to correct ebulliometer readings for the effect of sugar and to correct for the influence of other constituents on the boiling point. One method is to multiply the percentage of sugar by 0.05 and subtract it from the alcohol concentration. Procopio (18) proposed to use the following formula for the Malligand ebulliometer:

$$\text{Ethanol }(\%) = E\% \frac{100 - (S\%)(0.62)}{100}$$

where $E\%$ is the ethanol determined by the ebulliometer (vol %) and $S\%$ is the sugar content (wt %). The tables of Love (19) and Churchward (20) may also be useful.

Separation by Distillation

Once the volatile components have been separated from the dissolved solids, the measurement of ethanol is greatly simplified.

There are a number of equipment setups described in the literature (5, 11, 13, 14, 21–23). Figure 11 shows a practical and simple alcohol macrodistillation setup. The distillation flask usually has a 500-mL capacity, and ground-glass joints are preferred. Some type of foam breaker is essential, especially when young wines are being distilled. Antifoam agents are not recommended unless absolutely necessary. The condenser should be of adequate length and cooling capacity, and its outlet should reach nearly to the bottom of the receiving flask. One satisfactory microdistillation arrangement using steam distillation (Figure 12) is designed for the rapid distillation of 0.5–2 mL of wine.

The electric Kirk-type distillation (No. 51105-006, VWR Scientific, Box 3200, San Francisco, CA 94119) gave repeatability and reproducibility equal to those of the steam Scott-type apparatus (Scott Laboratories, San Rafael, CA 94901) in a collaborative study (24). The Kirk is now an alternative distilling apparatus in the official AOAC dichromate method for ethanol (25). An economical electronic distilling apparatus has been recommended (26). After transferring the wine, the flask is rinsed three times and is then added to the distilling flask. A CaO suspension (120 g/L) and an antifoam agent are placed in the flask and ≥ 75 mL is distilled. The apparatus automatically stops when ≥ 75 mL has been distilled.

Some of the common errors that occur are: failure to neutralize wine high in

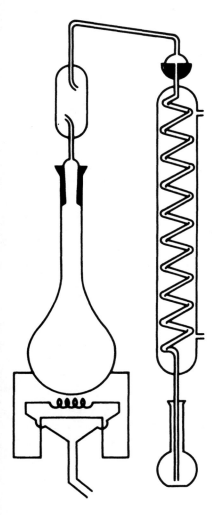

Figure 11. Macrodistillation equipment for alcohol separation.

sulfur dioxide or acetic acid; losses during distillation as a result of poor connections, foaming over, or insufficient cooling capacity of the condenser; evaporation losses between condenser and receiving flask; failure to bring to proper volume or temperature; and dirty equipment.

Procedure

Macrodistillation Apparatus. Fill the wine sample in a 100- or 200-mL volumetric flask held at 20°C, then fill the flask to volume with a few millimeters of the sample when it is temperature equilibrated. Pour the wine sample into the

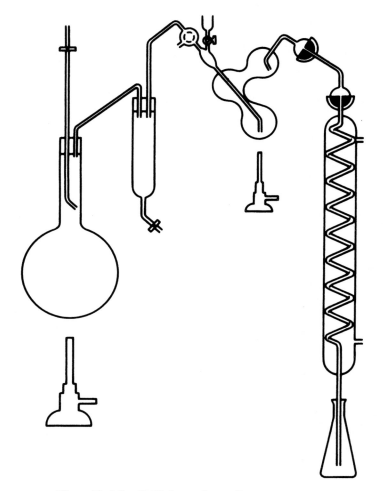

Figure 12. Microdistillation equipment for steam generator.

distillation flask and rinse the volumetric flask with a total of 50 mL of water in three equal portions. Add the rinse to the wine. Add a few boiling chips, place the same volumetric flask as the receiver under the condenser, and distill until 90–95% of the original volume of the wine sample is distilled over. Remove the flask from the condenser and rinse off the tip of the condenser into the flask with a few milliliters of water. Return the volumetric flask to the water bath at 20°C and, after equilibration, dilute to volume with water. The residue may be washed into a 100- or 200-mL volumetric flask, made to volume and temperature, and used for determining the extract, sugar, and other nonvolatile constituents.

If the volatile acidity—expressed as acetic acid—of the wine exceeds 1 mg/mL, neutralize the wine before distillation. From the total acidity, calculate the amount of base required and add 2 N sodium hydroxide solution accordingly. Also neutralize the wine if it contains a high amount of sulfur dioxide. If the wine foams excessively during the distillation, it is better to distill the unneutralized wine first and then neutralize the distillate and redistill it.

Microdistillation Apparatus. Place the wine sample in a 20°C bath and equilibrate at this temperature. Bring the steam generator to a boil with the vent open. Place the receiving flask under the condenser and turn the water on in the condenser. Pipet 1 mL of the sample into the distillation bulb and rinse the sample into the bulb with several milliliters of water until the distilling bulb is about one-third full. Close the filling vent and allow the steam to distill the alcohol.

Place a microburner below the distillation bulb and adjust the flame so that the liquid volume remains constant in the bulb. Collect the required volume of sample. Its use in the chemical determination of ethanol is discussed below. Remove the receiving flask, vent the steam generator, and remove the residue from the distilling bulb by an automatic siphon system. Rinse and drain the bulb to have it ready for the next sample.

The Cash distillation apparatus (normally used for volatile acidity determination) can also be used to steam distill alcohol samples ranging from 1 to 10 mL in volume (Figure 6). Adequate cooling, beyond that necessary for volatile acidity condensing, is required.

Determination of Ethanol in the Distillate

Once the ethanol has been separated from most of the interfering substances, a number of options are available for its determination. The most common methods correspond to general principles of density measurement. A newly developed method uses an oscillator whose rate of oscillation depends on the density of the liquid. A chemical method is also available.

Hydrometer

Ethanol determination by hydrometer is based on the Archimedean principle that an object placed in a liquid displaces a volume of liquid equal to its own weight. A special alcohol hydrometer (Figure 13) is immersed in the distillate. The hydrometer must be free of all residue, including traces of fatty or oily materials. Tey (17) recommends cleaning in 20% sodium hydroxide, then rinsing with dilute hydrochloric acid and distilled water. Depending on the specific gravity of the solution, the hydrometer will sink (or rise) to a depth at which it just displaces the volume of liquid that is exactly of the same weight as the hydrometer. Special hydrometers are available that are calibrated directly in vol % ethanol. Most hydrometers are calibrated at 15.56°C (60°F). The most useful hydrometers are in 5% ranges, with 0.05% or 0.1% divisions.

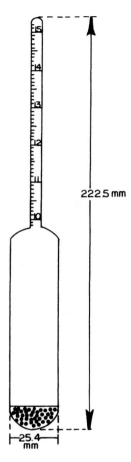

Figure 13. Alcohol hydrometer.

Procedure

Bring the sample of distillate to or near 15.56°C (60°F) (do not readjust volume). The hydrometer cylinder (32-mm i.d. × 220 mm) is rinsed with a few milliliters of the distillate. The cylinder and hydrometers must be free of grease and dirt and should be at or near 15.56°C (60°F). Hydrometers are very fragile and require careful handling. Hold the instrument by the stem, spin gently, and push downward: The hydrometer will seek its level, and the value can be read from the stem directly below the meniscus and can be recorded. Sufficient time must be allowed for the hydrometer to reach equilibrium. If the temperature of the sample differs from 15.56°C, see Table 15 for corrections.

Table 15. Temperature Corrections of Alcohol Hydrometers Calibrated at 15.56°C (60°F) in Volume % Ethanol

Observed alcohol content (vol %)	Add								Subtract															
	at 57°F 13.9°C	at 58°F 14.4°C	at 59°F 15.0°C	at 61°F 16.1°C	at 62°F 16.7°C	at 63°F 17.2°C	at 64°F 17.8°C	at 65°F 18.3°C	at 66°F 18.9°C	at 67°F 19.4°C	at 68°F 20.0°C	at 69°F 20.6°C	at 70°F 21.1°C	at 72°F 22.2°C	at 74°F 23.3°C	at 76°F 24.4°C	at 78°F 25.6°C	at 80°F 26.7°C
1	0.14	0.10	0.05	0.05	0.10	0.16	0.22	0.28	0.34	0.41	0.48	0.55	0.62	0.77	0.93	1.10	1.28	1.46
2	0.14	0.10	0.05	0.05	0.11	0.17	0.23	0.29	0.35	0.42	0.48	0.56	0.63	0.78	0.94	1.13	1.31	1.50
3	0.14	0.10	0.05	0.06	0.12	0.18	0.24	0.30	0.36	0.43	0.50	0.57	0.64	0.80	0.96	1.17	1.35	1.54
4	0.14	0.10	0.05	0.06	0.12	0.19	0.25	0.32	0.38	0.45	0.52	0.59	0.67	0.83	1.00	1.21	1.40	1.60
5	0.15	0.10	0.05	0.07	0.13	0.20	0.26	0.33	0.40	0.47	0.54	0.62	0.70	0.86	1.03	1.27	1.46	1.66
6	0.17	0.11	0.06	0.07	0.14	0.20	0.27	0.34	0.42	0.50	0.57	0.66	0.74	0.90	1.09	1.32	1.52	1.73
7	0.18	0.12	0.06	0.07	0.14	0.21	0.29	0.36	0.44	0.52	0.60	0.68	0.77	0.94	1.13	1.38	1.59	1.80
8	0.19	0.13	0.06	0.08	0.16	0.23	0.31	0.39	0.47	0.55	0.64	0.73	0.81	0.99	1.18	1.46	1.67	1.89
9	0.21	0.14	0.07	0.08	0.16	0.24	0.32	0.41	0.50	0.58	0.67	0.76	0.86	1.04	1.25	1.54	1.76	1.99
10	0.23	0.16	0.08	0.08	0.17	0.25	0.34	0.43	0.52	0.61	0.71	0.80	0.90	1.10	1.32	1.61	1.84	2.09
11	0.25	0.16	0.08	0.09	0.18	0.27	0.37	0.46	0.56	0.65	0.75	0.85	0.96	1.16	1.39	1.70	1.94	2.20
12	0.27	0.18	0.09	0.10	0.20	0.29	0.39	0.49	0.59	0.70	0.80	0.91	1.02	1.23	1.46	1.80	2.05	2.31
13	0.29	0.19	0.10	0.10	0.21	0.31	0.42	0.52	0.63	0.74	0.85	0.97	1.08	1.31	1.55	1.91	2.17	2.44
14	0.32	0.21	0.11	0.11	0.22	0.32	0.44	0.55	0.66	0.78	0.91	1.02	1.14	1.39	1.65	2.03	2.30	2.58
15	0.35	0.23	0.12	0.12	0.24	0.35	0.48	0.60	0.71	0.84	0.97	1.10	1.23	1.50	1.76	2.16	2.44	2.72
16	0.37	0.24	0.12	0.13	0.26	0.38	0.52	0.65	0.77	0.90	1.03	1.17	1.31	1.60	1.88	2.28	2.58	2.87
17	0.40	0.26	0.13	0.14	0.27	0.41	0.54	0.68	0.82	0.96	1.10	1.25	1.40	1.70	1.99	2.41	2.72	3.02
18	0.44	0.29	0.14	0.14	0.29	0.44	0.58	0.73	0.88	1.03	1.18	1.33	1.49	1.80	2.10	2.54	2.86	3.17
19	0.47	0.32	0.16	0.15	0.30	0.46	0.62	0.78	0.94	1.10	1.26	1.42	1.58	1.90	2.22	2.65	2.98	3.33
20	0.51	0.34	0.17	0.16	0.32	0.49	0.66	0.82	0.98	1.15	1.33	1.48	1.65	2.00	2.32	2.76	3.10	3.45
21	0.53	0.35	0.18	0.17	0.34	0.51	0.68	0.85	1.02	1.20	1.38	1.54	1.72	2.06	2.41	2.84	3.20	3.56
22	0.56	0.38	0.19	0.17	0.36	0.53	0.71	0.90	1.07	1.25	1.44	1.61	1.78	2.13	2.48	2.93	3.30	3.67
23	0.58	0.40	0.20	0.18	0.37	0.55	0.74	0.92	1.11	1.30	1.49	1.66	1.84	2.20	2.56	3.03	3.40	3.78
24	0.60	0.40	0.20	0.18	0.38	0.56	0.77	0.96	1.16	1.35	1.54	1.72	1.91	2.27	2.65			

To or from the observed

From Reference 22.

From the ethanol concentration observed and the volumes of the distillate, and the original wine sample (if different), the ethanol concentration of the latter can be calculated.

Some common sources of error are: (a) failure to neutralize wine that is high in sulfur dioxide or acetic acid, (b) losses during distillation (due to poor connections, foaming over, insufficient cooling capacity of the condenser, evaporation losses between condenser and receiving flask), (c) incorrect reading of hydrometer (parallax or direction), (d) failure to bring original distillate to proper volume or temperature, and (e) dirty equipment (the cylinder, the volumetric flask, and/or the hydrometer).

Pycnometer

The pycnometer method is considered to be the international standard (12, 13). It is generally used as the method of final comparison and is also an approved method of the AOAC (14). The weight of a certain volume of an alcohol distillate is compared to the weight of exactly the same volume of a water solution. The ratio of the weights of the two solutions gives the specific gravity of the distillate. From this value the alcohol can be calculated as vol %. A good analytical balance (sensitivity of ± 0.1 mg) is required as well as a vibrational-free, draft-free room to house the balance and a special volumetric glass container (the pycnometer) equipped with a thermometer (Figure 14). A repeatability of 0.1% and a reproducibility of 0.2 vol % can be attained 19 times in 20, for an alcohol content of 10–17% (24, 27), by skilled operators. It is not a method for amateurs. Considerable practice is necessary for acceptable results.

Procedure

Clean the pycnometer by soaking in cleaning solution, then thoroughly rinse with distilled water, with alcohol, and finally with acetone. Allow the pycnometer to dry, then wipe with a chamois skin. Weigh the empty pycnometer with the cap on the side-arm and with the ground-glass fitted thermometer in place. Fill the pycnometer with recently boiled distilled water 3–4°C cooler than the water bath temperature of 20°C (68°F). Insert the thermometer, and after the vessel and liquid have equilibrated to the water bath temperature, put on the cap. Use a piece of filter paper to remove any liquid overflow. Remove the assembled pycnometer from the water bath, dry with a cloth and then with the chamois, and weigh. The difference between the weights of the empty pycnometer and the full one is the weight of the water (W). Empty the pycnometer, carefully rinse with the alcohol distillate, and fill and repeat as with the water. The weight of the distillate (S) is the difference between the empty pycnometer weight and the weight of the pycnometer filled with the distillate. The specific gravity of the distillate is expressed as S/W. Correct the specific gravity for temperature by reference to the appropriate table in the AOAC *Official Methods of Analysis* (14). Correct the specific

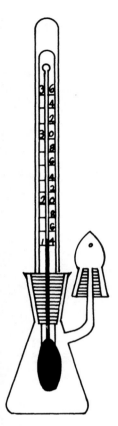

Figure 14. Pycnometer with thermometer.

gravity to *in vacuo* conditions by the following formula:

$$\text{Specific gravity } in \ vacuo = \frac{S - 0.00105W}{W + 0.00105W}$$

The exact volume of the pycnometer can be calculated from the weight of water held in the pycnometer at a known temperature. The weight of air held at that volume at that temperature can also be calculated. Knowing these two values, the density of the alcohol can be calculated (28). Corrections for mass of water and distillates to correct volumes at 20°C are given by Jaulmes et al. (27). For details of derivation of the international alcohol tables, see Jaulmes et al. (29). Table 16 is for various methods of expressing ethanol content by different systems and temperatures (30). Equations for correcting ethanol content to 20°C when measured at other temperatures have been derived (31).

Table 16. Interconversion of Alcohol Content at Various Temperatures

Percent by volume, 15.56°C (60°F) (U.S. Bur. Stds.)	Percent by volume, 20°C (IUPAC)	Percent by volume, 15°C (Osborne)	Percent by volume, 15°C (Gay-Lussac)	Percent by volume, 15.56°C (Windisch)	Percent by volume, 15.56°C (Tralles)	Percent by volume, 15.56°C (English)	Percent by volume, 15°C (Swedish)
1.00	1.00	1.00	0.96	1.00	0.98	1.00	1.01
2.00	2.01	2.00	1.95	2.01	1.98	2.00	1.99
3.00	3.01	3.00	2.93	3.01	2.98	3.01	2.94
4.00	4.01	4.00	3.95	4.01	3.98	4.01	3.93
5.00	5.02	5.00	4.97	5.02	4.98	5.02	5.00
6.00	6.03	6.00	5.97	6.02	5.98	6.02	6.01
7.00	7.03	7.00	6.94	7.00	6.97	7.00	7.09
8.00	8.03	8.00	7.94	8.00	7.97	8.00	8.11
9.00	9.04	9.00	8.92	8.97	8.96	8.99	9.12
10.00	10.04	10.00	9.91	9.97	9.97	9.99	10.06
11.00	11.04	10.99	10.91	10.99	10.99	11.01	11.01
12.00	12.05	12.00	11.92	12.02	12.03	12.04	12.04
13.00	13.05	13.00	12.92	13.02	13.03	13.04	13.09
14.00	14.05	14.00	13.93	14.06	14.05	14.05	14.12
15.00	15.05	14.99	14.92	15.08	15.08	15.06	15.08
16.00	16.05	15.99	15.91	16.10	16.12	16.05	16.05
17.00	17.06	16.99	16.91	17.13	17.12	17.08	17.13
18.00	18.06	17.99	17.88	18.13	18.18	18.08	18.13
19.00	19.06	18.99	18.88	19.13	19.13	19.08	19.13
20.00	20.06	19.99	19.82	20.10	20.11	20.05	20.12
21.00	21.07	20.99	20.81	21.10	21.09	21.05	21.02
22.00	22.07	21.99	21.81	22.10	22.09	22.05	22.02
23.00	23.07	22.99	22.82	23.10	23.09	23.05	23.09
24.00	24.07	23.99	23.85	24.10	24.10	24.07	24.12
25.00	25.07	24.99	24.87	25.10	25.10	25.07	25.10

From Reference 20.

Hydrostatic Balance

The density of the distillate can also be determined by the use of the Westphal hydrostatic balance (13). To secure the required sensitivity, the delicate equipment must be operated carefully. It is easily subject to damage or deterioration so is seldom used. The recent hydrostatic balance results for ethanol in 60 fermenting musts and wines differed by 0.05% or less when compared to those obtained by pycnometry (32). The instrument automatically allows for variation in temperatures.

Density Meter

A precise density meter (sometimes reported as the Mettler–Paar® density meter or as an oscillating U-tube digital density meter) has been developed (33). The theory of this instrument involves the development and solution of an equation describing simple harmonic motion. This theory is applicable to a glass U-tube containing the test liquid. The U-tube is caused electromagnetically to vibrate at its natural frequency. The density of the test liquid is proportional to the change in the U-tube vibrating frequency:

$$f = \frac{1}{2\pi} \left(\frac{c}{M_0 + eV} \right)^{1/2}$$

where f = vibrational frequency
c = elastic modulus of the glass
M_0 = mass of the U-tube
π = density of the liquid
V = volume of the liquid
$T = 1/f$ (T = period of oscillation)
$e = A(T^2 - B)$

Note that

$$A = \frac{4\pi^2 V}{c} \quad \text{and} \quad B = \frac{4\pi^2 M_0}{c}$$

Since A and B are characteristics of the apparatus, the only variable to determine is T, or a time measurement, which can be done electronically and with great precision. The temperature of the sample must be controlled precisely at 20°C (68°F). Density measurements on many liquids, accurate to four decimal places, have been reported (34), and the method compares favorably to the pycnometer method. The instrument appears to be equal or superior to the pycnometer for density determination of juice and wine (35–37). The time required for each measurement is less than 5 min. In fact, Tey (17) used a minimum of 2 min.

Joaskelainen (38) compared results from a pycnometer, from gas chromatography, and from a density meter. He found the gas chromatographic results to be the most reliable, followed by the density meter results and the pycnometer results, respectively. Strunk et al. (39) conducted a collaborative study of 14 laboratories with seven samples ranging from 0.5 to 85% ethanol. They found the density meter results to be 0.00 to 0.085% lower than the true value. The nonlinearity occurs when the specific gravity of ethanol solutions is plotted against the concentration. A graph or programmed calculator can be used to correct for this difference. They found it necessary to rinse the U-tube with acetone and to dry with steam after 10 samples or whenever erratic digital displays are noted. It has been shown (40) that is is difficult to operate the density meter at temperatures other than 15.56°C (60°F) and that results have to be corrected. A program was developed that read proof directly at 20°C (68°F). The calculator program results compared favorably with AOAC and NBS tables. The 20°C temperature of operation allows the density meter to operate more satisfactorily because of fogging of the viewing window at lower than ambient temperatures. A copy of the program with documentation and associated data may be obtained from L. E. Stewart at the National Laboratory Center of the Bureau of Alcohol, Tobacco, and Firearms, 1401 Research Blvd., Rockville, MD 20850. Figure 15 is a diagram of the essential parts of the density meter. The instruction manual for the Mettler-Paar Digital Density Meter, (41) should be carefully followed. It gives tables for the density of air and water at various temperatures. Recently, complete density–percent-alcohol tables and computer programs have been published (41a).

Refractive Index Measurement

The ethanol content of the distillate can also be estimated from the measurement of the refractive index. Ethanol causes the most significant contribution to

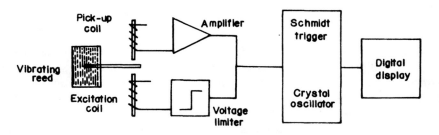

Figure 15. Schematic diagram of the Mettler-Paar digital density meter.

changes in refractive index; the other components that interfere in the distillate are primarily the secondary alcohols. Jaulmes et al. (42) have graphically compared the effects of various components on the refractive index. These are, in decreasing order of greatest effect on the ethanol: amyl alcohol, sulfur dioxide, isobutanol, acetic acid, ammonia, methanol, and carbon dioxide. A special refractometer is required for these measurements (refractive indices between 1.333 and 1.346 with a maximum error of 2×10^{-5}). Close ($\pm 0.25\,°C$) temperature control (and correction if necessary) is also required. Tables are available for dilution corrections (43) and for relating the refractive index to percent alcohol at several temperatures (44). Table 17 relates specific gravity to alcohol concentration. The errors ($\pm 0.1\%$ v/v) are comparable with those of the pycnometric method when a precision refractometer is properly operated.

Chemical Oxidation

The use of dichromate to oxidize ethanol to acetic acid was originally proposed in 1903 and since that time has been developed into a useful procedure (45–49). Collaborative studies show excellent results (50, 51) for winery laboratory use. Modifications have reduced the distilling time to several minutes, and the overall determination to 5 min (52); the method has also been automated for use with continuous systems (53).

The oxidation of ethanol in the presence of sulfuric acid and dichromate yields acetic acid:

$$2Cr_2O_7^{2-} + 3C_2H_5OH + 16H^+ \longrightarrow 4Cr^{3+} + 3CH_3COOH + 11H_2O$$

The proper hydrogen ion concentration is necessary; otherwise the ethanol may be oxidized to acetaldehyde or to a mixture of acetaldehyde and acetic acid. The dichromate that has not been used is reduced by titration with ferrous ammonium sulfate:

$$Cr_2O_7^{2-} + 6Fe^{2+} + 14H^+ \longrightarrow 2Cr^{3+} + 6Fe^{3+} + 7H_2O$$

The potential for the $Cr_2O_7^{2-} \rightarrow Cr^{3+}$ reaction in N hydrochloric acid is 1.09 V. The visual end point of the titration is not sharp. Two organic indicators have been successfully used: sodium diphenylamine sulfonate ($E = 0.85$ V) and 1,10-o-phenanthroline ($E = 1.06$ V).

Procedure

Prepare the dichromate solution in the following way. Add 325 mL of concentrated sulfuric acid to about 400 mL of water in a 1-L volumetric flask, mix, and cool to 80–90°C. Dissolve exactly 33.768 g of potassium dichromate primary standard in the mixture and dilute the solution to volume with water at 20°C.

Table 17. Proof, Percentage Alcohol by Volume and by Weight, and Specific Gravity at 15.56°C (60°F)[a]

Proof	Alcohol (vol %)	Alcohol (wt %)	Alcohol (g/100 mL)	Specific gravity
0.0	0.00	0.00	0.00	1.00000
1.0	0.50	0.40	0.40	0.99925
2.0	1.00	0.795	0.79	0.99850
3.0	1.50	1.19	1.19	0.99777
4.0	2.00	1.593	1.59	0.99703
5.0	2.50	1.99	1.98	0.99630
6.0	3.00	2.392	2.38	0.99558
7.0	3.50	2.80	2.78	0.99488
8.0	4.00	3.194	3.18	0.99418
9.0	4.50	3.60	3.58	0.99350
10.0	5.00	3.998	3.97	0.99281
11.0	5.50	4.40	4.37	0.99215
12.0	6.00	4.804	4.76	0.99148
13.0	6.50	5.21	5.16	0.99085
14.0	7.00	5.612	5.56	0.99021
15.0	7.50	6.02	5.96	0.98960
16.0	8.00	6.422	6.35	0.98898
17.0	8.50	6.83	6.75	0.98838
18.0	9.00	7.234	7.14	0.98778
19.0	9.50	7.64	7.54	0.98719
20.0	10.00	8.047	7.93	0.98659
21.0	10.50	8.45	8.33	0.98600
22.0	11.00	8.862	8.73	0.98542
23.0	11.50	9.27	9.13	0.98485
24.0	12.00	9.679	9.52	0.98427
25.0	12.50	10.08	9.92	0.98372
26.0	13.00	10.497	10.31	0.98316
27.0	13.50	10.90	10.71	0.98262
28.0	14.00	11.317	11.11	0.98208
29.0	14.50	11.72	11.51	0.98155
30.0	15.00	12.138	11.90	0.98102
31.0	15.50	12.54	12.30	0.98049
32.0	16.00	12.961	12.69	0.97994
33.0	16.50	13.37	13.09	0.97946
34.0	17.00	13.786	13.49	0.97892
35.0	17.50	14.19	13.89	0.97841
36.0	18.00	14.612	14.28	0.97791
37.0	18.50	15.02	14.68	0.97741
38.0	19.00	15.440	15.08	0.97691
39.0	19.50	15.84	15.47	0.97642

Table 17. (*Continued*)

Proof	Alcohol (vol %)	Alcohol (wt %)	Alcohol (g/100 mL)	Specific gravity
40.0	20.00	16.269	15.87	0.97593
41.0	20.50	16.67	16.26	0.97543
42.0	21.00	17.100	16.66	0.97493
43.0	21.50	17.51	17.06	0.97442
44.0	22.00	17.933	17.46	0.97392
45.0	22.50	18.34	17.86	0.97341

From Reference 44.
^aReferred to water at the same temperature. Multiply by 0.99908 to convert to specific gravity referred to water at 4°C (39.2°F).

Prepare the ferrous ammonium sulfate solution by dissolving 135.5 g of ferrous ammonium sulfate hexahydrate ($FeSO_4(NH_4)_2SO_4 \cdot 6H_2O$) in 500 mL of water in a 1-L volumetric flask, adding 30 mL of concentrated sulfuric acid, and diluting with water to volume at 20°C.

Prepare the 1,10-phenanthroline-ferrous sulfate indicator solution by adding 0.695 g of ferrous sulfate heptahydrate to about 50 mL of water in a 100-mL volumetric flask, then adding 1.485 g of 1,10-o-phenanthroline, and diluting to volume with water.

Carry out the microdistillation procedure, as described earlier (see Figure 12). Use a 50-mL flask, as the receiver, containing 25 mL of dichromate solution. First, do a blank distillation using water as the sample. Distill 15 mL over into the receiving flask, then remove the flask, stopper, and place in a 60°C water bath. After 20–25 min the oxidation is complete; now remove the flask and rinse the contents quantitatively with water into a 500-mL Erlenmeyer flask and titrate with ferrous ammonium sulfate solution to a green color under daylight conditions. Add several drops of indicator solution and continue the titration. When the sharp change from blue-green to brownish-purple occurs, the end point has been reached. Carry out blank titration daily because the titer of ferrous ammonium sulfate solution slowly changes.

Carry out the sample treatment using 1 mL of wine in the same way. Calculate the result as follows:

$$\text{Ethanol (g/100 mL)} = \left[25 - \left(25\frac{A}{B}\right)\right]\left(\frac{7.933}{10}\right)$$

where A = volume of ferrous ammonium sulfate used to titrate the remaining dichromate (mL)

B = volume of ferrous ammonium sulfate used to titrate the blank (mL)

The equation is valid only if the dichromate is prepared to the exact specifications. To calculate as vol %, the equation reduces to

$$\text{Ethanol (vol \%)} = 25 - 25\frac{A}{B}$$

The errors that can occur most frequently are those associated with poor handling and distillation techniques. Wines high in acetaldehyde cause small errors, usually less than 0.05% as ethanol. Standard deviations of ±0.04 and ±0.07% for wines containing 11 and 22% ethanol, respectively, have been reported (51). The effect of various volatile wine components on the dichromate determination has been studied (54). These components appear as ethanol equivalents and give a slightly higher value than methods such as gas chromatography (see Table 18). However, with pycnometer, hydrometer, and other density or refractive-index-type measurements, these compounds are also additive and in general the methods correlate well.

Pilone (55) conducted a collaborative study of the AOAC official dichromate-titrimetric method (11.008–11.011) and a spectrophotometric procedure. The official method gave better results, and the spectrophotometric method was not recommended.

Table 18. Effects of Other Volatile Components on the Apparent Ethanol Concentration by the Dichromate Oxidation Method

Component	"Ethanol equivalent" per 100 mg/L sample average (%, v/v)	Amount detected as 0.01% (v/v) ethanol: sample average (mg/L)
Methanol	0.024	41
1-Propanol	0.014	74
1-Butanol	0.004	250
2-Methyl-1-propanol	0.014	71
3-Methyl-1-butanol	0.014	71
Acetaldehyde	0.011	108
2,3-Butanediol	0.00094	1080
Ethyl acetate	0.0089	113
2-Butanone	0.0012	850
Glycerin	None detected	—
Potassium sorbate	None detected	—
Potassium metabisulfite	None detected	—

From Reference 57.

Rapid Oxidation Method

Rebelein developed a distillation method that is similar to the microdistillation technique given previously but that uses simple available equipment. The total time was 6–7 min for wine, with a mean error of ±0.05 g/100 mL (56). This method has been widely tested, and the validity of the method has been verified. The procedure is patented in Germany (57).

Procedure

Make up the following solutions. Add 67.445 g of potassium chromate to distilled water in a 1-L volumetric flask, dissolve, and bring to volume at 20°C with distilled water. Into a second 1-L volumetric flask add 770 mL of 65% nitric acid and bring the solution to volume at 20°C with distilled water. Weigh out 300 g of potassium iodide, dissolve with 100 mL of 1 N sodium hydroxide, and bring to 1-L volume at 20°C with distilled water. Accurately weigh out 86.194 g of potassium thiosulfate, dissolve in 100 mL of 1 N sodium hydroxide and 500 mL of distilled water in a 1-L volumetric flask; then bring to volume at 20°C with distilled water. Finally add 10 g of soluble starch to 500 mL of water in which 20 g of potassium iodide and 10 mL of N sodium hydroxide have been dissolved.

The equipment for the distillation appears in Figure 16. Into the 500-mL Erlenmeyer receiving flask add 10 mL of the chromate solution and 25 mL of the nitric acid solution. (This will oxidize up to 1 mL of a 11.0 vol % ethanol solution; if the ethanol concentration exceeds this, the wine sample must be diluted prior to addition.) The air-condenser exit tube should dip into this mixture. Put 12 mL of distilled water and the 1 mL of wine into the 100-mL distilling flask. Add some boiling chips and antifoam agent, if required, and connect the system.

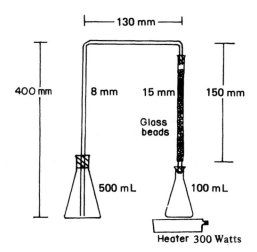

Figure 16. Diagram of Rebelein's apparatus for alcohol distillation.

Rapidly heat at boiling for 3 min. (Distillation should be complete.) Put 300 mL of distilled water into the receiving flask. Add 10 mL of the potassium iodide solution and 10 mL of the starch solution. Add a magnetic stirrer bar and titrate the solution with the thiosulfate solution to a blue, starch end point. The reactions are:

$$2CrO_4^{2-} + 3C_2H_5OH + 10H^+ \longrightarrow 2Cr^{3+} + 3C_2H_4O_2 + 5H_2O$$

(The use of nitric rather than sulfuric acid facilitates the oxidation by forming an ester with the ethanol.)

Then the excess chromate is reduced:

$$2CrO_4^{2-} + 9I^- + 16H^+ \longrightarrow 2Cr^{3+} + 3I_3^- + 8H_2O$$

The iodine produced is titrated with the thiosulfate:

$$I_3^- + 2S_2O_3^{2-} \longrightarrow 3I^- + S_4O_6^{2-}$$

$$\frac{22.1A}{22.1}\,(11.5\text{ g}/100\text{ mL}) = \text{g ethanol}/100\text{ mL}$$

where A = thiosulfate standard solution (mL). (To convert the answer to vol %, divide by 0.7933.)

Gas Chromatography (GC)

The determination of ethanol in wines by GC or HPLC is now a straightforward procedure. Acetone has been used as the internal standard, and good results were reported (58, 59). A precision of ±0.2 vol % for a wine containing 20% ethanol is indicated (60). Several others (61–63) have successfully determined ethanol with standard columns and equipment. Since other methods are influenced by other alcohols as well as ethanol, the results with GC and HPLC are usually less by about 0.06 vol % (17).

This procedure offers several advantages over other methods: (a) the method can be automated, (b) the precision is excellent, and (c) results are obtained extremely rapidly. In large wineries dealing with many samples, these factors can become extremely important.

A system that meets the qualifications above has been described (54). It consisted of a Varian® 1400 GC with a flame ionization detector equipped with a 6-ft × 1/4-in. copper column packed with 3% Carbowax® 600 on 40/60 mesh Chromosorb® T. Column temperature was 80°C; injector and detector temperatures were 120°C and 125°C, respectively. Helium was the carrier gas, and the flow was 110 mL/min. An autosampler was used with an injection of 0.5 μL. The wine or ethanol samples were diluted using a York 2111 automatic diluter (0.001 mL of wine to 25.0 mL) and an internal standard diluent (n-butanol). A Varian 480 electronic integrator was used to read the gas chromatographic response. Using standard ethanol solutions, it was determined that the response varied about 0.1% over a 10-day period. For 100 samples the precision was ±0.21 on duplicate determinations. For 100 samples the precision was 0.16 relative to the mean, including the error of dilution. In a collaborative study (64) the gas-liquid chromatography (GLC) procedure had a repeatability of 0.036 and a reproducibility of 0.142 for samples from 7.3 to 23.6% v/v. No difference between three columns (Poropak® QS, Carbowax 1500, and Carbowax 600) were noted. Of the 14 collaborating laboratories the results of two were eliminated as outliers. This procedure is now official AOAC method.

HPLC

A new (3) HPLC method uses a Varian (Vista® 500) with a Varian RI-3 refractometer and a Rheodyne® (model 7125) injector (10 µL) with a Spectra-Physics® 4270 recording integrator. The inverse silica column (Varian MCH 5N) of 15 cm × 4 mm was kept at 20.00 ± 0.01°C in a 20-L thermostat. A 100-µL syringe (710 SNR 80665) was used for the Rheodyne injector. A Millipore filter for 0.50-µm particles (FHLP 02500) was used for spirits; one for 0.45-µm particles (HATF 02500) was used for wines. Distilled water was the mobile phase (1.5 mL/min) and was prepared just before use and used at 20°C. The pressure was 214 bar. The attenuation of the refractometer was 20–100 ($\Delta RI \times 10^{-6}$). The samples were at 20°C for 20 min before use. The wine samples were filtered before use. Five standard calibration samples were prepared for the ranges 0.10–1.00, 1.00–10.00, 10.00–30.00, and 30.00–50.00% ethanol. The response is linear between 0.1 and 55% v/v of ethanol. The intensity of the signal at a retention time of 1.98 min is proportional to the ethanol content. The usual wine constituents (glucose, tartaric acid, glycerol, etc.) have lower and higher retention times than ethanol. For 12 wines the results were 11.68% by the new procedure and 11.69% by the pycnometer. It is claimed to be rapid and reproducible. Methanol can also be determined at the same time.

Enzymatic Analysis

Ethanol can be determined by measuring the change in nicotinamide adenine dinucleotide (NAD^+) to the reduced nicotinamide adenine dinucleotide (NADH) by absorption changes at 340 nm (65). The reaction is:

$$CH_3CH_2OH + NAD^+ \xrightarrow{\text{Alcohol dehydrogenase}} CH_3CHO + NADH + H^+$$

The solution is buffered toward the alkaline side, and semicarbazide is added to remove the acetaldehyde formed, thus forcing the reaction to completion. The test is very sensitive. It is very useful for solutions of less than 1% v/v ethanol.

The configuration of NAD^+ and NADH is given below.

NAD$^+$ NADH

Changes in the relative amounts of each configuration cause changes in the absorbance (Figure 17). The maximum for NADH (also NADPH) is at 340 nm. The molar extinction coefficient at 340 nm for NADH (also NADPH) at 25–37°C is 6.3×10^3 L $\times$ mol^{-1} $\times$ cm^{-1}.

The classic method of enzymatic analysis for ethanol is given below.

Procedure

Prepare the following solutions. Make up a buffer solution (pH 9.0) by dissolving 10 g of $Na_4P_2O_7 \cdot 10H_2O$, 2.5 g of semicarbazide hydrochloride, and 0.5 g of glycine in 250 mL of distilled water in a 300-mL volumetric flask. Bring to volume at 20°C with distilled water. (The solution is stable for 3 weeks at 5°C.) Dissolve 72 mg of NAD^+ in 6 mL of distilled water. (This solution also is stable for 4 weeks at +4°C.) Make up a solution of alcohol dehydrogenase, 30 mg/mL. Either use 1 mL of undiluted suspension or dissolve 50-mg of lyophilized enzyme ($\simeq$ 30 mg of enzyme protein) in 1 mL of distilled water. (The undiluted suspension is stable for 6 months at +4°C; the aqueous solution is stable for only 1 week at +4°C.) Prepare dilute ethanol standards daily.

Dilute samples so that they are in the range of 0.01–0.15 g/L. For example, a wine with about 10 g/100 mL ethanol concentration would have to be diluted 1000-fold to 0.10 g/L to get it in this range. Prepare a blank spectrophotometric sample by adding 2.50 mL of buffer, 0.10 mL of NAD^+ solution, and 0.50 mL of water to a test tube; mix and after 2 min read optical density (E_1), then add 0.02 mL of NADH solution and mix again. Seal the tube and put it into a 37°C water bath for 25 min. Do the same preparation for the sample except add 0.10 mL of the sample and only 0.40 mL of water. Read the optical density of the sample (E_2) at 340 nm using a 1-mm light-path cuvette. Read both samples with the same cuvette.

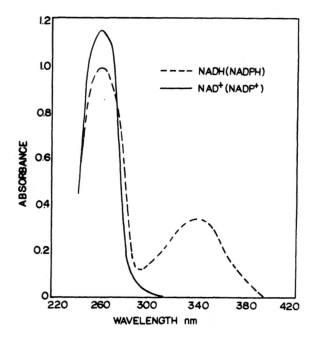

Figure 17. Absorption spectra of nicotinamide adenine dinucleotides.

The value of the alcohol is calculated as follows:

$E = E_s - E_b$
$E_s = $ sample $E_2 - $ sample E_1
$E_b = $ blank $E_2 - $ blank E_1

$$\text{Ethanol (g/100 mL)} = \frac{(V)(MW)(\Delta E)(F)}{(e)(d)(v)}$$

where V = final volume (3.12 mL)
  MW = molecular weight 46.07 (for ethanol)
  e = molar extinction coefficient at 340 nm
    (6.3 L × mmol^{-1} × cm^{-1})
  d = light path (cm) (0.1)
  v = sample volume (100 mL)
  F = dilution factor (1000 for dry wines)

It is best to decolorize the wines prior to determinations.
Kits developed for this method of analysis for wines are discussed by McCloskey and Replogle (66). The kit they used contained amino-oxyacetic acid

in place of semicarbazide to remove the acetaldehyde. Time was reduced to 7–15 min. An even more rapid kit system is available commercially (Boehringer GmbH, Mannheim, Germany) which uses aldehyde dehydrogenase to convert the acetaldehyde to acetic acid:

$$CH_3CHO + NAD^+ + H_2O \xrightarrow{\text{Aldehyde dehydrogenase}} CH_3COOH + NADH + H^+$$

This method only requires a total of 5 min of development time at 25°C. The slowness of the detailed method is due to the speed at which the semicarbazide binds with the acetaldehyde. This quicker method is similar in all other respects except only 0.5 mol of ethanol is oxidized per mol of NAD^+ used.

Bouvier (67) used an automatic enzymatic procedure with the Technicon Auto Analyser, a microcomputer, and other automated equipment. It processed 40–50 samples per hour of dry and sweet white, rosé, and red wines with errors not over ± 0.05 vol %, which is comparable to other methods. See also References 68 and 69.

Use of alcohol oxidase (E.C. 1.1.3.13) has been proposed (70, 71). The immobilized enzyme was in the middle layer of a three-layer membrane (70). Ethanol reaches the enzyme and produces hydrogen peroxide and acetaldehyde. The hydrogen peroxide diffuses through the inner cellulose acetate layer and is oxidized at the platinum anode producing a current directly proportional to the hydrogen peroxide content and hence to the ethanol percent. The Industrial Analyzer® (Model 27, Yellow Springs Instrument Co., Yellow Springs, OH) was used. The method is reported to be rapid, and the membranes last 10 days. The results on eight wines were within ± 0.16 of the AOAC (14) procedure. Alcohol oxidase has been used in another procedure (71). The half-life of the enzyme was 7 days when diluted beer was analyzed. A layer of cigarette paper mounted on a Clark oxygen electrode covered by a dialysis membrane was the method employed. The results were slightly lower than by pycnometer or refractometer. More recently (72) immobilized alcohol oxidase was used with only 1 or 2 mL of wine. A reading required only 1 min.

All these enzymatic methods are limited with respect to the accuracy to which the sample can be diluted. At best for autodilutors this is ± 0.15 vol % for table wines. Other secondary alcohols, including methyl, isoamyl, active amyl, isobutyl, and n-propyl alcohols, are also oxidized to some extent, causing positive errors.

METHANOL

Methanol is not a direct fermentation product, and it is present in very small amounts in grape wines. Occasionally it is found in excess in fruit wines. The major source is the pectinase activity, which demethylates the methyl ester of α-1,4-D-galacturonopyranose units in grape pectins. Addition of pectinase to freshly crushed grapes will increase the amounts of methanol found. The cause of this activity again is the pectin esterase. It is a normal and needed enzyme when polygalacturonases are used to split up pectins present in grapes. Without partial demethylation, the pectin-splitting properties of the usual enzyme complexes added are not efficient. Yeast rarely produces any natural esterase activity to cause pectin demethylation during fermentation. Grapes have some natural pectin esterase activity that is intensified by maceration of the skins (73).

$$\text{[pectin-COOCH}_3\text{]} \xrightarrow[H_2O]{\text{ESTERASE}} \text{[pectin-COOH]} + CH_3OH$$

Table 19 summarizes methanol contents in table wines as reported in recent reviews (74, 75).

The oral lethal dose for methanol is 340 mg/kg of body weight for humans (76). This alcohol is metabolized by the liver, similarly to ethanol. In excess it can damage the optic nerve and cause blindness. The California State Food and Drug Department has set a tolerance of 0.35 vol % for methanol in brandy. Italian law allows a maximum of 2000 mg/L in brandy made from wine and 8000 mg/L if made from pomace. France has a 4800 mg/L minimum from brandies distilled from pomace wines. The suggested European Economic Community limits are twice the Italian limits (77).

There are two general methods commonly used for the determination of methanol in the presence of ethanol: chemical oxidation to formaldehyde with the development of a colored product by derivative formation or GC.

Chemical Determination

Methanol is first separated from the nonvolatile constituents by a simple distillation as discussed for ethanol determination. It is then oxidized to formaldehyde, which reacts with either rose analine or chromotropic acid. The Denigès oxidation procedure uses potassium permanganate. During the distillation, ethanol also distills together with methanol. Various workers (78) have investigated the influence of the ethanol concentration on the reaction and have established the optimum concentration. Rebelein's method (79, 80) is based on these results.

Procedure

Reagents. Prepare a potassium permanganate–orthophosphoric acid solution in the following way. Dissolve 200 g of orthophosphoric acid in 1 L of water, then titrate a 10-mL aliquot with a 1 N sodium hydroxide solution. Calculate the

Table 19. Methanol Reported in White and Red Table Wines from Various Countries

Country	Number of samples	Methanol (mg/L) Range	Average
White table wine			
Albania	11	22–50	34
Australia	19	49–129	79
France	55	27–114	55
Germany	89	34–144	76
Italy	78	0–182	62
Japan	54	13–129	56
Portugal	26	0–119	66
Spain	23	41–240	104
United States	30	4–107	43
Rosé wine and red table wines			
Albania	51	43–136	68
Australia	20	156–188	165
France	76	44–200	165
Italy	110	0–245	110
Japan	57	38–269	150
Portugal	34	63–264	154
Spain	18	66–199	115
United States	23	48–227	140

From References 74 and 75.

concentration of orthophosphoric acid in the 990-mL solution and adjust it to 3 N by the addition of either water or orthophosphoric acid. Dissolve 52.67 g of potassium permanganate in 1 L of water. Add 60 mL of the permanganate solution and 100 mL of the 3 N orthophosphoric acid solution into a 250-mL volumetric flask and dilute to volume.

Prepare an oxalic acid–sulfuric acid solution by mixing 15.75 g of oxalic acid dihydrate with 100 mL/water, add 25 g of concentrated sulfuric acid, and dilute carefully to 250 mL.

Prepare chromotropic acid (1,8-dihydroxynaphthalene-3,6-disulfonic acid) solution in the following way. Dissolve 300 mg of the acid in 20 mL of water. If chromotropic acid is available as its salt, dissolve 10 g of the salt in 25 mL of water; add 2 mL of concentrated sulfuric acid, then 50 mL of methanol; heat just to boiling and filter. Cool and add 100 mL or more of isopropanol to precipitate the free chromotropic acid. If the available chromotropic acid needs purification, carry out the same procedure but without the addition of sulfuric acid.

Determination. Carry out the distillation as described for ethanol determination by obtaining a distillate whose volume is equal to the volume of the original sample. Determine the ethanol concentration (V vol %) of the distillate and calculate the volume of distillate to be used for methanol determination (mL) as $250/V$. Measure this amount of distillate from a buret calibrated to 0.05 mL into a ground-glass, stoppered, wide-mouth flask, then add water to bring the volume to 48 mL. Add 1 mL of a 1 N silver nitrate solution and 0.5 mL of a 30% potassium hydroxide solution. Boil for 30 min with a reflux condenser attached, then cool and wash down the inside of the reflux condenser several times with a total of 10 mL of water. Attach to a distilling apparatus and distill into a 50-mL volumetric flask. When the distillation is finished, dilute to volume; the ethanol concentration of this solution will be exactly 5 vol %. Pipet a 2-mL aliquot of this distillate into a 10-mL glass-stoppered graduated cylinder, then add 5 mL of the potassium permanganate–orthophosphoric acid solution. Stopper and shake. After 15 min, add 2 mL of the oxalic acid–sulfuric acid solution and loosen the stopper to allow escape of the carbon dioxide evolved. After 15 min the solution should become colorless and clear. If some manganese dioxide should be visible, close the stopper, mix gently once or twice, and loosen the stopper again. It is necessary to maintain a high concentration of carbon dioxide for accurate determination. Now immediately add a 1-mL aliquot of this solution to 1 mL of the chromotropic acid solution in a 25-mL glass-stoppered flask, then add 10 mL of 96% sulfuric acid using an all-glass dispensing buret, mix, and stopper. Place in a 60°C bath for 20 min, then remove to a 20°C bath and cool. Using a 10-mm cuvette, measure the absorbance at 570 nm against an air standard.

To prepare a methanol stock solution, dilute 1 mL of methanol in a 1000-mL flask to volume with water. Into individual 50-mL volumetric flasks add 10 mL of a 25 vol % ethanol solution and 5-, 10-, 15-, 20-, and 25-mL aliquots of the methanol stock solution, then dilute to volume. Pipet 2-mL aliquots of each solution and carry out the oxidation in glass-stoppered graduated cylinders, starting with the addition of the permanganate–orthophosphoric acid solution. Continue the determination as above and plot the results as milligrams of methanol per liter versus absorbance. The line will not go through zero but will be linear.

Determine the methanol concentration of the sample from the calibration curve and correct for the initial dilution.

The colorimetric procedure for methanol may be automated, especially if many samples have to be analyzed regularly. Details for 20 samples per hour are given in Reference 81. See also p. 115.

Gas Chromatography (GC)

The methanol content of wines can be determined by GC. Usually the distillate is injected directly into the instrument. Earliest workers used low-molecular-weight Carbowax as the liquid phase for the separation. Later the solid-phase gas chromatographic resins were used to attain better separation. The AOAC (14) recommends Carbowax 1500 at 23% on Chromosorb W for distilled spirits. This does not effectively separate ethanol and methanol for the lower alcohol distillates of wine, however. Porapak Q has been used to quantitate methanol in wine distillates (82). The method is given as follows.

Procedure

Thoroughly clean a 2-m, 0.2-cm-i.d., stainless steel column. Pack it with Porapak QS (silylated ethylvinylbenzene polymer). Condition the column for 24 hr at 210°C. Use nitrogen carrier gas with a flow of 40 mL/min. Keep injector port and detector at 210°C and 220°C, respectively. Operate the column at 115°C. Use 3-mL injections and use a flame ionization detector. Prepare standard solutions of reagent grade methanol of 0, 25, 50, 100, and 200 mg/L in water. Take 1000 mg (density = 0.7914 g/mL) of methanol and dilute with water to 1 L in a volumetric flask (1 mL = 1 mg). Then place 0, 2.5, 5, 10, and 20 mL of this stock solution into 100-mL volumetric flasks and bring to volume with water.

Inject these samples into GLC apparatus, record peak heights, and plot calibration curves. The wine samples can be injected directly, but it is necessary to replace the glass insert in the injector daily and to heat the column to 200°C overnight to remove organic matter. Otherwise a 1:1 distillate of the wine can be used with no organic buildup and no cleanup.

HIGHER ALCOHOLS (FUSEL OIL)

Longer-chain alcohols are formed during fermentation (83). Originally the pathway of formation was thought to be primarily by the Ehrlich mechanism: oxidation to the keto acid, decarboxylation, and finally reduction of the aldehyde to the corresponding alcohol. However, it was shown later that the Ehrlich pathway could account for only a fraction of the fusel oil formed and that most of the fusel oils are formed by way of the carbohydrate route (84–86). Figure 18 presents an abbreviated series pathway for the significant higher alcohols. The three main pathways are explained below:

1. The scheme of Ehrlich–Neubauer and Fromherz is as follows:

$$R_1HCNH_2COOH + R_2COCOOH \xrightarrow{Transaminase} R_1COCOOH$$
$$+ R_2HCNH_2COOH$$

$$R_1COCOOH \xrightarrow{Decarboxylase} R_1CHO \xrightarrow[NADH]{Dehydrogenase.} R_1CH_2OH$$

2. Another pathway is by condensation of the α-keto acids with acetyl-CoA to form α-keto acids with an additional $-CH_2$ at the C-3 position:

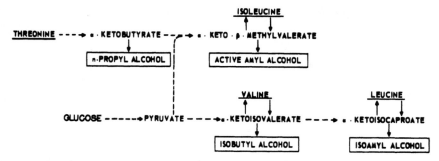

Figure 18. A general abbreviated scheme for the formation of the major higher alcohols.

$$R_1COCOOH \xrightarrow[\text{Acetyl CoA}]{\text{Acetotransferase,}} R_1COHCH_2COOH \xrightarrow{\text{Rearrangement}}$$
$$\qquad\qquad\qquad\qquad\qquad\;\; |$$
$$\qquad\qquad\qquad\qquad\qquad\; COOH$$

$$R_1HCCOOHCHOHCOOH \xrightarrow{\text{Dehydrogenase}} R_1HCCOOHCOCOOH \xrightarrow{\text{Decarboxylase}}$$
$$\qquad\qquad\qquad\qquad\qquad\qquad\qquad\qquad\qquad R_1CH_2COCOOH$$

From this point the α-keto acid is decarboxylated again and reduced from the aldehyde to the higher alcohol.

3. A third mechanism is the condensation with pyruvate or "active" aldehyde:

$$R_1COCOOH \xrightarrow[\text{or } CH_3CHO]{CH_3COCOOH} CH_3CO\overset{\overset{R_1}{|}}{\underset{\underset{COOH}{|}}{C}}OH \xrightarrow{\text{Rearrangement}}$$

$$\overset{CH_3}{\underset{OH}{\overset{|}{R_1C}\underset{|}{COCOOH}}} \xrightarrow{\text{Dehydrogenase}} \overset{CH_3}{\underset{OH}{\overset{|}{R_1CH}\underset{|}{COHCOOH}}} \xrightarrow{\text{Hydrolase}} \overset{CH_3}{\overset{|}{R_1HCCOCOOH}}$$

This method adds a methyl group to the C-3 position. From this point on, the formation is the same as in the previous two schemes.

The common higher alcohols are derived from all three schemes: *n*-propanol from schemes 1 and 2, isobutanol and active amyl alcohol from 1 and 3, and isoamyl alcohol from all three.

Table 20 lists the amounts of higher alcohols found in wines from various countries (85–93). Must, high in amino acids, yields wines with lower amounts

Table 20. Ranges and Average Values (mg/L) of Higher Alcohols Found in Wines from Various Countries

Country	Number of Samples	1-Propanol		2-Methyl-1-propanol (Isobutyl)		2-Methyl-1-butanol (Active Amyl)		3-Methyl-1-butanol (Isoamyl)		1-Hexanol		2-Phenyl-ethanol	
		Range	Average	Range	Average	Range	Average	Range	Average	Range	Average	Range	Average
Australia	39	15–50	27	15–56	36	—	—	89–328	201[a]	—	—	—	—
Australia	79	—	—	—	—	—	—	—	—	1.3–12.0	4	5–74	30
Austria	12	32–55	45	35–81	63	19–52	27	83–217	128	—	—	—	—
France	95	15–59	32	41–140	78	—	—	114–400	234[a]	—	—	—	—
France	10	—	—	—	—	—	—	—	—	—	—	10–74	33
Germany	89	11–38	21	16–104	40	—	—	155–344	229[a]	—	—	—	—
Italy	52	7–45	25	7–148	55	—	—	92–352	284	—	—	—	—
Japan	111	14–56	26	15–174	68	—	—	122–399	226[a]	—	—	—	—
Portugal	28	15–43	28	45–101	70	—	—	163–322	527[a]	—	—	—	—
Spain	41	13–36	25	25–94	59	—	—	132–237	180[a]	—	—	—	—
United States	60	16–68	34	22–57	46	29–96	52	95–288	183	—	—	—	—
United States	197	—	—	16–92	45	28–90	54	99–315	190	—	—	—	—

From References 75 and 85–93.

[a]Total values for active and isoamyl alcohols.

of higher alcohols associated with the individual amino acids (87). This is caused by increased amino acid synthesis by way of carbon skeletons from glucose. During the process a portion of the α-keto acid is decarboxylated and reduced before being transaminated to make the desired amino acid. Increased aeration can cause increase in higher alcohols during fermentation. Likewise, optimum growth conditions cause higher amounts of isobutanol and active and isoamyl alcohol (88). n-Propanol is formed in lesser amounts at optimum temperatures (88). Yeast strains cause variations in amounts of higher alcohols formed (73).

French regulations state that brandies can contain a minimum value of 1500 mg/L of total higher alcohols. Italian regulations allow a maximum value of 5000 mg/L (77).

Only in rare instances are the higher alcohols present in sufficient quantities to cause sensory effects in table wines. Threshold values for n-propyl, isobutyl, and active (+)isoamyl alcohols of 500, 500, and 300 mg/L, respectively, have been reported (93a). Occasionally the combined active and isoamyl alcohols exceed 300 mg/L, usually in a "late-harvested" red wine. However, in desert wines, especially if fortified with high-proof alcohol from pot stills, a detectable level is not unusual, although not desirable. These compounds undoubtedly have an additive effect to the wine aroma. In sound wines, 2-butanol (94) is found in concentrations ranging from 0.1 to 6.4 mg/100 mL. Three bacterially spoiled wines had 16–74 mg/100 mL. It was not recommended that 5 mg/L be the limit in wines intended for distillation. The spoiled wines were also high in allyl alcohol (8–24 mg/L). Sound wines had 0–1.0 mg/L.

In Table 20, 2-phenylethanol was found in wines in concentrations ranging from 5 to 74 mg/L (average 32 mg/L). In 268 wines (95) (Japanese and foreign) it averaged only 11 mg/L and it was considered unlikely that the 2-phenylethanol contributed to the odor of the wine. But at 74 mg/L it might (Beaujolias?). The concentration range of tyrosol (92) in 20 wines was found to be 16–45 mg/L (average 25 mg/L). Tryptophol appeared (92) in very small amounts, 0–0.8 mg/L (average 0.2 mg/L). Recently (96) 4.9–24.8 mg/L was reported. Studies using GC yielded γ-butyrolactone levels of only 0 and 13 mg/L, and it was not considered as contributing to the odor of wines (95).

The higher alcohols can be determined as a group or by chromatography individually. There are no procedures in the wine section of the AOAC (14), but sections 9.075–9.077 for spirits do apply to the determination of n-propanol, isobutanol, and total amyl alcohol. Sections 9.091–9.093 cover methanol.

Martin et al. (97) used a Hewlett-Packard® HPLC (model 5710A) with flame ionization detector for several alcohols and ethyl acetate. Each peak was confirmed by mass spectral data from a mass spectrometer. Agreement with standard procedures was only fair for methanol, n-propanol, and total amyl alcohols. It was better for ethyl acetate, isobutanol, and acetic acid.

Chemical Determination

The usual method involves the Kormarowsky reaction (98, 99). Higher alcohols, with the exception of 1-propanol, readily react with p-dimethylaminobenzaldehyde, salicyaldehyde, vanillin, and so on, in the presence of concentrated sulfuric acid, forming colored products. The sulfuric acid dehydrates the alcohols to unsaturated hydrocarbons, which react with aromatic aldehydes. The reactions are not stoichiometric. Other compounds that are dehydrated to unsaturated hydrocarbons and that would interfere are aldehydes, ketones, acetals, and terpenes. Pretreatment of the distillate with silver salt solves this problem. Phosphoric acid can be used instead of sulfuric acid, but the reactions are less complete.

The method given below was originally described for brandy distillate (100) but has been modified for the analysis of wines.

Procedure

Prepare the wine sample by carrying out the distillation using the macrodistillation equipment for ethanol (Figure 11). The distillate will usually be in the desired concentration range.

Prepare fusel oil standard solutions in the following way. Redistill reagent grade (water-free) isoamyl alcohol and collect the 131–132°C cut. Similarly redistill isobutyl alcohol and collect the 107.5–108°C cut. Mix the two pure fractions at a ratio of 4 vol of isoamyl and 1 vol of isobutyl alcohol. Weigh out exactly 1 g of the mixture and dilute to 1000 mL in a volumetric flask with water. This is the stock solution. Pipet 0-, 5-, 10-, 15-, 20-, 25-, and 35-mL aliquots of the stock solution into 1000-mL volumetric flasks. Add 7 mL of 95% ethanol purified by first refluxing 50% ethanol with silver sulfate, then distilling and collecting the 10–75% fraction.

Pipet 25 mL of the wine distillate into a 250-mL flask. Add 0.25 g of silver sulfate, 0.5 mL of 1:1 sulfuric acid, and a few boiling stones; place a reflux condenser on the flask, and reflux for 15 min. While still heating, add 5 mL of 6 N sodium hydroxide solution through the reflux condenser and a pinch of zinc granules. Reflux 30 min more. Cool, remove the reflux condenser, and rinse off its interior into the flask. Connect a ground-glass adaptor from the 250-mL flask to a coil condenser of the macrodistillation still, using a 50-mL volumetric receiving flask under the condenser, and distill off about 48 mL. Dilute the contents of the receiving flask to volume at 20°C with water. This step may not be required for wines that are low in aldehydes.

Carefully pipet 1 mL of the distillate into a 25 × 150-mm Pyrex test tube, then place it into an ice bath and add 20 mL of a cold p-dimethylaminobenzaldehyde–sulfuric acid solution. Prepare this solution fresh daily by dissolving 0.5 g of p-dimethylaminobenzaldehyde in 1 L of concentrated sulfuric acid; keep the solution in ice. During the mixing of the distillate and the reagent, agitate so that no local overheating occurs.

Prepare a set of tubes in the same manner using the standard solutions. All the contents of the tubes should be water white. Cover all tubes with aluminum foil caps and transfer to a boiling water bath. Heat for exactly 20 min, then cool in an ice bath. After 5 or 10 min bring to room temperature in a bath, and determine the absorbance at 525 nm against the reagent blank. Prepare a standard curve by plotting the concentration of the standards as milligrams per liter against absorbance. Calculate the fusel oil present according to the following formula:

$$\text{Fusel oil (mg/1000 mL of wine)} = \frac{(C)(50)}{v}$$

where C = fusel oil concentration in the standard solutions (mg/1000 mL), obtained from the calibration curve
v = volume of wine distillate (mL)

Gas Chromatography

The first report on the analysis of fusel oils in brandies (101) was followed by many investigations. The major problem encountered in the earlier work, according to these reports, was in separation. According to these reports, columns prepared with glycerol, diglycerol, and triethanolamine liquid phases would separate the amyl alcohols, but not isobutyl alcohol, from ethanol. However, Webb and Kepner (102) succeeded, by direct injection of a wine distillate, in separating all the common higher alcohols in wine on a diglycerol column.

Carbowax 1500 and 20M have been used for routine analysis of wine distillates (80, 103). These liquid phases fail to separate active amyl from isoamyl alcohol. The use of Carbowax 600 (75) has some advantages because acetaldehyde, ethyl acetate, and methanol can be determined simultaneously with the higher alcohols. Again the amyl alcohols are not separated. Some of the other liquid phases that also fail to separate the amyl alcohols, but do separate the other higher alcohols, are dodecyl phthalate (104), 1,2,3-tris(2-cyanoethoxy)propane (89), and UCON oil 50-4B-400 (91).

The method for separation of the amyl alcohols and good quantification of all the higher alcohols is a modified version of the method (105) using 2% 1,2,6-hexanetriol and 2% glycerol. This is given below.

Procedure

Prepare the following stock solutions:

1. 5 g of combined amyl alcohol diluted to 500 mL with 50% ethanol; 10 mg/mL (Eastman No. 18, for fusel oil assay).
2. 1 g of isobutyl alcohol diluted to 500 mL with 50% ethanol; 2 mg/mL (Eastman No. 303, for fusel oil assay).

3. 1 g of 1-propyl alcohol diluted to 500 mL with 50% ethanol; 2 mg/mL (reagent grade or better).

Prepare the following working standards from the stock solutions.

Standard number	n-Propyl		Isobutyl		Combined amyl	
	mL added	mg/100 mL	mL added	mg/100 mL	mL added	mg/100 mL
1	1.0	(2)	1.0	(2)	0.5	(5)
2	2.0	(4)	2.0	(4)	1.0	(10)
3	3.0	(6)	3.0	(6)	1.5	(15)
4	5.0	(10)	5.0	(10)	3.0	(30)
5	10.0	(20)	10.0	(20)	5.0	(50)

( ) Amount in the standards.

Dilute each standard to 100 mL with water plus sufficient 95% alcohol to make the alcohol content of each standard 12.5% v/v.

Prepare an internal standard by diluting 1 mL of reagent grade 3-pentanol to 100 mL using 95% ethanol. Add 1.0 mL of this to each working standard solution.

Determine the composition of the isoamyl alcohol by GC peak ratio. The Eastman No. 18 sample contains 76.2% isoamyl and 23.8% active amyl alcohols. Therefore the working standards contain the following:

Standard number	Isoamyl (mg/100)	Active-amyl (mg/100)
1	3.8	1.2
2	7.6	2.4
3	11.4	3.6
4	22.9	7.1
5	38.1	11.9

The active amyl–isoamyl composition of commercial amyl alcohols samples varies widely and must be determined for each.

Prepare a GC column by coating a 2% 1,2,6-hexanetriol plus 2% glycerol liquid phase on Gas Chrome R (100/120 mesh) and pack into a 6 ft × 1/8-in. o.d. thin-walled stainless steel tube. Weigh out 0.4 g each of the liquid phases, dissolve in methyl alcohol, and pour over 19.6 g of solid support. Higher-efficiency solid supports do not make as good a column as this regular firebrick. Remove the solvent using an infrared lamp or steam bath. Dry a thoroughly cleaned stainless steel tubing. Pack the column. Condition the column overnight at 80°C in the GC oven under a nitrogen flow of 20–25 mL/min.

The conditions for use are:

Oven temperature	65°C
Detector and injector	200°C
Nitrogen carrier	30 mL/min
Hydrogen	25 mL/min
Air	250 mL/min

Inject 2 µL of each working standard. Use an electrometer sensitivity of 1 × 32. Plot peak height against concentration of each component. (This requires four standard curves, one for each of four fusel oil constituents.)

To analyze the sample, add 0.1 mL of internal standard solution to 10 mL of a 1:1 wine distillate. Inject 2 µL and determine peak height of each alcohol. Express peak heights in concentration units from the standard curves.

Hexanol can be separated by GC. One method reported (92) used 15% diisodecyl phthalate on Chromosorb W, 60/80 mesh, at 135°C isothermal column temperature, with 2-phenyl-2-propanol as an internal standard. Ethyl lactate interferes when using a Carbowax liquid phase. Alcohols with higher boiling points such as tyrosol, tryptophol, and 2-phenylethanol are formed primarily by the following reactions to the corresponding amino acids (94).

$$RCNH_2COOH \xrightarrow{FAD} RCNHCOOH$$

$$RCNHCOOH \xrightarrow{H_2O} RCOCOOH + NH_3$$

$$RCOCOOH \xrightarrow{Decarboxylase} RCHO + CO_2$$

$$RCHO \xrightarrow{NADH} RCH_2OH$$

There are many other trace alcohols present that could be measured quantitatively by GC if necessary. Lists of these alcohols have been published by several authors.

GLYCEROL

Oura (106) has explained the conditions that will affect the relative production of the glycerol during yeast fermentation. Normally fermentation leads to an excess of energy (ATP) within the cell. This condition will activate pyruvate carboxylase, which consequently causes formation of succinate through the TCA cycle and, in turn, leads to an excess of reduced respiratory nucleotides:

$$\text{Pyruvate} \xrightarrow{\text{Pyruvate carboxylase}} \text{Acetyl-CoA}$$

$$NAD^+ \longrightarrow NADH$$

$$\text{Acetyl-CoA} \xrightarrow{TCA} \text{Citrate} \longrightarrow \text{Aconitrate} \longrightarrow \text{Isocitrate}$$

$$\text{Isocitrate} \longrightarrow \alpha\text{-Ketoglutarate} \longrightarrow \text{Succinate}$$

$$NAD^+ \diagup \diagdown NADH \quad NAD^+ \diagup \diagdown NADH$$

The excess NADH is used to form glycerol (see Figure 19). This allows the optimum redox state of the cell to be maintained. Production of other side-products such as 2,3-butanediols are also facilitated.

Other pathways, such as acetate formation, contribute minor amounts of reduced nucleotides to the pool.

Table 21 lists the ranges and average values of glycerol found in wines of several countries (107–117). Practices that serve to alter the amounts formed are fermentation temperature (107) and yeast strain (108, 109); pH, initial sugar concentration, and aeration conditions also can affect glycerol content. Sulfur dioxide, unless used in extreme excess, has little or no effect on the amounts produced.

Red wines usually have greater amounts of glycerol than do white wines. This is partially because a higher fermentation temperature is used in the production of red wines.

Glycerol can be determined in a number of ways, but until recently the procedure was time-consuming. The old method was to separate and purify and actually weigh the glycerol. Now rapid chemical, enzymatic, GC or HPLC procedures make the analysis simpler.

Chemical Determination

The most widely used chemical method is that of Rebelein (118). Sufficient clarification and purification is done to eliminate most interfering substances.

$$CH_2OH\text{-}CO\text{-}CH_2PO_4H_2 \xrightleftharpoons{NADH \leftrightarrow NAD^+} CH_2OH\text{-}CHOH\text{-}CH_2PO_4H_2 \longrightarrow CH_2OH\text{-}CHOH\text{-}CH_2OH$$

DIHYDROXYACETONE PHOSPHATE GLYCEROL

Figure 19. Pathway of formation of glycerol.

Table 21. Concentrations of Glycerol Found in Wines from Various Countries

Country	Number of samples	Glycerol (g/L) Range	Average
Austria	72	1.36–9.94	6.5
France	71	6.59–23.00[a]	8.6[a]
Germany	146	2.0–20.0	7.7
Italy	38	4.12–10.8	7.2
Italy	89	3.26–6.77	6.3
Japan	42	4.90–12.8	7.4
Japan	80	3.57–9.59	7.1
USSR	13	1.11–4.32	2.9
United States	100	1.9–14.7	7.2

From References 95 and 107–117.
[a] Includes botrytized samples.

Glycerol is oxidized with periodic acid and the formaldehyde formed is reacted with phloroglucinol to give a colored product that is measured on a spectrophotometer. The method as outlined in the official handbook of the OIV (13) is given below.

Procedure

Prepare a 0.05 M periodic acid solution in the following way. Into a 1-L volumetric flask, add 11.5 g of potassium iodate, 800 mL of water, and 15 mL of concentrated sulfuric acid. Heat gently on a water bath until dissolved. Cool and dilute to volume at 20°C. This solution is good for about 2 weeks.

Prepare a standard glycerol solution by diluting 20 mL of distilled glycerol to 100 mL with water. Determine the glycerol concentration using a refractometer, then dilute an aliquot to prepare a standard solution with a glycerol concentration of 0.1 g/L.

Separation of Glycerol and 2,3-Butanediol from Interfering Substances. Into a 100-mL rubber-stoppered flask, place 5 g of barium hydroxide (which should not be finely powdered) and 5 g of sand, and pipet 10 mL of the wine down the side of the flask; if the wine has more than 12% sugar, use only 5 mL. Agitate vigorously for 30 sec. Let set and agitate from time to time. Add 50 mL of redistilled acetone and again agitate for 30 sec. Place the flask in a 45°C water bath and agitate strongly for 5 min. Vacuum filter the hot solution through a medium porcelain crucible filter into a vacuum flask. Wash the precipitate with 40 mL of water and 5 mL of 1 N sodium hydroxide solution, then three times with redistilled acetone, resuspending the precipitate each time by mixing with a spatula; be sure that the precipitate is as dry as possible. Transfer the collected filtrate and washings with acetone into a distilling flask equipped with a thermometer. Distill until the distilling temperature reaches 100°C, then distill off 20 mL of water. Turn off the heat and add 5 mL of N sulfuric acid. Cool, quantitatively transfer the contents of the distilling flask into a 50-mL volumetric flask, and dilute to volume at 20°C. Filter the contents of the flask and collect the filtrate. Use this filtrate for the determination of glycerol (below) and 2,3-butanediol.

Determination of Glycerol. Pipet a 5-mL aliquot of the filtrate into a 100-mL volumetric flask and dilute to volume with water. Pipet 20 mL of this solution into a 60-mL glass-stoppered flask, add 10 mL of the 0.05 M periodic acid solution, shake, and let stand 5 min. Add quickly, down the side of the flask, 10 mL of 1 N sodium hydroxide solution, agitate, then add 10 mL of a 2% phloroglucinol solution and mix. Place a portion into a cuvette with a 10-mm light path and determine the absorbance at 480 nm against a reagent blank. The color reaches a maximum rapidly, is stable for only 50 or 60 sec, then deteriorates rapidly. Prepare the reagent blank by using 20 mL of water instead of the sample and adding all the reagents, starting with the 10 mL of periodic acid solution.

To establish a standard curve, dilute 25, 50, 75, 100, 125, 150, and 175 mL of the standard glycerol solution to 200 mL with water. Take 20-mL aliquots of these solutions and carry out the determination as above, starting with the addition of the 10 mL of 0.05 M periodic acid. If the respective concentrations of the solutions used for the standard curve are taken as 60, 120, 180, 240, 300, 360, and 420 mg/L, then, using a 10-mL wine sample, the glycerol concentration can be read directly from the curve. For 5-mL wine samples, multiply this result by 2. Multiply this value be the factor based on the sugar concentration of the wine sample using the factors listed in the accompanying tabulation.

Sugar concentration (g/100 mL)	Factor	Sugar concentration (g/100 mL)	Factor
0.5	1.020	8.0	1.123
1.0	1.030	9.0	1.134
2.0	1.045	10.0	1.145
3.0	1.060	11.0	1.156
4.0	1.073	12.0	1.168
5.0	1.086	13.0	1.180
6.0	1.099	14.0	1.191
7.0	1.111	15.0	1.202

Correction for Mannitol. The value obtained as glycerol concentration also includes sorbitol and mannitol. To eliminate the latter, carry out the following procedure with the 20-mL aliquot ready for the addition of periodic acid solution: In a 50-mL evaporating dish evaporate the 20-mL aliquot in the presence of 0.1 g of sodium sulfate and a few milliliter of 20% sulfuric acid. Continue evaporation on a 90°C water bath to a syrupy consistency. Dissolve the residue in 5 mL of absolute ethanol, transfer to a 40-mL centrifuge tube, wash the dish with several small portions of absolute ethanol, and bring the total volume in the centrifuge tube to 10 mL. To this, add 3–5-mL portions of anhydrous ethyl ether, mixing after each addition. Centrifuge and decant the supernatant liquid into a 125-mL Erlenmeyer flask. Wash the precipitate three times with 10-mL portions of a 2:3 mixture of ethanol and ethyl ether and collect all the washings in the 125-mL Erlenmeyer flask. Add 20 mL of water and distill off the organic solvents in a 100°C watern bath, then proceed with the sample as above, starting with the addition of 10 mL of 0.05 M periodic acid solution.

The method above gave precise and reproducible results (119). The results agreed well with those obtained by the enzymatic method.

Enzymatic Analysis

Drawert (120) suggested the following approach for the enzymatic determination of glycerol (abbreviations: ATP = adenosine triphosphate, ADP = adenosine diphosphate, NAD^+ = nicotinamide adenine dinucleotide, NADH = NAD^+ in the reduced form, PEP = phosphoenolpyruvate):

$$\text{Glycerol} + \text{ATP} \xrightarrow[\text{Mg}^{2+}]{\text{Glycerol kinase}} \text{Glycerol-1-phosphate} + \text{ADP}$$

$$\text{ADP} + \text{PEP} \xrightarrow{\text{Pyruvate kinase}} \text{ATP} + \text{Pyruvate}$$

$$\text{Pyruvate} + \text{NADH} + \text{H}^+ \xrightarrow{\text{Lactate dehydrogenase}} \text{Lactate} + \text{NAD}^+$$

The change in the NADH concentration is measured spectrophotometrically.

A more direct enzymatic system may be used (121, 122). Instead of going through the PEP → pyruvate step, the glycerol-1-P was reacted directly with NAD^+ as shown:

$$\text{Glycerol-1-phosphate} + \text{NAD}^+ \xrightarrow{\text{Glycerolphosphate dehydrogenase}}$$

$$\text{Dihydroxyacetone-P} + \text{NADH} + \text{H}^+$$

The recoveries of 95–100% are claimed. This procedure has been made continuous (123) by using an automatic sampler and a spectrophotometer at 340 nm. It was capable of 20 samples per hour.

The details of the procedure by the first scheme follow.

Procedure

Prepare a pH 7.6 buffer by dissolving 10 g of glycine and 0.25 g of magnesium sulfate heptahydrate in 80 mL of distilled water. Adjust the pH to 7.6 using about 2.5 mL of 5 M sodium hydroxide. Dilute the buffer to 100 mL in a volumetric flask. The buffer is stable for 3 months at 4°C. Dissolve 42 mg of NADH-Na$_2$, 120 mg of ATP-Na$_2$H$_2$, 60 mg of PEP-Na, and 300 mg of sodium bicarbonate in 6 mL of distilled water. The solution is stable for only 2 weeks at 4°C. Use undiluted suspension of 3 mg/mL of pyruvate kinase and 1 mg/mL of lactic dehydrogenase. Also use a 1 mg/L suspension of glycerol kinase. Both these solutions are stable for up to a year at 4°C.

Prepare a blank sample by addition of 1 mL of buffer, 0.10 mL of NADH-ATP-PEP solution, 2.00 mL of water, and 0.01 mL of PK/LDH suspensions to a 1-cm light-path cuvette. Mix well and read absorbance E_1 at 340 nm after 3 min. Add 0.1 mL of glycerol kinase suspension, mix, and read absorbance E_2 at 340 nm after 5–10 min. If reaction is still proceeding, read every 2 min until reaction stops. Do exactly the same for a sample of diluted wine that contains between 2 and 50 mg of glycerol (usually a 1:25 dilution of the wine), but add 0.10 mL of diluted sample and 1.90 mL of water instead of 2 mL of water.

$$\Delta E_B = \Delta E_1 - \Delta E_2$$
$$\Delta E_S = \Delta E_1 - \Delta E_2$$
$$\Delta E = \Delta E_S - \Delta E_B$$
ΔE_B = absorption of blank
ΔE_S = absorption of sample

$$\text{Glycerol (g/L)} = \frac{(V)(\text{MW})(F)(\Delta E)}{(E)(d)(v)(1000)}$$

where V = final volume (mL)
v = sample volume (mL)
MW = molecular weight of glycerol
d = light path (cm)
E = extinction coefficient of NADH at 340 nm (6.31 nmol^{-1} cm^{-1})
F = dilution factor (25 for wine)

In short, for these conditions glycerol (g/L) = 11.39ΔE.

Because of the high cost of enzymes, considerable savings can be obtained by immobilizing the enzymes (124). Glycerol dehydrogenase was immobilized on the inner face of a nylon tube. This catalyzes the reaction

$$\text{Glycerol} + \text{NAD}^+ \xrightarrow{\text{GDH}} \text{Dehydroxyacetone} + \text{NADH} + \text{H}^+$$

The method of preparing the tube is given in Reference 125. The system was

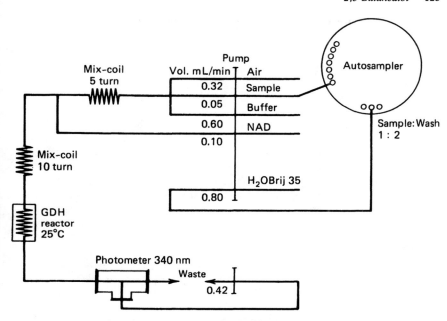

Figure 20. Flow diagram for automated determination of glycerol (124).

set up with a Technicon Auto-Analyzer (Technicon Instrument Co., Tarrytown, N.Y.). The flow diagram is shown in Figure 20.

The reagents required are a buffer solution (0.1 mol/L glycylglycine, pH 10.5, containing 0.03% of Brij 35 surfactant), NAD solution (45 mmol/L adjusted to pH 7.5 with KOH), and a wash solution (0.03% Brij in deionized water). Wines were diluted 1:200 and beers 1:100. The automated procedure allowed 60 samples per hour and a linear response of 1–50 mg/L of glycerol. The results compared favorably with those obtained in the usual manual enzymatic procedure with GK, PK, and LDH at 10% the cost.

In a collaborative study by 10 laboratories (126) for glycerol in grape juice by the enzymatic procedure, satisfactory results were obtained. The coefficient of variation for repeatability was 0.7%, and the reproducibility was 2.3%. The reason for the determination in grape juice is because it is an indicator of undesirable and excessive mold growth on grapes. Swiss law limits the glycerol in musts to 1 g/L. See also Junge (127).

Gas Chromatography

With the proper column and conditions, glycerol can be chromatographed and quantified. Using microporous beads, several workers (107, 128) have directly measured glycerol in wine. More recently a silylation technique, using a 5% Dexil®-300 on 80/100 mesh Chromosorb W was developed (116, 129). A promising column appears to be 10% SP-1000 on Chromosorb W, 100/200 mesh (130); this procedure is given here.

Procedure

Weigh out 20 g of glycerol into a 100-mL flask and bring to volume with 80 vol % ethanol solution. Put 0.0, 0.5, 1.0, 1.0, 3.0, 5.0, and 10.0 mL of this stock solution into 100-mL volumetric flasks. Make an internal standard solution of 50 mL of 1,4-butanediol in a 100-mL volumetric flask and bring to volume with water at 20°C. Pipet 1 mL of this internal standard into each of 100-mL volumetric flasks and bring each to volume with 80 vol % ethanol. Prepare a glass column 6 ft × ¼ in. and fill with 10% SP-1000 coated onto 100/200 mesh Chromosorb W. Cure the column overnight at 220°C. Use carrier glass flow of about 60 mL/min, and use column, injector, and detector temperature of 195, 200, and 210°C, respectively.

For wines containing 0.2–4.0 g/100 mL of glycerol, pipet 50 mL into a 100-mL volumetric flask (add 1 mL of internal standard solution, mix well, and bring to volume with 30 vol % ethanol).

Inject 0.2 µL of diluted sample or standard solutions into the GLC. Calculate the peak height ratios (or areas) of the standard solutions to the internal standard and plot a standard curve (R_s/R_i vs. glycerol in the diluted stock solutions of 0, 0.1, 0.2, 0.6, 1.0, and 2.0 g/100 mL). Determine the value of the sample by comparison to the standard curve, and multiply the value by the dilution factor.

Direct injection of diluted wine may be made in a glass column packed with Ucon LB 550X (131). The results compared well with the enzymatic procedure and were better than by the chemical procedure. Recently (132), HPLC with a cation exchange resin, HPX 87H, as the stationary phase and dilute sulfuric acid as the mobile phase was used. The results tended to be slightly lower than those with the enzymatic method.

Fluorometric Analysis

A comparison of methods indicated (129) that a fluorometric method was best suited for glycerol analysis. Recoveries of 99.3–100.9% ± 2.9–3.2% were noted. The method lends itself to automation.

2,3-BUTANEDIOL

Three isomers of 2,3-butanediol may be formed through fermentation by microorganisms (133):

$$\begin{array}{ccc}
\text{CH}_3 & \text{CH}_3 & \text{CH}_3 \\
| & | & | \\
\text{H—C—OH} & \text{HO—C—H} & \text{H—C—OH} \\
| & | & | \\
\text{H—C—OH} & \text{H—C—OH} & \text{HO—C—H} \\
| & | & | \\
\text{CH}_3 & \text{CH}_3 & \text{CH}_3 \\
meso\text{-2,3-Butanediol} & \text{D(−)-2,3-Butanediol} & \text{L(+)-2,3-Butanediol} \\
 & (levo) & (dextro)
\end{array}$$

Wine yeast fermentations give rise to primarily the meso and levo forms (134). They are formed by the reduction of acetoin by NADH. Acetoin is formed from pyruvic acid:

$$2\text{CH}_3\text{—C(=O)—COOH} \longrightarrow \text{CH}_3\text{—CH(OH)—C(=O)—CH}_3 \longrightarrow \text{CH}_3\text{—CH(OH)—CH(OH)—CH}_3$$

Pyruvic acid   Acetoin   2,3-Butanediol

Consequently, only half as much NAD^+ is returned to the system when the acetoin is converted to 2,3-butanediol. Under aerobic yeast-growing conditions in wine, acetoin accumulates and no 2,3-butanediols are produced (135). More acetoin is formed at high sugar level in the must and at high fermentation temperatures (91, 134). See p. 120 concerning basic reasons. Table 22 summarizes some of the data available on amounts in wine. The meso form accounts for about 23–30% (134, 136) of the totals.

The determination of 2,3-butanediol in wine is similar to the determination of glycerol.

Chemical Determination

The chemical method involves the oxidation of the 2,3-butanediol to acetaldehyde and either the direct chemical determination of the acetaldehyde or the formation of a colored product with piperidine and nitroprusside by subsequent

Table 22. Concentrations of 2,3-Butaneodiol Found in Wines

Country	Number of samples	2,3-Butanediol (mg/L)	
		Range	Average
Germany	25	300–1200	551
Italy	38	250–1250	578
Italy	89	184–603	566
Japan	42	280–930	603
USSR	13	16–267	139
United States	107	165–1615	475

From References 90, 95, and 115–118.

spectrophotometric measurements (118). The determination is carried out from the filtrate prepared for the glycerol determination (13).

Procedure

Pipet a 5-mL portion of the filtrate prepared as for glycerol and 5 mL of a 27% aqueous sodium acetate solution into a 60-mL flask, add 10 mL of the 0.05 M periodic acid solution, agitate, and allow to stand for 2 min. Add, down the side of the flask, 5 mL of a 2% aqueous sodium nitroprusside solution and mix; then add 10 mL of a 10% aqueous piperidine solution and mix. Immediately transfer an aliquot into a 10-mm light-path cuvette and determine the absorbance in a spectrophotometer at 570 nm. The color is at a maximum at 30–40 sec, then rapidly fades. Zero the spectrophotometer with a blank containing 5 mL of water instead of the prepared wine and including all the reagents.

Prepare a standard solution of 2,3-butanediol containing 2.000 g in 1 L of water. Dilute this solution 1:10 with water. Take 0-, 10-, 20-, 30-, 40-, 50-, 60-, and 70-mL aliquots of this solution and dilute each in 100-mL volumetric flasks with water to volume. Take 5-mL aliquots of each of these solutions and develop the color as described above. Plot the absorbance against the concentration of 2,3-butanediol in the stock solutions, and use this calibration curve to establish the apparent 2,3-butanediol concentration in the sample. Correct this value according to the reducing sugar concentration using the values given in Reference 118. (For wines with less than 12% sugar this gives the answer in grams per liter. In wines with more than 12% sugar, twice this corrected value is the answer.)

Gas Chromatography

The separation and measurement of the two isomers of 2,3-butanediols in wine were performed successfully (134) by the direct injection of a wine into a 6-ft $\times$ $\frac{1}{4}$-in. column containing 10% UCON oil 75-H-90M on acid-washed Chromosorb W, 60/80 mesh. The instrument was operated under the following conditions: injection port, 200°C; flame ionization detector, 200°C; column, 125°C; and carrier gas, helium at 60 mL/min. The meso and levo isomers were separated, with the levo isomer eluting first. If dessert wine or wine with appreciable sugar is used, ghost peaks and a rising baseline can be a problem. This can be overcome by raising the oven temperature at 180°C for a few minutes, after several samples.

Another system (130) uses the same column and similar conditions described in the glycerol section. A separate standard curve is necessary. This system does not separate the isomers of the 2,3-butanediols. GC has been used directly with the wine diluted 1:1 (131). *Meso-* and *levo-*2,3-butanediols were separated. The results tended to be slightly higher than by the chemical procedure. The levo form tends to predominate (69%).

Enzymatic Analysis

An enzymatic method has been suggested (137) for measurement of the 2,3-butanediols. It is possible to prepare dehydrogenase(s) from *Sarcina hansenii* capable of oxidation of both the isomers to acetoin.

SORBITOL AND MANNITOL

Sorbitol, a sugar alcohol, is present in many fruits. The concentration levels found in grapes are much lower than those found in apples. This property has afforded a means of testing to determine whether apple wines have been used to dilute the grape wines. Yeasts do not alter the amount of sorbitol present in wine from that present in the original must. Table 23 (138–144) gives amounts reported in wines. Apple wines can exceed these values by 10-fold or more. However, wines made from botrytized grapes obviously have much more than those from sound grapes. Thus a limit of 70 mg/L for sorbitol does not apply to wines made from such grapes.

$$\begin{array}{cc}
\text{CH}_2\text{OH} & \text{CH}_2\text{OH} \\
\text{H}-\text{C}-\text{OH} & \text{HO}-\text{C}-\text{H} \\
\text{HO}-\text{C}-\text{H} & \text{HO}-\text{C}-\text{H} \\
\text{H}-\text{C}-\text{OH} & \text{H}-\text{C}-\text{OH} \\
\text{H}-\text{C}-\text{OH} & \text{H}-\text{C}-\text{OH} \\
\text{CH}_2-\text{OH} & \text{CH}_2-\text{OH} \\
\text{Sorbitol} & \text{Mannitol}
\end{array}$$

The presence of high amounts of mannitol is either an indication of a bacterially contaminated wine (often from a stuck fermentation) or of a wine made from grapes that were highly infected with botrytis. Lactic acid-producing bacteria will convert fructose to mannitol:

$$\text{Fructose-1-PO}_4 \xrightarrow[\text{Mannitol-PO}_4 \text{ dehydrogenase}]{\text{NADH} \quad \text{NAD}^+} \text{Mannitol-PO}_4 \xrightarrow{\text{Mannitol phosphatase}} \text{Mannitol} + \text{PO}_4$$

This is rare nowadays because of use of yeast cultures and temperature control during fermentation. A range of 1–30 g/L was reported (17). The role of *Botrytis cinerea* as the source of high glycerol and other polyalcohols is well known (138). After only 2 days' inoculation of berries the mannitol content rose from

Table 23. Concentrations of Sorbitol and Mannitol Reported in Wines (138–144)

Country	Number of samples	Range	Average
Sorbitol			
European	500	9–169	61
France			
White	16	29–113	70
Red	35	74–194	122
Sauternes	10	116–394	293
Germany	18	50–209	116
Q.b.A.	12	30–97	
Kabinett and Spätlese	63	14–220	
Auslese, Beerenauslese, and			
Trockenbeerenauslese	48	84–989	
Italy	10	5–88	34
Mannitol			
France			
White	7	84–323	184
Red	29	90–394	248
Moldy red	8	452–735	602
Germany	18	126–1401	602
Q.b.A.	12	110–552	
Kabinett and Spätlese	63	3–1672	
Auslese, Beerenauslese, and			
Trockenbeerenauslese	48	377–12,884	
Italy	10	430–1510	960

82 to 536 mg/L. See also the data in Table 23 for Auslese, Beerenauslese, Trockenbeerenauslese, and Sauternes.

Using HPLC and trimethylsilyl derivatives (with ribitol as an internal standard) followed by mass spectroscopy it was possible (143) to quantitatively determine the sugar alcohols erythritol, xylitol, arabitol, mannitol, sorbitol, and *meso*-inositol, as well as malic, tartaric, citric, gluconic, and mucic acids. In another HPLC procedure (144), glycerol, inositol (0.4–0.6 g/L), trehalose, arabinose, glucose, fructose, and sucrose were measured.

Polarimetric Determination

The polarimetry of the molybdic acid–polyalcohol complex as used with wine (145) for determining sorbitol has been shown to be in error (139). The main problem is the presence in all wines of mannitol, which also forms a complex with molydbic acid. Thus these two compounds are measured polarimetrically at the same time. A combined procedure (139, 146–148) involving thin-layer separation and enzymatic detemination of the sorbitol by sorbitol dehydrogenase and the polarimetric measurement of the total of mannitol and sorbitol can be employed, but it is time-consuming and of questionable practicability. At low acidities, only the mannitol phosphomolybdic complex is optically active (149). Suggested configurations of the optically active forms of the five molybdate complexes are given. Sugars do not significantly interfere unless the concentration exceeds 1 g/L total.

Enzymatic Analysis

By a rather indirect method, sorbitol and xylitol can be measured enzymatically (150). Sorbitol and xylitol react as follows:

$$\text{D-Sorbitol} + \text{NAD}^+ \xrightarrow{\text{Sorbitol dehydrogenase}} \text{Fructose} + \text{NADH} + \text{H}^+$$

$$\text{Xylitol} + \text{NAD}^+ \xrightarrow{\text{Sorbitol dehydrogenase}} \text{Xylulose} + \text{NADH} + \text{H}^+$$

The NADH is reconverted to NAD^+ by lactic dehydrogenase and pyruvate. Fructose is determined as shown earlier for enzymatic determination of fructose with hexokinase.

The total sorbitol + xylitol is determined by having the NADH formed react with iodine nitrotetrazolium chloride to give a formazan. The absorption is measured at 492 nm. Reducing substances such as ascorbic acid interfere. The xylitol is determined by the difference between the total and the sorbitol.

Gas Chromatography

Acetylation of the polyalcohols after removal of the sugars and alcohol has been suggested (143, 145). The acetylated polyalcohols are then chromatographed on a fluorosilicone column and subjected to flame ionization detection. One method allows for measurement of a number of polyalcohols (sorbitol, glycerol, erythritol, adonitol, arabitol, xylitol, mannitol, mesoinositol) and is given below (140).

Procedure

Prepare Amberlite IRA 400 (80/100 mesh) anion resin by washing with 1 N sodium hydroxide, then rinse with 1 N hydrochloric acid and rinse again with water unit Cl^- is absent by the silver nitrate test. Finally wash with 1 N sodium hydroxide and then wash with neutral water until the eluate is negative to phenolphthalein. Wash 10 mL of the resin into a 10 × 200-mm glass column.

Prepare an internal standard by adding 4 g of perseitol (a 7-carbon polyalcohol) to a liter of water. Add 2 mL of this to 20 mL of wine and mix. This is done before ion exchanging the sample.

Place 5 mL of wine with internal standard added onto the column and wash with 100 mL of water. Collect the eluate, which contains the polyalcohols; the acids and sugars remain on the column. The column can be regenerated by washing with 100 mL of 1 N sodium hydroxide and testing for neutrality of the eluate as before. After four cycles of use, regenerate by use of acid, then base, as in preparation.

Concentrate the eluate to dryness under vacuum at 60°C. When sample nears dryness, add 5-10 mL of carbon tetrachloride to azeotropically distill off the remaining water. Add 2 mL of acetic anhydride and 0.4 mL of pyridine. Heat under reflux at 100°C for 1 hr.

Prepare a $\frac{1}{8}$-in. i.d. × 3-m stainless steel column. Coat Gas Chrome Q with a 7% load of QF-1. Condition the column overnight under about 30 mL/min nitrogen gas flow at 230°C. Conditions of operation are as follows: injector and detector, 270°C; oven, 210°C isothermal. Flame ionization is the method of detection.

Inject 5 mL of the acetylated sample onto the column. Sorbitol elutes after about 18-20 min, and the internal standard at 35 min.

Prepare a standard curve for sorbitol. To make up the standard solution of 0, 50, 100, 150, 200 mg/L in water, add 1 g of sorbitol to a 1-L flask and bring to volume with water at 20°C. Take 0, 5, 10, 15, and 10 mL of this stock solution, add to 100-mL volumetric flasks, and bring to volume at 20°C. Take 20 mL of each, add to 100-mL volumetric flasks, and bring to volume at 20°C. Take 20 mL of each, add 2 mL of the internal standard, and treat as with wine, putting it through all the preparatory and derivatizing steps. Plot a standard curve comparing the original concentrations to the peak height ratios of sorbitol to internal standard. Compare wine values to standard curve.

Other polyalcohols such as mannitol and mesoinositol readily lend themselves to analysis by the same system.

HPLC

Sorbitol can be determined in juice, especially apple juice, by the same HPLC system reported on p. 42 for sugars. The necessary variations are given as follows.

Procedure

Make up a standard solution by adding 2.000 g of sorbitol to a 1-L volumetric flask and bring to volume with water. Divide into 1-mL ampules and freeze for future use. All the conditions and materials as noted in the HPLC method for

Table 24. Concentration of Erythritol and Arabitol (mg/L) in Table Wines[a]

Wine	Erythritol			Arabitol		
	Number of samples	Range	Average	Number of samples	Range	Average
White	7	33–100	91	7	13–59	56
Red	29	64–116	91	29	32–111	56
Moldy red	8	160–272	205	8	165–329	242
Q.b.A.	12	34–94	—	12	12–77	—
Kabinett and Spätlese	63	7–227	—	63	6–216	—
Auslese	8	93–234	—	8	13–370	—
Beeren and Trockenbeerenauslese	40	70–692	—	40	175–2,353	—
Red	19	53–114	—	19	14–315	—
Red	16	52–182	—	16	49–148	—

From References 138 and 143.

[a] The first three wines are French (138), the wines on the next five lines are German, and the last wine is Spanish (143).

sugars are retained. Inject a 10-μL sample of the standard. Repeat to check for repeatability. If satisfactory, determine the area/mg sorbitol.

Prepare sample as suggested for the sugar determination. No dilution should be required. The retention time for sorbitol is approximately 18.0 min. Determine the area for the sorbitol in the test sample.

To calculate the concentration of sorbitol, use the following equation:

$$\text{Peak area (sorbitol)} \times \text{mg (sorbitol)}/\text{area (std)} = \text{mg sorbitol}/\text{L}$$

OTHER POLYALCOHOLS

Mesoinositol, a cyclic sugar alcohol, is one of the growth factors for yeast. However, it is present in such large amounts in grape juice that it is seldom, if ever, growth-limiting in wine fermentation. Yeast can synthesize it if required. Mesoinositol is very stable in wine and is not removed during the usual stabilization or fining operations. One report (151) gives a range of 220–730 mg/L (average 436 mg/L). There are chemical (152), enzymatic (153), microbiological (151), and GC (154, 155) methods available for determining mesoinositol.

Other sugar alcohols present are erythritol and arabitol. Both are found in increasing amounts in fruit made from mold-infected grapes. The amounts found in various wines are given in Table 24. After separation from acids and sugars the polyalcohols are now usually determined by HPLC.

Xylitol in the German wines had the following concentrations: Q.b.A., 0–22 mg/L; Kabinett and Spätlese, 0–27 mg/L; Auslese, 0–25 mg/L; Beeren- and Trockenbeerenauslese, 0–95 mg/L; the German red, 0–23 mg/L; and the Spanish red, 0–42 mg/L.

TERPENE COMPOUNDS

With the recognition that this class of compounds is of importance to the odor of muscat and other aromatic varieties of grapes, a number of procedures have been developed for their separation, identification, and, in some cases, quantitative determination. Gas chromatography with mass spectrometry has been widely used in these studies. The precursors of the terpene compounds are the polyhydroxylated monoterpene polyols (156). Both free terpenes and their glycosides are reported. The glycosides, after acid hydrolysis, yield nerol, α-terpineol, linalool, and geraniol. These generally decrease during fermentation and storage (157). Terpineol, hotrienol, and linalool oxides increase under the same conditions. However, under acid conditions, hotrienol rearranges to nerol oxide, which accounts for the very small amount of hotrienol found in aged wines (157). Some aged wines were found to contain rather high concentrations of α-terpineol and nerol (158). The conversion of linalool and geraniol to α-terpineol may explain the increase.

Since the terpenes are currently the subject of widespread research, methods for their identification and determination are sure to change rapidly. Gas chromatography with mass spectrometry has been recommended (159). A colorimetric method that might be used for control purposes (amount of muscat aroma in grapes delivered to the winery) may be helpful (160). The volatiles are separated into free volatile terpenes (FVT) and potentially volatile terpenes (PVT).

REFERENCES

1. M. A. Amerine, H. W. Berg, R. E. Kunkee, C. S. Ough, V. L. Singleton, and A. D. Webb, *The Technology of Wine Making*, 4th ed. Avi Publishing Co., Westport, CT, 1980.
2. M. Denis, *Ind. Aliment. Agric.* **100**, 527–530 (1983).
3. E. Martin, V. Iadaresta, J.-C. Giacometti, and J. Vogel, *Mitt. Geb. Lebensmittelunters Hyg.* **77**, 528–534 (1986).
4. J. Dujardin, *Recherches retrospectives sur l'art de la distillation historique de l'alcool, l'alambic et de l'alcoometrie*. Dujardin-Salleron, Paris, 1955.
5. J. Ribéreau-Gayon, E. Peynaud, P. Sudraud, and P. Ribéreau-Gayon, *Traité d'Oenologie*, Vol. 1. Dunod, Paris, 1976.
6. L. A. Williams and R. Boulton, *Am. J. Enol. Vitic.* **34**, 234–242 (1983).
7. C. S. Ough and M. A. Amerine, *Hilgardia* **34**, 585–600 (1963).
8. A. C. Houtman and C. S. du Plessis, *Wynboer* **564**, 104–105 (1978).
9. G. L. Marsh, *Am. J. Enol.* **9**, 53–58 (1958).
10. M. A. Amerine and E. B. Roessler, *Wines: Their Sensory Evaluation*, revised and enlarged ed. Freeman, New York, 1983.
11. P. Jaulmes, *Analyse des Vins*, 2nd ed. Librairie Coulet, Dubois et Poulain, Montpelier, 1951.
12. E. Vogt, revised by L. Jakob, E. Lemperle, and E. Weiss, *Der Wein; Bereitung, Behandlung, Untersuchungen*. Ulmer, Stuttgart, 1984.
13. *Recueil des Méthodes Internationales d'Analyse des Vins*. Office International de la Vigne et du Vin, Paris, 1978.
14. *Official Methods of Analysis*, 14th ed., Association of Official Analytical Chemists, Arlington, VA, 1984, pp. 220–230 and other sections.
15. O. de Marco and B. Leoci, *Riv. Vitic. Enol.* **28**, 145–161 (1963).
16. U.S. Internal Revenue Service, *Wine*, Part 240 of Title 26, IRS Publ. No. 146, Code of Federal Regulations. U.S. Gov. Printing Office, Washington, DC, 1970.
17. Y. Tey, *Rev. Fr. Oenol.* **22**, 35–39 (1982).
18. M. Procopio, *Riv. Vitic. Enol.* **3**, 139–145 (1950).
19. R. F. Love, *Ind. Eng. Chem., Anal. Ed.* **11**, 485–550 (1939).
20. C. R. Churchward, *Aust. Chem. Inst. J. Proc.* **7**, 18–30 (1939).
21. M. A. Amerine and M. A. Joslyn, *Table Wines, The Technology of Their Production*, 2nd ed. Univ. of California Press, Berkeley and Los Angeles, 1970.

22. U.S. Internal Revenue Service, *Wine*, Regul. No. 7. U.S. Gov. Printing Office, Washington, DC, 1945.
23. R. Franke and C. Junge, *Weinanalytik, Untersuchungen von Wein und ähnlichen alcoholischen Erzeugnissen sowie von Fruchtsäften*. Carl Heymanns Verlag, Cologne, 1970-1983.
24. C. Junge, *J. Assoc. Off. Anal. Chem.* **68**, 141-143 (1985).
25. R. H. Dyer, *J. Assoc. Off. Anal. Chem.* **68**, 255 (1985).
26. A. Gargano, A. Tamborni, and R. Bechetti, *Ind. Bevande* **14**, 351-353 (1985).
27. P. Jaulmes, T. Terret, and S. Brun, *Rev. Fr. Oenol.* **13**(47), 5-14 (1972).
28. Y. Tep and S. Brun, *Ann. Falsif. Expert. Chim.* **66**, 37-40 (1973).
29. P. Jaulmes, S. Brun, and Y. Tep, *Actual. Anal. Org. Pharm. Bromatol.* **24**, 1-26 (1976).
30. A. Caputi, Jr., *Am. J. Enol. Vitic.* **32**, 256 (1981).
31. A. Casanova, H. Cosson, and A.-M. Saugnac, *Ann. Technol. Agric.* **29**, 1-12 (1980).
32. Y. Tey, M. T. Cabanis, and J. Roffy, *Ind. Bevande* **13**, 84-86 (1984).
33. O. Kratky, H. Leopold, and H. Stabinger, *Z. Angew. Phys.* **4**, 273-277 (1969).
34. J. P. Edler, *Am. Lab.* April, pp. 75-81 (1978).
35. J. Schneyder, *Mitt. Hoeheren Bundeslehr- Versuchsanst. Wein- Obstbau, Klosterneuberg* **24**, 48-56 (1974).
36. M. Perscheid and F. Zürn, *Wein-Wiss.* **31**, 32-44 (1976).
37. G. Henning and E. Heike, *Z. Anal. Chem.* **265**, 97-104 (1973).
38. K. Joaskelainen, *Mallas Olut* **6**, 203-209 (1977).
39. D. H. Strunk, J. W. Hamman, J. C. Aicken, and A. A. Andreasen, *J. Assoc. Off. Anal. Chem.* **64**, 550-553 (1981).
40. L. E. Stewart, *J. Assoc. Off. Anal. Chem.* **66**, 1400-1404 (1983).
41. O. Kratky, H. Leopold, and H. Stabinger, *Paar Digital Density Meter, DMA 40, Instruction Manual*. Anton Paar, Graz, 1975.
41a. G. Gidaly, *Mitt. Klosterneuburg* **37**, 1-12 (1987).
42. P. Jaulmes, S. Brun, and J.-P. Laval, *Ann. Falsif. Expert. Chim.* **58**, 304-310 (1965).
43. P. A. Reeves, *J. Assoc. Off. Agric. Chem.* **48**, 476-478 (1965).
44. U.S. Bureau of Standards, *Standard Density and Volumetric Tables*, Circ. No. 19. U.S. Gov. Printing Office, Washington, DC, 1924.
45. H. W. Zimmerman, *Am. J. Enol. Vitic.* **14**, 205-213 (1963).
46. M. Flanzy, P. Benard, C. Flanzy, and A. Schober, *Ann. Technol. Agric.* **18**, 367-374 (1969).
47. J. F. Guymon and E. A. Crowell, *J. Assoc. Off. Agric. Chem.* **42**, 383-389 (1959).
48. R. L. Morrison and T. E. Edwards, *Am. J. Enol. Vitic.* **14**, 185-193 (1963).
49. A. Caputi, Jr., M. Ueda, and T. Brown, *Am. J. Enol. Vitic.* **19**, 160-165 (1968).
50. A. Caputi, Jr. and D. Wright, *J. Assoc. Off. Anal. Chem.* **52**, 85-88 (1969).
51. A. Caputi, Jr., *J. Assoc. Off. Anal. Chem.* **53**, 11-12 (1970).
52. H. Rebelein, *Allg. Dtsch. Weinfachztg.* **107**, 590-594 (1971).
53. J. Sarris, J. N. Morfaux, P. Dupuy, and D. Hertzog, *Ind. Aliment. Agric.* **86**, 1241-1246 (1969).
54. B. Stackler and E. N. Christensen, *Am. Enol. Vitic.* **25**, 202-207 (1974).
55. G. Pilone, *J. Assoc. Off. Anal. Chem.* **68**, 188-189 (1985).
56. H. Rebelein, *Chem., Mikrobiol., Lebensm.* **2**, 112-121 (1973).
57. H. Rebelein and H. Steinert, *Offenlegungsschrift (Fed. Repub. Ger.)* **2**, 403 (1973).

58. R. J. Bouthilet, A. Caputi, Jr., and M. Ueda, *J. Assoc. Off. Agric. Chem.* **44**, 410-414 (1961).
59. R. L. Morrison, *Am. J. Enol. Vitic.* **12**, 101-106 (1961).
60. B. V. Lipis, N. K. L. Grinberg, and T. A. Manuilova, *Tekhnol. Pishch. Prod. Rast. Priozkhozhd.* pp. 116-118 (1966); *Chem. Abstr.* **67**, 89739b (1967).
61. H. P. Pietsch, R. Oehler, and D. Kaspuck, *Nahrung* **12**, 885-886 (1968).
62. C. S. Lie, A. D. Haukeli, and J. J. Gether, *Brygmesteren* **27**, 281-291 (1970).
63. J. Farakaš and A. Príbela, *Kvasny Prum.* **14**, 57-61 (1968).
64. A. Caputi, Jr. and D. P. Mooney, *J. Assoc. Off. Anal. Chem.* **66**, 1152-1157 (1983).
65. E. Bernt and L. Gutmann, in *Methods of Enzymatic Analysis*, Vol. 3, H. A. Bergmeyer, Ed. Verlag Chemie, Weinheim, 1974, pp. 1499-1502.
66. L. P. McCloskey and L. L. Replogle, *Am. J. Enol. Vitic.* **25**, 194-197 (1974).
67. J. C. Bouvier, *Rev. Fr. Oenol.* **22**, 47-48 (1982).
68. J. Sarris, J. N. Morfaux, P. Dupuy, and D. Hertzog, *Ind. Aliment. Agric.* **86**, 1241-1246 (1969).
69. W. van Steebergen-Horrocks and J. H. S. Williams, *Z. Lebensm.-Unters. -Forsch.* **168**, 112-114 (1979).
70. M. Mason, *Am. J. Enol. Vitic.* **34**, 173-175 (1983).
71. A. Scheter-Graff, H. Huck, and A.-S. Schmidt, *Z. Lebensm.-Unters. -Forsch.* **177**, 356-358 (1983).
72. T. Hirata, R. Izumi, K. Takahashi, and K. Yoshizawa, *Nippon Jozo Kyokai Zasshi* **80**, 206-207 (1985).
73. M. Flanzy and Y. Loisel, *Ann. Technol. Agric.* **7**, 311-321 (1958).
74. B. Gnekow and C. S. Ough, *Am. J. Enol. Vitic.* **27**, 1-6 (1976).
75. T. Shinohara and M. Watanabe, *Agric. Biol. Chem.* **40**, 2475-2477 (1976).
76. *The Toxic Substance List*, U.S. Department of Health, Education and Welfare, Washington, DC, 1973.
77. P. Sudraud, *Ann. Falsif. Expert. Chim.* **67**, 289-295 (1974).
78. G. F. Beyer, *J. Assoc. Off. Anal. Chem.* **34**, 745-748 (1951).
79. H. Rebelein, *Dtsch. Lebensm.-Rundsch.* **61**, 211-212 (1965).
80. D. Hess and F. Koppe, in *Alkoholische Genussmittle*, Vol. 7. W. Diemair, Ed. Springer-Verlag, Berlin, 1968, pp. 311-495.
81. J. Le Roux, *Agrochemophysica* **3**, 11-16 (1971).
82. C. Y. Lee, T. E. Acree, and R. M. Butts, *Anal. Chem.* **47**, 747-748 (1975).
83. J. F. Guymon, *Dev. Ind. Microbiol.* **7**, 88-96 (1966).
84. J. F. Guymon, J. L. Ingraham, and E. A. Crowell, *Arch. Biochem. Biophys.* **95**, 163-168 (1961).
85. J. L. Ingraham and J. F. Guymon, *Arch. Biochem. Biophys.* **88**, 157-166 (1960).
86. J. L. Ingraham, J. F. Guymon, and E. A. Crowell, *Arch. Biochem. Biophys.* **95**, 169-175 (1961).
87. C. S. Ough and A. A. Bell, *Am. J. Enol. Vitic.* **31**, 122-123 (1980).
88. C. S. Ough, J. F. Guymon, and E. A. Crowell, *J. Food Sci.* **31**, 620-625 (1966).
89. F. Prillinger and H. Horwatitsch, *Mitt. Hoeheren Bundeslehr- Versuchanst. Wein- Obstbau, Klosterneuberg, Gartenbau, Schoenbrunn, Ser. A.* **15**, 72-79 (1965).
90. C. S. Ough and M. A. Amerine, *Am. J. Enol. Vitic.* **18**, 157-164 (1967).

91. B. C. Rankine and K. F. Pocock, *Vitis* **8**, 23-27 (1969).
92. J. C. Sapis and P. Ribéreau-Gayon, *Ann. Technol. Agric.* **18**, 221-229 (1969).
93a. B. C. Rankine, *J. Sci. Food Agric.* **18**, 583-589 (1967).
93. S. D'Agostino, E. Carruba, B. Pastena, and C. Alagna, *Riv. Vitic. Enol.* **32**, 496-513 (1979).
94. W. Postel, *Dtsch. Lebensm.-Rundsch.* **78**, 211-215 (1982).
95. T. Shinohara and M. Watanabe, *Nippon Jozo Kyokai Zasshi* **71**, 888-889 (1976).
96. C. Gil and C. Gómez-Cordovés, *Food Chem.* **22**, 59-65 (1986).
97. G. E. Martin, J. M. Burggraff, R. H. Dyer, and P. C. Buscemi, *J. Assoc. Off. Anal. Chem.* **64**, 186-190 (1981).
98. W. B. D. Penniman, D. C. Smith, and E. I. Lawshe, *Ind. Eng. Chem., Anal. Ed.* **9**, 91-95 (1937).
99. H. W. Coles and W. E. Tournay, *Ind. Eng. Chem., Anal. Ed.* **14**, 20-22 (1942).
100. J. F. Guymon, *Analytical Procedures for Brandy, A Laboratory Manual.* University of California, Davis, 1959.
101. R. J. Bouthilet and W. Lowery, *J. Assoc. Off. Agric. Chem.* **42**, 634-637 (1959).
102. A. D. Webb and R. E. Kepner, *Am. J. Enol. Vitic.* **12**, 51-59 (1961).
103. R. L. Brunelle, *J. Assoc. Off. Anal. Chem.* **51**, 915-921 (1968).
104. I. Otvös, I. Szeb, I. Bikfalvi, L. Pásztor, and G. Pálvi, *Elelmiszervizsglalati Kozl.* **22**, 142-151 (1976).
105. J. H. Kahn, F. M. Trent, P. A. Shipley, and R. A. Vordenberg, *J. Assoc. Off. Anal. Chem.* **51**, 1330-1333 (1968).
106. E. Oura, *Process Biochem.* **12**(3), 19-21, 35 (1977).
107. C. S. Ough, D. Fong, and M. A. Amerine, *Am. J. Enol. Vitic.* **23**, 1-5 (1972).
108. B. C. Rankine and D. A. Bridson, *Am. J. Enol. Vitic.* **22**, 6-12 (1971).
109. C. Otoguro, S. Ogino, and M. Watanabe, *Nippon Jozo Kyokai Zasshi* **78**, 214-219 (1983).
110. E. Vogt, revised by L. Jakob, E. Lemperle, and E. Weiss, *Der Wein; Bereitung, Behandlung, Untersuchungen.* Ulmer, Stuttgart, 1984.
111. M. A. Amerine, *Adv. Food Res.* **5**, 353-510 (1954).
112. L. R. Mattick and A. C. Rice, *Am. J. Enol. Vitic.* **21**, 213-215 (1970).
113. F. Drawert and G. Kupper, *Z. Lebensm.-Unters. -Forsch.* **123**, 211-217 (1973).
114. K. Möhler and S. Looser, *Z. Lebensm.-Unters. -Forsch.* **140**, 149-154 (1969).
115. M. Castino and L. Usseglio-Tomasset, *Riv. Vitic. Enol.* **21**, 465-480 (1968).
116. S. Spagna Musso, B. Mincione, and V. Coppola, *Riv. Vitic. Enol.* **32**, 47-64 (1979).
117. M. A. Pimenova, I. D. Belousova, and S. P. Avakiants, *Vinodel. Vinograd. SSSR* **6**, 57 (1983).
118. H. Rebelein, *Z. Lebensm.-Unters. -Forsch.* **105**, 296-311 (1957).
119. K. Wagner and P. Kruetzer, *Wein-Wiss.* **33**, 109-113 (1978).
120. F. Drawert, *Vitis* **3**, 237-238 (1963).
121. K. Mayer and I. Busch, *Mitt. Geb. Lebensmittelunters. Hyg.* **54**, 297-303 (1963).
122. J. J. Boltralik and H. Noll, *Anal. Biochem.* **1**, 269-273 (1960).
123. J.-L. Battle and Y. Collon, *Connaiss. Vigne Vin* **13**, 45-50 (1979).
124. W. Hinsch, A. Antonijevic, and V. Sundaram, *Z. Lebensm.-Unters. -Forsch.* **171**, 449-450 (1980).
125. W. Hinsch and P. V. Sundaram, *Clin. Chim. Acta* **104**, 87-94 (1980).

126. E. Walter and P. Kohler, Z. *Lebensm.-Unters. -Forsch.* **180**, 121-125 (1985).
127. C. Junge, *Lebensm. Chem. Gerichtl. Chem.* **33**, 67-70 (1979).
128. M. F. Feil and L. Marinelli, *Proc. Am. Soc. Brew. Chem.* **1969**, 29-34 (1969).
129. P. Avellini, P. Damiani, and G. Burini, *J. Assoc. Off. Anal. Chem.* **60**, 536-540 (1947).
130. G. E. Martin, R. H. Dyer, and D. M. Figert, *J. Assoc. Off. Anal. Chem.* **58**, 1147-1149 (1975).
131. W. Postel, L. Adam, and M. Rustler, *Dtsch. Lebensm.-Rundsch.* **78**, 170-172 (1982).
132. P. Pfeiffer and F. Radler, *Z. Lebensm.-Unters. -Forsch.* **181**, 24-27 (1985).
133. G. A. Ledingham and A. C. Neish, in *Industrial Fermentations*, Vol. 2, L. A. Underkoffer and R. J. Hickey, Eds. Chem. Publ. Co., New York, 1954, pp. 27-93.
134. J. F. Guymon and E. A. Croswell, *Am. J. Enol. Vitic.* **18**, 200-209 (1967).
135. C. S. Ough and M. A. Amerine, *Am. J. Enol. Vitic.* **23**, 128-131 (1972).
136. A. C. Neish, *Can. J. Res., Sect. B* **28**, 660-661 (1950).
137. H. Muraki and H. Masuda, *J. Sci. Food Agric.* **21**, 345-350 (1976).
138. M. O. Dubernet, A. Bertrand, and P. Ribéreau-Gayon, *C. R. Hebd. Seances Acad. Sci., Ser. D* **279**, 1561-1564 (1974).
139. H. Thaler and G. Lippke, *Dtsch. Lebensm.-Rundsch.* **66**, 96 (1970).
140. A. Bertrand and R. Pissard, *Ann. Falsif. Expert. Chim.* **69**, 571-579 (1976).
141. A. Patschky, *Allg. Dtsch. Weinfachztg.* **110**, 1336-1337 (1974).
142. P. Damiani and M. Brogioni, *Ind. Aliment. (Pinerolo, Italy)* **10**, 81-84 (1971).
143. W. R. Sponholz and H. H. Dittrich, *Vitis* **24**, 97-105 (1985).
144. W. Flak, *Mitt. Klosterneuburg* **31**, 204-208 (1981).
145. M. Frèrejacques, *C. R. Hebd. Seances Acad. Sci.* **200**, 1410-1412 (1935).
146. H. Tanner, *Schweiz. Z. Obst- Weinbau* **103**, 610-617 (1967).
147. H. Tanner and M. Duperrex, *Fruchtsaft-Ind.* **13**, 98-114 (1968).
148. H. G. Williams-Ashman, in *Methods of Enzymatic Analysis*, H. U. Bergmeyer, Ed. Verlag Chemie, Weinheim, 1963, pp. 167-170.
149. J. Dobladalova and R. P. Upton, *J. Assoc. Off. Anal. Chem.* **56**, 1382-1387 (1973).
150. H. O. Beutler and J. Becker, *Dtsch. Lebensm.-Rundsch.* **73**, 182-187 (1977).
151. E. Peynaud and S. Lafourcade, *Ann. Technol. Agric.* **4**, 381-396 (1955).
152. M. K. Gaitonde and M. Griffiths, *Anal. Biochem.* **15**, 532-535 (1966).
153. F. C. Charalampous and P. Abrahams, *J. Biol. Chem.* **225**, 575-583 (1957).
154. C. C. Sweeley, R. Bentley, M. Makita, and W. W. Wells, *J. Am. Chem. Soc.* **85**, 2497-2507 (1963).
155. Y. C. Lee and C. E. Ballow, *J. Chromatogr.* **18**, 147-149 (1965).
156. C. R. Strauss, P. J. Williams, B. Wilson, and E. Dimitriadis, *Proc. ALKO Symp. Flavour Res. Alcohol. Beverages, 1984*, Vol. 3, pp. 51-60 (1984).
157. P. J. Williams, C. R. Strauss, and B. Wilson, *J. Agric. Food Chem.* **28**, 766-771 (1980).
158. R. F. Simpson and G. C. Miller, *Vitis* **22**, 51-63 (1983).
159. R. Di Stefano and C. Ciolfi, *Riv. Vitic. Enol.* **36**, 126-143 (1983).
160. E. Dimitriades and P. J. Williams, *Am. J. Enol. Vitic.* **35**, 66-71 (1984).

Four
CARBONYL COMPOUNDS

A large number of aldehydes, ketones, and related compounds have been found in fermented beverages. Some of these have little sensory importance but a few do.

The relation of alcohols, aldehydes, keto acids, and amino acids is shown in Table 25.

Acetaldehyde is the carbonyl compound that has received the most attention. It is unwelcome as an odor in table wines but is looked for in fino sherries. Likewise, hydroxymethylfurfural is unwanted in table wines and even in many dessert wines—port, for example. But in wines that have been baked—Marsala and some sweet sherries for example—it is a normal constituent and its odor is appreciated. 2,3-Butanedione (diacetyl), when present in small amounts [below 4 mg/L (1)], seems to go undetected, although amounts as low as 0.5 mg/L have been considered as harmful to quality (2). Its presence indicates bacterial activity and/or oxidation.

Acetoin (3-hydroxy-2-butanone) is found in wines that have been altered by chemical or bacterial action. The same is true for 2,3-pentandione. Both of these compounds are found in wines that have undergone a malolactic fermentation. For acetoin, 1 mg/L has been suggested as an upper limit (2). Neither of these is considered desirable in above threshold amounts. Some of these compounds are interrelated:

$$\text{2,3-Butanediol} \rightleftharpoons \text{Acetoin} \rightleftharpoons \text{Diacetyl}$$

Several aliphatic aldehydes have been reported in fermented beverages (e.g., butyraldehyde and isovaleraldehyde). The stale odor of bottled beer has been traced to such compounds (3). Increases in acetoin, *trans*-2-propenal, heptanal, nonal, and *trans*-2-nonenal were reported in beers with 30 days of aging (4). In Italian wines, compounds such as methanal, acetaldehyde, propanal, 2-methylpropanal, pentanal, and hexanal were found as aldehydes; propanone, butanone, 3-hydroxybutanone, and 2-pentanone were found as ketones (5).

Acetals are equilibrium products of carbonyls and alcohols. Reducing sugars

Table 25. Some Alcohol- and Amino Acid-Related Carbonyl Compounds Found in Fermented Products

Alcohols	Aldehydes	Keto acids	Amino acids
Ethanol	Acetaldehyde	Pyruvic acid	Alanine
Glycol	Glyoxal	Hydroxypyruvic acid	Serine
Propanol	Glyoxal	α-Ketobutyric acid	α-Aminobutyric acid
Butanol	Butyraldehyde	—	—
Isobutanol	Isobutyraldehyde	α-Ketoisovaleric acid	Valine
Isoamyl alcohol	Isovaleraldehyde	α-Ketoisocaproic acid	Leucine
Active amyl alcohol	Active valer-aldehyde	α-Keto-β-methylvaleric acid	Isoleucine
Hexanol	Hexanal	—	—
Heptanol	Heptanal	—	—
—	—	Oxalacetic acid	Aspartic acid
—	—	α-Ketoglutaric acid	Glutamic acid
Phenethyl alcohol	Phenethyl-aldehyde	Phenylpyruvic acid	Phenylalanine
Tyrosol	—	Hydroxyphenylpyruvic acid	Tyrosine
Tryptophol	—	—	Tryptophan

are also carbonyls but are considered in Chapter 1. The carbonyls range from highly volatile liquids such as acetaldehydes to high-boiling-point liquids such as hydroxymethylfurfural. Most of the carbonyls are highly reactive and readily produce derivatives with many compounds. For further information, References 6–12 may be consulted.

ACETALDEHYDE

Acetaldehyde, acetoin, and diacetyl are closely linked metabolically (12). Acetaldehyde is a precursor for acetate (13, 14) as well as for ethanol. It is formed from pyruvate through the glycolytic pathway enzymes (15). There is considerable variation in the amount present in wines because of increased fermentation temperature, aeration resulting in higher acetaldehyde, higher pH, and lack of pantothenate and thiamine (16). High levels of SO_2 in musts before and during fermentation result in higher levels of fixed acetaldehyde in the wines. Use of less SO_2 in musts produces wines of relatively higher free acetaldehyde. Table 26 lists some acetaldehyde concentrations reported in wines (17–24). The amounts present in table wines have decreased in recent years, mainly because of the more rational use of sulfur dioxide.

There are several chemical methods for the determination of acetaldehyde.

Table 26. Reported Acetaldehyde Concentrations in Wines

Country	Wine type	Number of samples	Acetaldehyde (mg/L) Range	Acetaldehyde (mg/L) Average	Reference
General	Table	764	3–494	54.4	17
	Dessert	415	15–264	86.8	17
	Sherry[a]	25	90–500	218	17
Brazil	Table	19	55–783	162	18
France	Table	40	14–71	24.6	19
Germany	Table	4	33–79	58.2	20
	Table	6	3–45	21.0	21
	Table	7	40–238	105.4	22
	Table	17	7–67	29	23
United States (California)	Table	4	32–91	69.3	24
	Dessert	2	40–85	62.7	24
	Sherry[a]	2	104–248	126	24

[a] Primarily California-style sherry.

It can be measured directly by iodiometric titration or by derivative formation of colored compounds and subsequent spectrophotometric measurement. In addition, acetaldehyde can be determined by gas chromatography.

Chemical Determination

Jaulmes and Espezel (25) determined the acetaldehyde content of wines by distilling the acetaldehyde into a neutral bisulfite solution. The excess bisulfite is titrated at pH 2.0. The acetaldehyde–bisulfite complex, $CH_3CH(OH)SO_3^-$, does not dissociate at this pH. Subsequently the pH is adjusted to 9.0, and the released bisulfite is titrated quantitatively with iodine. The following equations are pertinent:

$$OH^- + CH_3CH(OH)SO_3^- \xrightarrow[\text{heat}]{\text{pH 8.0}} CH_3CHO + SO_3^{2-} + H_2O \quad \text{(distill)}$$

$$H^+ + CH_3CHO + SO_3^{2-} \xrightarrow[\text{fast}]{\text{pH 7.1}} CH_3CH(OH)SO_3^- \quad \text{(receiving flask)}$$

$$CH_3CH(OH)SO_3^- + H^+ \xrightarrow[\text{very slow}]{\text{pH 2.0}} CH_3CHO + H_2SO_3$$

$$SO_2 + I_3^- + 2H_2O \xrightarrow[\text{fast}]{\text{pH 2.0}} SO_4^{2-} + 3I^- + 4H^+ \quad \text{(removal of excess } HSO_3^-\text{)}$$

$$OH^- + CH_3CH(OH)SO_3^- \xrightleftharpoons[\text{fast}]{\text{pH 8.8-9.5}} CH_3CHO + SO_3^{2-}$$

$$+ H_2O \text{ (release bound } HSO_3^-\text{)}$$

$$SO_3^{2-} + I_3^- + 2OH^- \xrightleftharpoons[\text{fast}]{\text{pH 8.8-9.5}} SO_4^{2-} + 3I^-$$

$$+ H_2O \text{ (titration of freed } HSO_3^-\text{)}$$

When analyzing wines for total acetaldehyde content, a small percentage (3-4% in wines containing 20% ethanol; less than 1% in table wine containing 12% ethanol) is bound as acetal. This is not recovered in the usual procedures. The procedure given below is that of Jaulmes and Hamelle, as tested in the United States (26), and is an official method of the AOAC (27). However, one report (28) indicates that the results are 1-20% higher than those obtained by enzymatic or colorimetric procedures (22) because these are more specific for acetaldehyde.

Modifications to consider the acetal concentration (29) can be made. The air oxidative changes taking place during the alkaline titration step are prevented by addition of a chelating agent (EDTA) to bind the copper that is present. Copper is a catalyst for the oxidation reaction (30). Addition of a small amount of isopropyl alcohol also inhibits the oxidation (31).

Procedure

Potassium Metabisulfite Solution. Dissolve 15 g of $K_2S_2O_5$ in water, add 70 mL of concentrated hydrochloric acid, and dilute to 1 L with water. The titer of 10 mL of this solution should not be less than 24 mL of 0.1 N iodine solution.

Phosphate-EDTA Solution. Dissolve any of the following combinations and 4.5 g of the disodium salt of EDTA in water and dilute to 1 L:

200 g of $Na_3PO_4 \cdot 12H_2O$
188 g of $Na_2HPO_4 \cdot 12H_2O$ + 21 g of NaOH
72.6 g of $NaH_2PO_4 \cdot H_2O$ + 42 g of NaOH
71.7 g of KH_2PO_4 + 42 g of NaOH

Sodium Borate Solution. Dissolve 100 g of boric acid plus 170 g of sodium hydroxide in water and dilute to 1 L.

Use the macrodistillation equipment shown in Figure 11. Pipet 50 mL of the wine sample containing less than 30 mg of acetaldehyde into the distilling flask and add 50 mL of a saturated borax solution to bring the pH to about 8. Distill 50 mL into a 750-mL Erlenmeyer flask containing 300 mL of water and 10 mL each of the potassium metabisulfite and the phosphate-EDTA solutions; the pH of the solution in the Erlenmeyer flask prior to distillation should be in the range 7.0-7.2.

After the distillation add 10 mL of 3 N hydrochloric acid and 10 mL of a freshly prepared 0.2% starch solution to the Erlenmeyer flask. Mix and immediately titrate with 0.1 N iodine solution just to a faint blue end point. Add 10 mL of the sodium borate solution and rapidly titrate the liberated bisulfite with 0.02 N iodine solution, using a 25-mL buret, to the same blue end point. Avoid direct sunlight. The pH of the solution should be 8.8–9.5.

$$\text{Acetaldehyde (mg/L)} = \frac{(V)(N)(22.0)(1000)}{v}$$

where V = volume of iodine solution used for titration after the addition of the sodium borate solution (mL)
N = normality of the iodine solution
v = volume of wine sample (mL)

If the first titration with iodine, to remove the excess bisulfite, takes only a few drops, either the bisulfite solution was not up to proper strength or the wine contains excessive amounts of acetaldehyde; if the second titration consumes excessive amounts of iodine and yellow iodoform formation is noted, the buffer was improperly prepared and the solution is too alkaline.

A colorimetric method for acetaldehyde uses 3-methylbenzothiol-2-one hydrazone to react with the acetaldehyde after it is separated from the wine. Ferric chloride is added to oxidize the excess reagent. The oxidized reagent reacts with the aldehyde-reagent compound to form a highly colored product. The reaction is stopped by addition of acetone and absorbance is determined at 630–666 nm within a few minutes (20, 24, 32). The following reactions take place during the determination (32):

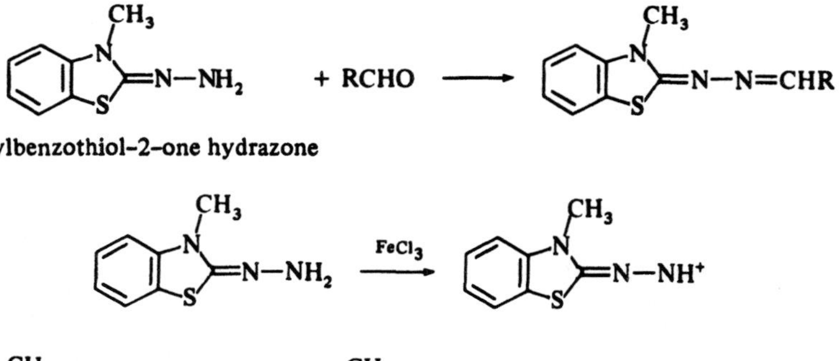

3-Methylbenzothiol-2-one hydrazone

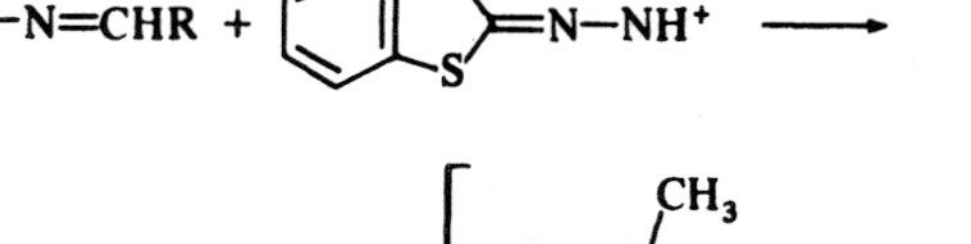

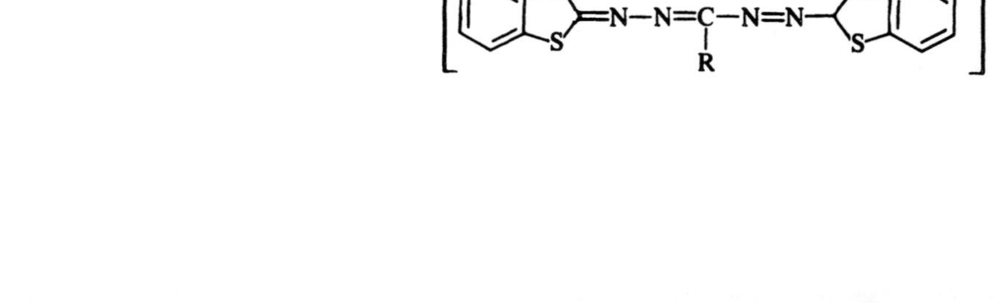

Rebelein (22) suggested a colorimetric method that is rapid and simple and gives results comparable to those with the standard chemical method. The wine is decolorized, filtered, and treated with piperidine and nitroprusside solutions to develop a colored reaction product proportional to the amount of sulfur dioxide-bound acetaldehyde [also called the bisulfite complex, or $CH_3CH(OH)SO_3^-$] present. The procedure follows.

Procedure

Nitroprusside Solution. In a 250-mL volumetric flask, dissolve 1.0 g of $Na_2Fe(CN)_5NO_3$ in water and dilute to volume (prepare daily).
Piperidine Solution. Dilute 2 mL of piperidine with 20 mL of water.
Aqueous 5% Sulfur Dioxide Solution. Saturate water with gaseous sulfur dioxide at 20°C.

To a 25-mL aliquot of the wine sample in a 50-mL Erlenmeyer flask, add 2 g of activated charcoal, mix well, and filter clear after 2 min. Pipet a 2-mL aliquot of this filtrate into a 25-mL Erlenmeyer flask containing 5 mL of the nitroprusside solution and 5 mL of the piperidine solution. Mix, transfer a portion to a 10-mm cuvette, and measure the absorbance immediately at 570 nm in a spectrophotometer. The maximum absorbance is reached in 50 sec. The reaction takes place with the bound acetaldehyde [$CH_3CH(OH)SO_3^-$].

If there is more free acetaldehyde present in the wine than would normally be in equilibrium with an excess of sulfur dioxide, add additional sulfur dioxide to the wine prior to the analysis, then allow to stand for 1 hr.

Calibration. Weight out approximately 1 g of freshly distilled acetaldehyde and dissolve it in 500 mL of water in a 1-L volumetric flask. Add 30 mL of the 5% aqueous sulfur dioxide solution to the flask and dilute to volume with water. Determine the exact concentration of the acetaldehyde in this solution in the following way.

To a 50-mL aliquot add 1:3 hydrochloric acid, 10 mL of a 0.2% starch solution, and 100 mL of water, then titrate to a blue end point with 0.1 N iodine solution to remove excess bisulfite solution. Then add 10 mL of the sodium borate solution (see preceding procedure), and titrate the freed sulfur dioxide with 0.1 N iodine solution to a faint blue end point.

$$\text{Acetaldehyde (mg/L)} = (V)(44)$$

where V = volume of 0.1 N iodine solution used for the titration (mL)

Pipet 2-, 5-, 10-, 15-, and 20-mL aliquots of the standard acetaldehyde solution into individual 100-mL volumetric flasks and dilute each to volume with water. From each flask, take a 2-mL aliquot, place it into a separate 25-mL Erlenmeyer flask, and add, to each flask, 5 mL of the nitroprusside solution and 5 mL of the piperidine solution. Measure the absorbance as described above. Beer's law is followed from 0 to 200 mg/L.

Gas Chromatography

Acetaldehyde is easily separated from the other constituents of wine on almost any semipolar or polar column. A typical liquid phase used is diethylene glycol succinate (DEGS) or Carbowax 20M.

HPLC

More recently, HPLC has been used to determine acetaldehyde in wines (33).

Procedure

React the wine at 60°C for 30 min with acetylacetone, ammonium acetate, and acetic acid. Add, after cooling, 5 mL of the lutidine derivative of propionaldehyde in chloroform as an internal standard. Discard the aqueous phase and dry the organic phase over anhydrous Na_2SO_4. Place a 50-μL volume in the HPLC. Treat the column of silica gel with 3-aminopropyltriethoxysilane or with $5NH_2$ Nucleosil®. Pack stainless steel columns (250 mm × 4 mm i.d.) using the balance density method and a 10-mL stainless steel packer at the rate of 500 kg/cm^2 (Kyowa Seimitsu Type® KHW-20 ultra-high-pressure pump). The mobile phase was a mixture of hexane and ethanol (25:1).

In five commercial wines (Japanese?), only concentrations of 0.2–1.2 mg/L were found. For GC with beer see Delcour et al. (34). For distilled beverages, HPLC separation with diode-array spectroscopic identification has been successful in separating and identifying 14 carbonyls and quantitatively measuring 10 of them (35). With suitable concentration this procedure might be adaptable to wines.

A somewhat more straightforward procedure (23) is to adjust the pH of the wine to 8 and distill an aliquot into acidified 2,4-dinitrophenylhydrazine. The precipitate (containing the volatile carbonyls) is fine-screen filtered and washed well with water. Mix the precipitate with 2-ketoglutaric acid. The dried precipitate is placed in a glass tube at 300°C leading to the HPLC. The aldehydes are volatilized and their peaks are measured. The amounts found in 17 German wines (mg/L) were: acetaldehyde, see Table 26; propanal, 0.06–2.27 (average 0.77); 2-methylpropanal, 0.06–3.6 (average 0.50); propenal, 0.06–3.8 (average 0.95); butanal, 0.01–1.41 (average 0.22); 2- + 3-methylbutanal, 0.08–2.55 (average 0.47); pentanal, 0.05–14.36 (average 1.43); butenal, 0–0.72 (average 0.23); and hexanal, 0–0.24 (average 0.04). It is of interest that lactic-spoiled wines had lower amounts of acetaldehydes and higher amounts of other aldehydes than did normal wines.

Colorimetric Procedure

A recent colorimetric procedure (36) uses 10% piperidine and 2% freshly prepared sodium nitroprusside for development of the color. The diluted wine (2–

5×) is distilled into a 50-mL flask containing 10 mL of 5% metabisulfite (cold). Bring to volume. To a 5-mL aliquot add 5 mL of the piperidine reagent and 5 mL of the nitroprusside reagent. The absorbance is determined at 570 nm. A standard curve is prepared with known amounts of acetaldehyde. The concentrations of acetaldehyde produced by 14 yeast strains varied from 16 to 72 mg/L (ethanol 9.7–12.9%).

A similar and direct method designed for wines uses semicarbazide and spectrophotometric determination at 223 nm (37).

Enzymatic Determination

Although alcohol dehydrogenase (ADH) is not specific for acetaldehyde it has been used for its determination. The reaction is

$$\text{Acetaldehyde} + \text{NAD}^+ + \text{H}_2\text{O} \xrightarrow{\text{Aldehyde dehydrogenase}} \text{Acetate} + \text{NADH} + \text{H}^+$$

Since the reaction takes place at pH 8.7, free and fixed (1-hydroxyethanesulfonic acids) acetaldehyde are determined together. The McCloskey–Mahaney (38) procedure is given here.

Procedure

Prepare reagent A buffer by dissolving 7.9 g of glycylglycine, 3.73 g of KCl, and 2 g of polyvinylpyrrolidone in 450 mL of deionized water. Adjust the pH to 8.7 with 5 N NaOH and make to 500 mL. This is stable for 6 months at 0–4°C. Reagent A NAD$^+$ is made by dissolving 250 mg of NAD$^+$ in 6 mL of deionized water. This is stable for a year at 0–4°C. Reagent B is aldehyde dehydrogenase (E.C. 1.2.1.5). Dissolve 2 mg (7 IU) in 1.0 mL of deionized water. This is stable for 24 hr at 4°C. The powdered enzyme is stable for 4 months at below 0°C (desiccated). A stock acetaldehyde solution is prepared in a 1-L volumetric flask to which 500 mL of deionized water is added, followed by ~1.5 mL of acetaldehyde; then bring to volume with water. This solution is stable for 4 months at 0°C. **Caution:** Handle pure acetaldehyde in a fume hood.

To calibrate the acetaldehyde stock solution, pipet 50 mL to a 250-mL Erlenmeyer flask containing 1 mL of sodium bisulfite (125 mg/mL), add 10 mL of 3 N HCl and 2 mL of 1% starch solution, and then titrate the excess bisulfite with 0.100 N iodine to a blue end point. Pipet 10 mL of sodium borate solution (10 g of boric acid plus 17 g of NaOH made to 100 mL), then titrate with iodine to the blue end point. Acetaldehyde (mg/L = $V \times 44$, where V is the volume of 0.100 N iodine used in the second titration. The stock solution must be restandardized monthly. To prepare acetaldehyde standards, pipet 1-, 2-, 3-, 5-, and 10-mL aliquots into 100-mL volumetric flasks, then bring to volume with water. These are stable for 3–4 weeks at 0–4°C.

Dilute wines with over 125 mg/L of acetaldehyde before using. Red wines need decolorizing, which may result in loss of acetaldehyde. Warm the buffer A to 26–30°C and add 3 mL to each cuvette; then add 0.10 mL of NAD$^+$ solution and 0.100 mL of sample. Mix and read the absorbance at 240 nm and record as E_1. Then add 0.025 mL of reagent B, mix, and incubate for 10 min at 26–30°C. Read absorbance and record as E_2. The $E_2 - E_1$ is the E_s. Repeat with only water. The absorbance is E_b. $E_s - E_b = E_{std}$. Prepare a standard curve with the acetaldehyde standards, then plot versus absorbance. Read the acetaldehyde content of the wine from this curve. Obviously, automatic or other accurate pipets must be used for small volumes. Essentially the same procedure has been used for beer.

The aromatic aldehydes vanillin, 4-hydroxybenzaldehyde, cinnamaldehyde, benzaldehyde, piperonal, ethyl vanillin, syringaldehyde, and coumarin may be separated by thin-layer chromatography (39).

α-DIKETONES AND α-HYDROXYKETONES

Diacetyl (2,3-butanedione) and 2,3-pentanedione (both vicinal ketones) and their precursors 2-acetolactate, 2-acetohydroxybutyrate, acetoin, and 3-hydroxy-2-pentanone, along with traces of related compounds, have been reported in wines.

Acetoin can be formed by several routes, the most obvious being the addition of thiamine pyrophosphate–pyruvate complex to pyruvate to form α-acetolactate, followed by decarboxylation to acetoin. Reaction of the thiamine pyrophosphate–pyruvate complex with acetyl-CoA forms diacetyl, which may be reduced to acetoin. This appears to be the main pathway (40–42). Acetoin can be reduced to 2,3-butanediol by microbial action. It is generally agreed that brewer's yeast does not form diacetyl but that its precursor, α-acetolactate, is directly converted to diacetyl. At the same time, some of the α-acetolactate is converted to acetoin. It appears that as long as yeast cells are present in a fermenting medium, diacetyl cannot be detected because of the high diacetyl- and 2,3-pentanedione-reducing activity of the yeast cells (43, 44).

$CH_3-CO-CO-CH_3 \longleftrightarrow CH_3CHOH-CO-CH_3 \longleftrightarrow CH_3-CHOH-CHOH-CH_3$
Diacetyl   Acetoin   2,3-Butanediol

$CH_3-CH_3-CO-CO-CH_3 \longleftrightarrow CH_3-CH_2-CHOH-CO-CH_3 \longleftrightarrow$
2,3-Pentanedione   3-Hydroxy-2-pentanone

$CH_3-CH_3-CHOH-CHOH-CH_3$
2,3-Pentanediol

Under aerobic conditions, yeasts reduce acetoin to 2,3-butanediol and also reduce 2,3-pentanedione to 2,3-pentanediol.

At higher fermentation temperatures, bacteria and certain yeasts (45) result

Table 27. Reported Acetoin and Diacetyl Concentrations (mg/L) in Wines

Country	Type of wine	Acetoin			Diacetyl			Reference
		Number of samples	Range	Average	Number of samples	Range	Average	
Australia	White table	15	0.7– 4.3	1.8	15	0.1 –1.3	0.67	46
	Red table	70	1.5–44.0	10.6	70	0.3 –4.6	1.4	46
	Red table				466	0.1 –7.5	2.4	47
Finland	White table	8	8–44	12				40
	Red table	9	6–53	46				40
	Miscellaneous				16	0.16–1.20	0.5	48
France	Miscellaneous	20	2–84	10.0	30	0.4 –1.8	0.6	19
Germany	Table[a]	2	1.5– 5.9	3.7	11	0.2 –0.6	0.3	49
	Table[b]	14	3.0–31.8	11.9	14	0.7 –4.3	1.4	46
	Miscellaneous	100	0–29	7.8				17
	Red table	20	5.9–38.2	15.0	20	0.26–4.06	1.46	41
	White table	20	1.9–31.7	5.9	20	0.08–3.40	0.42	41
Japan	White table	17	1–138	7.3				45
	Red table	12	T–43.1	11.4				45
Soviet Union	Table	11	8.0–36.2	15.1	11	0.14–2.92	0.68	11
United States	Dry table	3	5.5–18.5	10.3	2	0.2 –2.2	0.5	50
(California)	Fortified sweet	13	37.5–236	86.0	5	0.2 –0.5	0.3	50
	Baked sherry	3	5.6–28.4	18.8				50
	Flor sherry[c]	4	74–350	196	4	0.2 –0.4	0.3	50

[a] No malolactic fermentation.
[b] With malolactic fermentation.
[c] Spanish-type sherry made with special film yeast.

in higher amounts of acetoin and diacetyl. Natural vinegars are higher in acetoin than are imitation vinegars. Table 27 lists data for the acetoin and diacetyl content of various wines (46–50). The 2,3-butanediol in four of these wines amounted to 0.38–0.68 g/L. Few quantitative data on the other components are reported, and the amounts are small (11): 0.02–0.28 mg/L for 2,3-pentanedione, 0.45–2.8 mg/L for 3-hydroxy-2-pentanone, and 7.0–18.0 mg/L for 2,3-pentanediol.

ACETOIN AND DIACETYL

In wine spirits the acetals 3,3-diethoxybutan-2-one and 1,1,3-triethoxypropane have been identified. They are derived from diacetyl and acrolein. When used for fortifying wines of pH >4.2 they may produce off-odors (51). For a method form determining the vicinal ketones as well as 2-acetolactate and 2-acetohydroxybutyrate see References 51 and 52.

Acetoin, diacetyl, and 2,3-pentadione have been measured by the same chemical reactions by most investigators. They can also be separated and measured by gas chromatography. They can be derivatized and determined by other means. In an enzymatic micromethod for acetoin (53), NADH is used to reduce the acetoin to 2,3-butanediol. The change in absorbance of the NADH is measured spectrophotometrically. The isomers were equally reactive with the 2,3-butanediol dehydrogenase obtained from *Sarcina hansenii*. The dehydrogenase was extremely stable. Maximum reaction occurred at pH 5.0, but the activity lasted longer at pH 6.5. No wine interference was noted. Red wine was decolorized. Recovery of added acetoin was in excess of 90%.

The most common method is based on the reaction of diacetyl and acetoin with α-naphthol and creatine. This is the Voges–Proskauer reaction. The usual procedure is to react the wine distillate with the reagents and read the absorbance within about 5 min to determine diacetyl. The rate of reaction of acetoin is slower, and the reading after 1 hr gives the total response of diacetyl and acetoin. If the acetoin concentration exceeds 20 mg/L, such as in fortified dessert wines or aerated sherries, the method probably does not give valid information for diacetyl. Details on the method can be found in References 54 and 55.

Procedure

α-Naphthol Solution. Dissolve 5 g of α-naphthol in 100 mL of isopropanol, add 0.5 g of decolorizing charcoal, mix for 30 min, then filter. Store in an amber bottle.

Potassium Hydroxide–Creatine Solution. In a 100-mL volumetric flask, dissolve 40.00 g of potassium hydroxide in a small amount of water, dilute to about

90 mL, add 0.5 g of creatine, dissolve, then dilute to volume. Store in a polyethylene bottle under refrigeration.

Diacetyl Stock Solution. Dissolve 1 g of pure diacetyl in 1 L of water, then dilute a 1-mL aliquot to 100 mL.

Acetoin Stock Solution. Dissolve 1 g of pure acetoin in 1 L of water, then dilute a 20-mL aliquot to 100 mL.

To 100-mL volumetric flasks add 0, 2, 5, 10, 20, and 50 mL of the diluted diacetyl stock solution and dilute each to volume. Also add similar amounts of the diluted acetoin stock solution into separate 100-mL volumetric flasks and dilute them to volume with water. To 10-mL aliquots of the diacetyl solutions, add 2 mL of the potassium hydroxide–creatine solution and 2 mL of the α-naphthol solution, mix, stopper, and read the absorbance at 530 nm after 1 min. Repeat this procedure with 10-mL aliquots of the acetoin solutions but let the color develop for 1 hr, then read the absorbance at 530 nm. Use a reagent blank containing 10 mL of water instead of the sample that was held for the same time period. Prepare calibration curves for both diacetyl and acetoin.

Place a 5-mL aliquot of the wine sample into the steam microdistillation apparatus (Figure 13) and steam distill until slightly less than 50 mL has been collected. The tip of the condenser outlet should dip into a small amount of water in the receiver. Dilute to 50-mL volume and use a 10-mL aliquot to carry out the determination as outlined above, starting with the addition of 2 mL of the potassium hydroxide–creatine solution.

Compare the absorbance at 1 min to the curve developed from the diacetyl standards. Compare the 1-hr absorbance reading to the curve developed for the acetoin standards. Since the absorbance of diacetyl will be included in the 1-hr reading, subtract from the latter the absorbance read after 1 min.

Gas Chromatography

It is difficult to determine diacetyl in the usual gas chromatographic system because it is present in relatively small amounts and its peak is overlapped on most columns by the ethanol peak. This problem can be overcome by using an electron capture detector that has no response for ethanol but that is sensitive for ketones (47). The headspace over the wine sample has been analyzed using acetone as the internal standard (47). It is also possible to neutralize to pH 7 with calcium hydroxide or to utilize a post column coated with 8% Carbowax 20M on Chromosorb P-AW plus 3% boric acid (56). For beer, a sensitive broad-spectrum gas chromatographic procedure (57) separated not only diacetyl and 2,3-pentadione but also their precursors α-acetolactic and α-acetohydroxybutyric acids. Diacetyl and α-acetolactate were also determined in beer by gas chromatography (43). A gas chromatographic method for wine (41) measures diacetyl, 2,3-pentanedione, and acetoin. In the first step, diacetyl and 2,3-pentanedione are measured. Then acetoin is oxidized to diacetyl and determined by difference. The procedure is as follows (47).

Procedure

Into a 60-mL bottle add 10 mL of the wine sample, 1 mL of a 1:9 acetone–water solution, add 2 mL of water, then close the bottle with a serum cap. Completely immerse the bottle in a 25°C water bath for 30 min, then raise until the septum is out of the water and dry the septum. Withdraw a 2–5 μL aliquot of the headspace with a gas-tight syringe and inject it into the gas chromatography operated under the following conditions:

Column	6 ft × $\frac{1}{8}$-in. o.d., filled with Porapak Q
Temperature	
Injection port	175°C
Column	150°C
Detector	110°C
Carrier gas and flow rate	Nitrogen, 30 mL/min (column), 110 mL/min (makeup at the detector)
Detector	Electron capture

Under these conditions the retention times of acetone and diacetyl are 4.0 and 9.5 min, respectively. Water and ethanol emerge at 0.5 and 2.2 min, respectively.

Acetoin can be determined by direct injection of the wine sample into the gas chromatograph operated under the following conditions (58):

Column	6 ft × $\frac{1}{4}$-in. o.d. packed
Column packing	3% UCON oil 75H-90M on Chromosorb G AW 100/200 mesh
Temperature	
Injection port	150°C
Column	140°C
Detector	180°C
Carrier gas and flow rate	Helium, 60 mL/min
Detector	Flame ionization

Acetoin can be converted to diacetyl by acidic steam distillation, and the diacetyl formed can be determined by analysis of the headspace using an electron capture detector (48). Before the conversion, acetoin is separated from diacetyl and other diketones and aldehydes.

Procedure

Distill a 5-mL wine sample as described above for the Voges–Proskauer method until 50 mL has been collected. To the distillate, add a 0.25% solution of dinitrophenyl hydrazine in 2 N hydrochloric acid. Mix overnight with a magnetic stirrer at 4°C, then filter off the precipitated diketones and aldehydes. To the filtrate add activated carbon, such as Darco G-60 (obtainable from most chemical supply houses). Mix for 1 hr, filter, and wash well with 100 mL of 95% ethanol and 100 mL of water to remove butanediol, if present. The α-hydroxyketone-phenylhydrazone is adsorbed on the carbon. Transfer the carbon precipitate to a steam microdistillation apparatus (e.g., Figure 13); add 4 mL of 1:1 sulfuric acid, and steam distill until 50 mL has been collected. Take a 10-mL aliquot of the distillate, place it in a 60-mL bottle, and continue the determination as given for diacetyl, starting with the addition of the acetone-water solution.

Acetoin may also be determined by a new procedure (52). In a colorimetric method (59) diacetyl reacts with thiosemicarbazide and the extinction is measured at 323 nm.

HYDROXYMETHYLFURFURAL

Hydroxymethylfurfural (HMF) is formed under severe dehydration conditions, or prolonged heating or aging of wines containing fructose, according to the following reaction:

$$\underset{\text{Fructose}}{\text{HOH}_2\text{C}\diagup\text{O}\diagdown\text{CH}_2\text{OH}} \xrightarrow{\Delta} \underset{\text{Hydroxymethylfurfural}}{\text{HOH}_2\text{C}\diagup\text{O}\diagdown\text{CHO}} + 3\text{H}_2\text{O}$$

This reaction does not take place to a measurable extent in grapes on the vines, even though they may be overripe (60). In wines the sensory threshold is about 100 mg/L.

California sherries and Madera-type wines are high in hydroxymethylfurfural; Amerine (61) reported that HMF ranged 0–300 mg/L in California sherries, whereas another report on commercial sherries gave the range as 0–680 mg/L (62). The amounts were not well correlated with the sugar levels, but the lower values were from dry sherry wines while the higher values were from sweet sherries. Soviet "Madera" was found to contain 340 mg/L, a sherry 116 mg/L, and four Soviet brandies 35–135 mg/L (63). In 71 Hungarian table wines (64) the hydroxymethylfurfural was 1–8 mg/L (or 2.5 mg/L average). In 123 Tokaji wines the HMF content was 2–87 mg/L (or 29.4 mg/L average). In 87 Czechoslovakian table wines (65, 66) the HMF ranged from not detectable to 5.2 mg/L (average 2.0 mg/L). In 30 Tokay wines from the same country (1981–1983) the HMF ranged from 3.5 to 37.5 mg/L (average 15.4 mg/L). In 24 Japanese, 8 German, 5 French, and 5 U.S. table wines the HMF ranged from a trace to 27.3 mg/L (67). However, Spanish finos (6) contained between 1.4 and 8.2 mg/L, but ammontillados and olorosos (5) had 23.5–93.2 mg/L. Ports

(68) of known provenance contained 0–27 mg/L (average 7 mg/L). Commercial ports of unknown production and age had 3–36 mg/L (average 25 mg/L). The conclusion was that some of the latter ports may have been blended with reduced musts or caramel, or that the wines had undergone heat madeirization.

Hydroxymethylfurfural can be estimated colorimetrically and easily determined by paper or thin-layer chromatography, or spectrophotometrically. The early methods were primarily estimating the color by extraction of HMF from the wine and reacting with Fiehe's solution. Winkler's method (69, 70) has been used extensively for the quantitative determination of HMF in wines. It involves reaction with barbituric acid and p-toluidine:

$$\text{barbituric acid} + HOH_2C\text{-furan-}CHO + p\text{-toluidine} \longrightarrow$$

$$\text{barbituric acid}=CH-C(OH)=CH-CH=C(CH_2OH)-NH-C_6H_4-OH + H_2O$$

The method is subject to interference (71) from sulfur dioxide because the sulfur dioxide–HMF complex does not react. This can be overcome by removing the sulfur dioxide by iodine titration or addition of excess acetaldehyde (72).

Procedure

p-Toluidine Solution. In a 100-mL volumetric flask dissolve 10 g of p-toluidine in 40 mL of isopropanol, add 50 mL of glacial acetic acid, and dilute to volume with isopropanol. Prepare fresh daily.

Barbituric Acid Solution. Into a 100-mL volumetric flask, add 500 mg of barbituric acid to 90 mL of water. Warm on a water bath to dissolve the material, cool, and dilute to volume with water. The solution is stable for a week.

Be sure that the wine or other sample liquid is brilliant; if not, filter first.

If free sulfur dioxide is present in a concentration under 10 mg/L, pipet a 2-mL sample into each of two glass-stoppered reagent bottles. To each reagent bottle add 5 mL of the p-toluidine solution; to one of the bottles to be used as the blank add 1 mL of water while to the other test bottle add 1 mL of the barbituric acid solution. Mix and measure the absorbance of the second solution at 550 nm using a 10-mm cuvette against the blank. The absorbance is maximum after 2 or 3 min from the time of addition of the barbituric acid.

If the free sulfur dioxide concentration is more than 10 mg/L, pipet 15 mL of the filtered or brilliant sample into a 25-mL volumetric flask, add 2 mL of a 1% aqueous acetaldehyde solution, mix, and dilute to volume with water. Take a 2-mL aliquot of this solution and proceed as above.

Prepare a standard curve in the following way. Prepare a stock solution by dissolving 1 g of HMF in 1000 mL of water. Take 0.5-, 1.0-, 2.0-, 3.0-, and 4.0-mL aliquots of this stock solution and dilute each to 100 mL in volumetric flasks. Pipet 2-mL aliquots of each of these solutions and develop the color by addition of the reagents as described for the less than 10 mg/L sulfur dioxide samples. Plot absorbancy versus concentration.

For wine samples with less than 10 mg/L free sulfur dioxide concentration, the HMF concentration can be read directly from the standard curve. For samples containing more than 10 mg/L free sulfur dioxide, multiply the values read from the calibration curve by 5/3 because of the dilution.

Other methods (60) are available, including a gas chromatographic procedure (65). HMF has also been determined in fruit juices and alcohol-free beverages by high-performance thin-layer chromatography using spectrophotometry (72).

To determine HMF (and furfural), GC was used (67). A dual hydrogen flame ionizing detector was employed. To 1 mL of a sample (or standards for calibration—2, 3, 10, 15, and 20 μg/mL), 10 μg of n-octanol and 20 μg of methyl heptadecanoate were added as internal and supplementary standards plus 30 mL of deionized water and 30 mL of dichloromethane. The mixture was stirred for 15 min and the solvent layer was separated and concentrated in a rotary evaporator. A portion of the concentrate was injected on the glass column (2 m × 3-mm i.d.) and peaked with 20% diethylene glycol succinate (DEGS) on Chromosorb W (AW-DVCS) 60–80 mesh. The oven temperature was programmed from 100°C to 210°C at 4°/min. The injection and detection temperatures were set at 300°C, and the flow rate of the carrier gas (N_2) was set at 30 mL/min. A plot of the ratio of peak height (for furfural or HMF) was linear up to 20 μg/mL.

Reverse-phase HPLC and spectrophotometric detection has also been employed (68).

OTHER CARBONYLS

Off-flavors have been reported in spirits as being due to acrolein (2-propenal) and to crotonaldehyde (2-butenal). Both of these compounds are unsaturated aldehydes. A bitter taste in wine has been traced to acrolein (73). An HPLC method, using both a headspace technique and a spectrophotometric detector at 220 nm was sensitive to 100 pg for acrolein and 50 pg for crotonaldehyde (74).

Caramel was reported present in California fruit wines, in concentrations ranging from 0.2 to 4.5 mg/L (67).

REFERENCES

1. T. Takasawa, *Hakko Kogaku Kaishi* **59**, 225-229 (1981).
2. S. T. Tiurin, G. T. Ageeva, L. N. Blagonravova, I. N. Okolelov, and I. N. Eramova, *Vinod. Vinograd. SSSR* **376**, 23-26 (1983).
3. N. Hashimoto and Y. Kuroiwa, *Proc. Am. Soc. Brew. Chem.* **33**, 104-111 (1975).
4. A. Piondl, H. Westner, and E. Geiger, *Brauwissenschaft* **34**, 300-307 (1981).
5. A. Borea Carnacini, P. Capella, A. Amati, and U. Pallotta, *Am. J. Enol. Vitic.* **31**, 219-226 (1980).
6. M. A. Amerine and M. A. Joslyn, *Table Wines, The Technology of Their Production*, 2nd ed. Univ. of California Press, Berkeley and Los Angeles, 1970.
7. M. A. Amerine, H. W. Berg, R. E. Kunkee, C. S. Ough, V. L. Singleton, and A. D. Webb, *The Technology of Wine Making*, 4th ed. Avi Publishing Co., Westport, CT, 1980.
8. U.S. Internal Revenue Service, *Wine*, Regul. No. 7. U.S. Gov. Printing Office, Washington, DC, 1945.
9. J. Ribéreau-Gayon, E. Peynaud, P. Sudraud, and P. Ribéreau-Gayon, *Analyse et contrôles des vins*. Dunod, Paris, 1976.
10. J. G. Hanna and S. Siggia, in *Treatise on Analytical Chemistry*, Part II, Sect. B-2, Vol. 13, I. M. Kolthoff and P. J. Elving, Eds. Wiley (Interscience), New York, 1966, pp. 131-222.
11. A. V. Kavadze, A. K. Rodapulo, and I. A. Egorov, *Prikl. Biokhim. Mikrobiol.* **13**, 199-204 (1977).
12. M. A. Amerine, *J. Food Sci. Technol.* **1**, 87-98 (1964).
13. K. Nordström, *Acta Chem. Scand.* **20**, 474-478 (1966).
14. K. Nordström, *J. Inst. Brew.* **74**, 192-195 (1968).
15. A. L. Lehninger, *Biochemistry. The Molecular Basis of Cell Structure and Function*. Worth, New York, 1970.
16. K. Wucherpfenning and G. Semmler, *Z. Lebensm.-Unters. -Forsch.* **148**, 77-82, 138-145 (1972).
17. M. A. Amerine, *Adv. Food Res.* **5**, 353-510 (1954).
18. C. E. Daudt and A. C. Meller, *Rev. Cent. Cienc. Rurais (Univ. Fed. St. Maria)* **5**, 97-102 (1975).
19. J. Ribéreau-Gayon, E. Peynaud, P. Sudraud, and P. Ribéreau-Gayon, *Traité d'Oenologie*, Vol. 1. Dunod, Paris, 1972.
20. K. Heintze and F. Braun, *Z. Lebensm.-Unters. -Forsch.* **142**, 40-46 (1970).
21. E. Bayer, *J. Gas. Chromatogr.* **4**, 67-73 (1966).
22. H. Rebelein, *Dtsch. Lebensm.-Rundsch.* **66**, 6-11 (1970).
23. W.-R. Sponholz, *Z. Lebensm.-Unters. -Forsch.* **174**, 458-462 (1982).
24. J. L. Owades and J. M. Dono, *J. Assoc. Off. Anal. Chem.* **51**, 148-151 (1968).
25. P. Jaulmes and P. Espezel, *Ann. Falsif. Fraudes* **28**, 325-335 (1935).
26. J. F. Guymon and D. L. Wright, *J. Assoc. Off. Anal. Chem.* **50**, 305-307 (1967).
27. *Official Methods of Analysis*, 14th ed. Association of Official Analytical Chemists, Arlington, VA, 1984, pp. 220-280.
28. A. Joyeux and S. Lafon-Lafourcade, *Ann. Falsif. Expert Chim. Toxicol.* **72**, 321-324 (1979).
29. P. Jaulmes and P. Espezel, *Ann. Falsif. Fraudes* **47**, 9-14 (1954).

30. P. Jaulmes and G. Hamelle, *Ann. Falsif. Fraudes* **54**, 338-347 (1961).
31. L. F. Burroughs and A. H. Sparks, *Analyst (London)* **86**, 381-385 (1961).
32. R. Then and F. Radler, *Z. Lebensm.-Unters. -Forsch.* **138**, 163-169 (1968).
33. M. Okamoto, K. Ohtsuka, J. Imai, and F. Yamada, *J. Chromatogr.* **219**, 175-178 (1981).
34. J. A. Delcour, J. M. Caers, P. Dondeyne, F. Delvaux, and E. Robberechts, *J. Inst. Brew.* **88**, 384-386 (1982).
35. E. Puputti and P. Lehtonen, *J. Chromatogr.* **353**, 163-168 (1986).
36. R. Di Stefano and G. Ciolfi, *Riv. Vitic. Enol.* **35**, 474-480 (1982).
37. L. Ballesta, M. R. Olea Serrano, and R. Garcia-Villanova, *An. Bromatol.* **32**, 367-373 (1980).
38. L. P. McCloskey and P. Mahaney, *Am. J. Enol. Vitic.* **32**, 159-162 (1981).
39. E. Laub and R. Woller, *Mitteilungsbl. GDCh-Fachgruppe Lebensmittelchem. Gerichtl. Chem.* **9**, 268-272 (1974).
40. M. A. Amerine and C. S. Ough, *CRC Crit. Rev. Food Technol.* **2**, 407-515 (1972).
41. W. Postel and U. Güvenc, *Z. Lebensm.-Unters. -Forsch.* **138**, 35-44 (1976).
42. H. Soumalainen and P. Ronkainen, *Nature (London)* **220**, 792-793 (1968).
43. A. D. Haukeli and S. Lei, *Brygmesteren* **35**, 23-26 (1978).
44. W. Postel and B. Meier, *Z. Lebensm.-Unters. -Forsch.* **176**, 113-115 (1983).
45. S. Hu, H. Masuda, and H. Muraki, *Nippon Jozo Kyokai Zasshi* **72**, 449-454 (1977).
46. J. C. M. Fornachon and B. Lloyd, *J. Sci. Food Agric.* **16**, 710-716 (1965).
47. B. C. Rankine, J. C. M. Fornachon, and D. A. Bridson, *Vitis* **8**, 129-134 (1969).
48. P. Ronkainen, S. Brummer, and H. Soumalainen, *Am. J. Enol. Vitic.* **21**, 136-139 (1970).
49. H. H. Dittrich and E. Kerner, *Wein-Wiss.* **19**, 528-535 (1964).
50. P. Ronkainen and H. Soumalainen, *Mitt., Rebe Wein, Obstbau Fruechteverwert.* **19**, 102-108 (1969).
51. P. J. Williams and C. R. Strauss, *J. Sci. Food Agric.* **26**, 1127-1136 (1975).
52. W. Postel and B. Meier, *Z. Lebensm.-Unters. -Forsch.* **173**, 85-89 (1981).
53. H. Masuda and H. Muraki, *J. Sci. Food Agric.* **26**, 1027-1036 (1975).
54. J. F. Guymon and E. A. Crowell, *Am. J. Enol. Vitic.* **16**, 85-91 (1965).
55. D. I. Murdock, *Food Technol. (Chicago)* **21**, 157-161 (1967).
56. C. Barcelo, M. Gassiot, and M. Ferrer, *J. Chromatogr.* **147**, 463-469 (1978).
57. P. W. Gales, *J. Am. Soc. Brew. Chem.* **34**, 123-127 (1976).
58. C. S. Ough and M. A. Amerine, *Am. J. Enol. Vitic.* **18**, 157-164 (1967).
59. R. Garcia Villanova and M. R. F. Olea Serrano, *An. Bromatol.* **30**, 241-246 (1978).
60. C. Charpentier, *Ann. Falsif. Expert. Chim.* **62**, 271-282 (1969).
61. M. A. Amerine, *Food Res.* **13**, 264-269 (1948).
62. E. Meidell and F. Filipello, *Am. J. Enol. Vitic.* **20**, 164-168 (1969).
63. Z. A. Mamakova, T. A. Manuilova, L. A. Spektor, and M. V. Kalika, *Sadovod. Vinograd. Vinodel. Mold.* **31**(2), 30-32 (1976).
64. Z. Jeszenszky and P. Szalka, *Borgazdasag* **23**(1), 22-26 (1975).
65. F. Malík, A. Navara, and E. Minárik, *Mitt. Klosterneuburg* **35**, 45-47 (1985).
66. A. Navara, F. Malík, and E. Minárik, *Mitt. Klosterneuburg* **36**, 28-33 (1986).
67. J. Schmizu and M. Watanabe, *Agric. Biol. Chem.* **43**, 1365-1366 (1979).

68. M. A. Williams, R. C. Humphrey, and H. P. Reader, *Am. J. Enol. Vitic.* **34**, 57–60 (1983).
69. W. Postel, *Dtsch. Lebensm.-Rundsch.* **64**, 318–322 (1968).
70. J. F. Drilleau and C. Priquolt, *Ind. Aliment. Agric.* **88**, 699–704 (1971).
71. E. Carrubba, *Riv. Vitic. Enol.* **20**, 35–36 (1967).
72. H. Greve and J. Rebelein, *Fluess. Obst* **45**, 10–14 (1978).
73. W. C. Serjak, C. Day, J. M. Van Lanen, and C. S. Boruf, *Appl. Microbiol.* **2**, 9–13 (1954).
74. P. Lehtonen, R. Laakso, and E. Puputti, *Z. Lebensm.-Unters. -Forsch.* **178**, 487–489 (1984).

Five

ESTERS

The fruity character of *V. vinifera* wines is due, in part, to the volatile esters formed during fermentation. Lower temperatures of fermentation, removal of heavy solids before fermentation, and the use of good fermentation practices (sulfur dioxide and pure yeast) seem to be the main influences on the formation and retention of esters. Earlier work (1) indicated that temperatures of about 12–15°C were optimal for final maximum ester content in wines. In addition it was found (2) that the higher temperatures caused earlier and increased production of some of the esters but greater losses later in fermentation.

The addition of 2-phenyl alcohol to grape juice prior to fermentation was shown (3) to cause increases in ethyl laurate, ethyl valerate, ethyl caproate, and ethyl caprylate as well as causing favorable sensory effects. The addition of higher fatty acids (C_{14}, C_{16}, and C_{18}) increased ester formation during fermentation, whereas the addition of C_{18} unsaturated acids decreased ester formation (4). With the use of greater than normal yeast mass (5), three to four times as much esters were produced. Esters that increase the fruity sensory response are produced in greater amounts if the juice is settled and racked and fermented at low temperature (6). During the sherry-making process, ester content decreases (7, 8). The importance of yeast type is critical (1, 9–11). *Saccharomyces* produce greater amounts of the more favorable esters. Other yeast species tend to produce more of the longer-chain esters or ethyl acetate and lesser amounts of the fruity esters. Increased nitrogen levels in grape juice cause increased fruity volatile ester formation in the wine (12, 13).

The formation of esters during fermentation was investigated by Nordström (15). He concluded that the normal nonenzymatic formation of esters is much too slow and that the equilibrium much too far to the left to account for the major ester formation during fermentation at the normal pH of wine. The concentration of acetic acid apparently has little influence on the amounts of acetate esters formed.

$$R_1COOH + R_2OH \rightleftharpoons R_1COOR_2 + H_2O$$

During aging, the acetate esters decrease in amount as do the ethyl esters (14). Kinetic data have shown this relationship (15–17). The decrease rates to equilibrium are pseudo-first-order. The succinic, malic, and tartaric ethyl esters increase with age (18). Some rates of formation data have been given (19, 20). Parkkinen (21), working with Bakers' yeast, separated five esterases from the yeast. These preferentially hydrolyzed ethyl esters with acyl side-chains of lengths C_5–C_{12} (22, 23). The amounts found in equilibrium were consistent with those found in wines. *Brettanomyces* has been shown (24) to contain considerable esterase activity toward acetate esters, unlike *Saccharomyces*.

Propionic, isobutyric, and isovaleric acids originate from the fusel oil pathway and do not form ethyl esters. Small amounts of certain esters result from the must and are carried through the fermentation. The rest of the esters depend on syntheses. The acid portion of the esters mainly consists of even-numbered carbon atoms (from the fatty acid synthesis pathway); the alcohol portion is dominated by ethanol, but some is also formed via the fusel oil pathways. The mechanism shown in Figure 21 demonstrates the accepted biochemical route of ester formation during fermentation. The alcohol acetyltransferase responsible for acetate ester formation has been isolated (25). The enzymic activity is inhibited by unsaturated fatty acids (26). The most common ester is ethyl acetate, and the next most common is isoamyl acetate. Ethyl acetate (Table 28) (27–32) varies somewhat between white and red wines. This variation is probably a result of slight increases in the activities of *Acetobacter* and wild yeast during red wine fermentation and barrel storage. If wines are made carefully and not

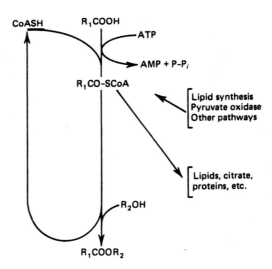

Figure 21. Formation of acetate and other esters by yeast.

Table 28. Concentrations of Ethyl Acetate (mg/L) in Wines of Various Countries

Country	White			Rosé			Red		
	Number of samples	Range	Average	Number of samples	Range	Average	Number of samples	Range	Average
Australia	19	19–84	46	—	—	—	20	40–146	85
France	36	28–261	97	9	46–86	62	50	71–168	114
Germany	89	11–96	33	—	—	—	—	—	—
Germany	77	46–99	73	—	—	—	—	—	—
Italy	63	72–128	101	—	—	—	—	—	—
Japan	54	28–130	60	9	20–63	44	48	42–232	98
Portugal	11	28–153	76	7	49–106	73	10	75–119	92
South Africa	9	15–51	29	—	—	—	—	—	—
South Africa	3	46–88	63	—	—	—	1	46	46
Spain	23	45–102	67	6	73–109	81	12	66–148	92
United States	8	35–79	54	—	—	—	—	—	—
United States	11	16–47	29	—	—	—	14	28–121	61
United States	199	23–135	67	—	—	—	40	26–125	67.1

From References 27–32.

stored in barrels, the level of ethyl acetate seems to be fairly constant, as suggested from published data (31, 32).

In *V. labruscana* and some crosses, methyl (and ethyl) anthranilate are natural constituents of the grape and can be found in wine at concentrations equal to isoamyl acetate. Table 29 gives the quantitative amounts of the methyl anthranilate in grape juice and wine (33–38). The cooler the climate and the riper the fruit, the higher the amount of methyl anthranilate (38).

Maximum cell growth can be correlated to maximum total ester formation (39). In addition, increase or decrease in nitrogen nutrients causes more or less yeast growth, thus exerting an effect on the production of esters (32).

Certain of the esters (longer-chain acid portions) are tightly bound to the yeast (40). The ethyl caprate was about equal to the isoamyl acetate, and the ethyl caproate and caprylate were there in relative amounts greater than those reported in wines. When distilling a sugar–yeast fermentation, about six times as much ethyl caprate was distilled if yeast cells were included (41). No ethyl laurate was found unless the yeast was added to the distillation. Ethyl caprylate, ethyl caprate, and ethyl laurate were distilled in significantly greater amounts if yeast cells were included, whereas isoamyl acetate, ethyl caproate, and 2-phenylethyl acetate in wine distillation did not depend on the presence of yeast. This is one reason for distilling material being put through the pot or column still with the yeast cells included. The odd-length chain fatty acids, esters, and methyl esters are in very small amounts. In distillates for spirits or brandies they are less than 0.1 mg/100 mL of ethanol (42).

Ethyl acetate can be formed in excess by aerobic yeasts, such as *Hansenula* and *Pichia*. Likewise, *Actobacter* sp. infections result in formation of large amounts of ethyl acetate. The malolactic fermentation also contributes considerable ethyl lactate to a wine. In the first two instances, control is necessary because the product is undesirable in wine. In the last instance the ethyl lactate has little odor and is of minor sensory consequence. The amounts of ethyl lac-

Table 29. Methyl Anthranilate in *Vitis labruscana* Beverages

Beverage	Variety	Range (mg/L)
Grape soda	Artificial	3.0–8.8
Grape juice	Concord	0.14–3.50
Wine	Concord	0.4–0.97
	Niagra	0.6–3.1
	Delaware	0.0–0.1
	Ives	0.4
	Catawba	0.2
	Elvera	Trace

From References 33–38.

Table 30. Concentrations of Some Volatile Esters (mg/L) Found in Wines

Country	n-Propyl acetate	Isobutyl acetate	Isoamyl acetate	n-Hexyl acetate	Phenethyl acetate	Ethyl butyrate	Ethyl caproate	Ethyl caprylate	Ethyl caprate	Ethyl succinate
Japan										
Number of samples	—	—	24	24	24	—	24	24	24	—
Range	—	—	tr[a]–1.2	0.0–0.1	0.0–0.2	—	tr–0.7	tr–1.0	0.0–0.5	—
Mean	—	—	0.3	0.0–0.1	0.0–0.2	—	tr–0.7	tr–1.0	0.0–0.5	—
South Africa										
Number of samples	—	—	9	9	9	—	9	9	9	9
Range	—	—	1.8–5.9	0.1–1.0	0.07–1.14	—	0.3–1.8	0.5–2.1	0.2–0.9	0.3–0.8
Mean	—	—	3.4	0.4	0.4	—	1.4	1.4	0.5	0.4
South Africa										
Number of samples	—	—	4	3	4	—	4	4	4	4
Range	—	—	0.3–5.6	0.08–0.5	0.03–1.07	—	0.2–1.2	0.3–1.8	0.1–0.6	0.3–6.3
Mean	—	—	3.2	0.25	0.6	—	0.9	1.3	0.4	1.9
United States										
Number of samples	10	11	12	—	—	—	—	—	—	—
Range	0.04–0.8	0.1–0.5	0.9–9.3	—	—	—	—	—	—	—
Mean	0.3	0.3	3.7	—	—	—	—	—	—	—
United States										
Number of samples	—	3	3	3	3	3	3	3	3	3
Range	—	0.0–0.2	1.4–6.2	0.0–0.06	0.1–0.9	0.2–0.3	0.4–0.9	0.5–1.3	0.1–0.6	0.2–0.3
Mean	—	0.1	3.1	0.03	0.4	0.3	0.6	1.0	0.4	0.3
United States										
Number of samples	4	4	4	4	4	4	4	4	4	—
Range	0.25–0.30	0.14–0.17	2.2–6.1	0.22–0.57	0.38–0.67	0.18–0.44	0.32–0.74	0.30–0.61	0.12–0.0	—
Mean	0.28	0.15	4.8	0.40	0.53	0.35	0.56	0.47	0.14	—

From References 1, 2, 4, 36, and 45.

[a] tr, trace.

tate found (43) in over 200 table wines varied from a trace to 534 mg/L. Sparkling wines have levels in the same range.

There are more qualitative than quantitative data on esters in wines. Reports on *V. vinifera* show only 11 esters in grapes but 83 different esters in wines (44). Table 30 (45) is a summary of data on the quantitative values reported for esters in wines (mainly white wines). The data are consistent between workers. More data are needed for red wines. A report (46) on analysis of wine samples for nine esters seldom quantified showed ethyl, isoamyl, and phenethyl esters of various acids other than acetate and butyrate. They were found in amounts ranging from a few micrograms per liter to several hundred micrograms per liter.

Ethyl acetate has reported thresholds in wine of 150-170 mg/L (47,48). This value usually exceeds the amounts shown to be present in wine. Some red, barrel-aged wines do exceed the threshold slightly, and this is considered by some to add complexity to aroma. There are few valid data for ester thresholds in wines. However, considerable data are available on the main esters from other sources (Table 31) (49-51). Because the alcohol content of wine is greater than that of beer or cider, we expect the threshold values to be greater. The volatility of esters decreases with increasing ethanol concentrations (52). This phenomenon is a decreasing log function from zero to about 80% v/v ethanol-

Table 31. Threshold Data for Volatile Esters in Several Alcoholic Beverages

	Threshold values (mg/L)		
	Beer		
Ester	Type 1[a]	Type 2[b]	Cider[c]
Ethyl acetate	33	25	35
n-Propyl acetate	30	—	25
Isobutyl acetate	0.5	—	0.9
Isoamyl acetate	1.6	2.0	2.0
n-Hexyl acetate	3.5	—	0.2
Phenethyl acetate	3.8	0.2	2.6
Ethyl butyrate	0.4	0.5	0.2
Ethyl caproate	0.2	0.2	0.2
Ethyl caprylate	0.9	1.0	0.8
Ethyl caprate	1.5	—	1.1

[a] From Reference 49.
[b] From Reference 50.
[c] From Reference 51.

Table 32. Toxicity of Some Wine Esters (55)

Ester	LD$_{50}$ Oral dose for rats (mg/kg body weight)
Ethyl acetate	5620
Isobutyl acetate	8000 mg/L[a]
Isoamyl acetate	200 mg/L[a]
Hexyl acetate	6160
Phenethyl acetate	1680
Ethyl butyrate	2200
Ethyl caproate	1600
Diethyl succinate	500
Methyl anthranilate	2910

From Reference 55.
[a]Inhaled by the rat—lowest published lethal concentration in air.

water solutions. From this consideration and reference to the ester concentration table of wines, it is evident that few of the individual esters ever reach threshold values in wine. Isoamyl acetate is a possible exception. Esters can have an additive sensory impact (53). Yet even if all esters are below individual threshold they may, together, be detectable or exhibit some sensory importance by synergistic action (54).

At the concentration determined in wines, the esters are far below toxic levels. Table 32 (55) gives LD$_{50}$ toxicity levels reported for the common esters. Most esters are hydrolyzed readily *in vivo* (56). Certainly the very acid conditions of the stomach would cause some hydrolysis. The moieties of the hydrolysis should be insignificant amounts compared to their toxicity levels and the amounts already present.

TOTAL VOLATILE ESTERS

The common chemical method of analysis for total volatile esters is given by the AOAC in the distilled beverage section of *Official Methods of Analysis* (56a). A 1:1 wine distillate is made, and hydroxamic acid is formed by reacting the distilled esters with hydroxylamine:

$$R-\overset{O}{\underset{\|}{C}}-OR_2 + NH_2OH \xrightarrow{NaOH} R_1-\overset{O}{\underset{\|}{C}}-NHOH + R_2OH$$

The hydroxamic acid complexes with ferric ion to form a yellow product:

$$3R_1-\overset{O}{\underset{}{C}}-NHOH + Fe^{3+} \xrightarrow{H^+} [\text{Fe complex}] + 3H^+$$

Procedure

Use purified ethyl acetate as a standard. Purify by distilling, and check purity on a gas–liquid chromatograph. For the standard, make up 10% ethanol solution saturated with potassium acid tartrate. Make up a 1 g/L ethyl acetate stock solution. From this and the standard ethanol solution make up 0-, 25-, 50-, 75-, and 100-mg/L standards. (Take 0, 25, 50, 75, and 100 mL and make to 1 L each.) Take 100 mL of these standards, add 50 mL of water, and distill. Collect 90 mL in a volumetric flask immersed in ice, and bring to volume with water. Mix and pipet 3 mL of each standard into a 15-mL-capacity test tube, add 2 mL of a 2 M $NH_2OH \cdot HCl$ solution (13.9 g/100 mL), and also add 2 mL of 3.5 N sodium hydroxide. Stopper with a neoprene stopper, shake, and let stand for 10 min. Then add 2 mL of 4 N hydrochloric acid and 2 mL of ferric chloride reagent ($FeCl_3 \cdot 6H_2O$, 10 g in 100 mL of 0.1 N HCl). Shake, and read directly at 510 mm in the spectrophotometer.

Treat wine or must samples in a similar manner and read results from the standard curve.

INDIVIDUAL VOLATILE ESTERS

Determination by the chemical means just outlined probably is as accurate as the more specific methods in that ethyl acetate makes up most of the volatile esters. Results of a large number of analyses by gas chromatography for ethyl acetate and by the chemical method for total volatile esters were almost identical (57).

Ethyl acetate separates rather easily and completely on several liquid phases with gas chromatography. A gas chromatographic method for wine (58) using a polyethyleneglycol liquid phase has been suggested. Carbowax 600, 4000, 20M or Porapak Q also can be used to separate the ester from interfering substances.

For separation and determination of individual esters by GLC, a capillary open tubular column coated with SP-2100 or SE 30 has been used (2). For brandy distillate (40) a Carbowax 20M packing in a standard $\frac{1}{8}$-in. stainless steel column proved adequate for separation.

For quantitative work with the individual esters, internal standards must be included in the samples prior to extraction and concentration and cleanup. The Freon extraction method as suggested by Rapp et al. (59) and modified (60) works very well. Modifications to suit the type of gas chromatographic column can be made in temperature programming, injection size, split ratio, and so on.

Procedure

Cool wine to 0°C to minimize emulsion formation. Add 20 mL of Freon 11 to the extraction equipment (59). Add 250 mL of the cooled wine. Add 2 mL of ethyl nanoate (3.20 mg/L wine) and 2 mL of *t*-amyl acetate (25 mg/L wine) as internal standards prior to addition to the apparatus. Attach a 25-mL receiving flask containing 20 mL of Freon 11. The condenser water should be at or very near 0°C. The lower portion of the apparatus should be cooled to prevent the Freon from boiling. Hold the receiving flask at about 35°C to achieve good reflux and extraction. The tip of the funnel that catches the refluxing Freon should dip below surface of the wine and be no more than 0.2 mm in diameter. Extract as long as it takes (for esters, 2–3 hr). Concentrate the extract to 1 mL using a Vigreux condenser (25 × 1 cm) at 35°C (takes about 45 min). Keep at 4°C in a closed container until ready to use. Injection port and flame ionization detectors should be at 200°C and 250°C, respectively. The component retention times and mass spectra should be used to verify the various compounds measured. Make standard curves using appropriate internal standards, and relate response factors to area ratios to calculate amounts.

METHYL ANTHRANILATE

Methyl anthranilate can be extracted from juice or wine with Freon® 11 and can then be concentrated and gas chromatographed (35). The limits of detection were about 0.1 mg/L, with about 4% average error. Others (33) used a liquid chromatograph, and the sensitivity was about 0.05 mg/L. No other compounds of similar retention time were fluorescent emitters at 432 nm. A report of a fairly rapid steam distillation method with fluorescence detection and measurement that has sensitivity in the required range has been published (34).

Procedure

Make up 0.5 M citric acid by dissolving 52.3 g of citric acid ($C_6H_8O_7 \cdot H_2O$) in water and bring to 500 mL. Dissolve 70.98 g of Na_2HPO_4 in water and bring to 500 mL. Prepare a pH 7.0 McIlvaine's buffer at 5 times normal concentration. Mix 176.5 mL of McIlvaine's solution *a* with 823.5 mL of solution *b*. Dissolve 18 g of methyl anthranilate ($NH_2C_6H_4COOCH_3$) in 10 mL of concentrated hydrochloric acid and recrystallize twice from absolute ethanol or until a melting point of 181°C is reached. Dry and dissolve 0.0124 g of methyl anthranilate hydrochloride in 1 liter of water. Take 50 mL and dilute to 500 mL with water. The solution contains 1 mg/L methyl anthranilate (as the free base). Store in dark glass. A 1000 mg/L solution is stable for several weeks at room temperature.

To standardize the fluorometer, put 10 and 20 mL of 1 mg/L methyl anthranilate solution into 50-mL volumetric flasks, add 10 mL of buffer, and bring to volume with water. The stronger solution should cause about an 85% scale deflection with the buffer blank.

For the sample, pipet 10 mL into a Kjeldahl microdistillation flask (Figure 12). Rinse with a few milliliters of water. Steam distill into a 50-mL volumetric flask containing 10 mL of pH 7.0 buffer and collect about 38 mL, so that a total of 48 mL is in the flask. Put the flask in a water bath and mix thoroughly, bring to volume, and read the fluorometer. If the sample, either grapes or wines, contains more than 2 mg/L of methyl anthranilate, dilute the sample accordingly before distilling. The ethanol enhancement of the fluorescence of methyl anthranilate is linear (61). A multiple regression equation to compensate for this effect can be calculated from model solutions:

$$Z = A + BX - CY$$

where Z = methyl anthranilate in the distillate (mg/L)
X = fluorometer reading
Y = ethanol in the distillate (% v/v)
A, B, C = multiple regression coefficients

Then the concentration of the methyl anthranilate in the wine M is

$$M(\text{mg/L}) = Z \times f$$

where f = dilution factor

(The ethanol effect can also be compensated for by adjusting all samples to 20% ethanol before distillation and making the standard samples up to 16% v/v ethanol; alternatively, use 20% ethanol solution instead of water for all dilutions.)

NONVOLATILE ESTERS

Volatility is only relative; however, some esters can be classified as nonvolatile. Among these are the monocaffeoyl, mono-*p*-coumaroyl, and monoferuloyl esters of tartaric acid. These three esters have been determined in musts of 11 different *V. vinifera* varieties (62). Pinot noir, Grenache, and White Riesling were consistently high in these esters. Table 33 gives the average values for the esters. They also suggest the presence of hydroxycinnamic acid–tartaric acid–glucose esters. Others (63) have followed the changes in these esters during fermentation and related the changes to browning capacity of the grapes.

Table 33. Concentrations (mg/L) of Some Nonvolatile Tartaric Acid Esters

Tartaric acid ester	Number of samples	Range	Average
Monocaffeoyl	10	70.9–233.8	120.7
Mono-p-coumaroyl	10	8.3– 33.8	18.1
Mono-feruloyl	10	1.6– 15.9	5.6

From Reference 62.

Table 34. Diethyl Esters of Nonvolatile Table Wines and Sherries

Country	Number	Range (mg/L)		
		Diethyl succinate	Diethyl malate	Diethyl tartrate
Table wine				
Japan	24	tr^a–30	tr–53	<0.5
Germany	8	2–15	13–43	<0.5
France	4	26–55	1–35	<0.5
Hungary	1	23	41	<0.5
USSR	—	5.2–6.6	—	—
Sherry				
Spain	10	12–38	2–16	0.5–100
USSR	—	2.0–2.2	—	—

From References 45, 64, and 65.
^{a} trace.

The determination of these esters is best done with high-performance liquid chromatography. Methods for these analyses are, for the most part, in the developmental stages.

The gas chromatographic determination of the other nonvolatile acids are given in Table 34 (45, 64, 65).

REFERENCES

1. C. E. Daudt and C. S. Ough, *Am. J. Enol. Vitic.* **24**, 130–135 (1973).
2. E. Killian and C. S. Ough, *Am. J. Enol. Vitic.* **30**, 301–305 (1979).
3. S. S. Karpov, G. G. Valuiko, A. A. Nalimova, and A. I. Keptine, *Sadovod. Vinograd. Vinodel. Mold.* **37**(2), 31–33 (1982).
4. K. Yoshizawa, *Nippon Nogei Kagaku Kaishi* **50**, 115–119 (1976).
5. A. G. Reva, *Vinodel. Vinograd. SSSR* (7), 17–18 (1978).
6. A. Bertrand, *Ann. Technol. Agric.* **27**, 231–233 (1978).

7. G. I. Kozub, Z. A. Mamakova, E. A. Skorbanova, and A. S. Maksimova, *Sadovod. Vinograd. Vinodel. Mold.* **39**(5), 26–29 (1984).
8. N. D. Chichashvili, *Prikl. Biokhim. Mikrobiol.* **15**, 909–914 (1979).
9. E. Soufleros and A. Bertrand, *Connaiss. Vigne Vin* **13**, 181–198 (1979).
10. R. M. Soles, C. S. Ough, and R. E. Kunkee, *Am. J. Enol. Vitic.* **33**, 94–98 (1982).
11. T. Shinohara and M. Watanabe, *Agric. Biol. Chem.* **45**, 2645–2651 (1981).
12. A. A. Bell, C. S. Ough, and W. M. Kliewer, *Am. J. Enol. Vitic.* **30**, 124–129 (1979).
13. C. S. Ough and T. H. Lee, *Am. J. Enol. Vitic.* **32**, 125–127 (1981).
14. J. Marias, *Vitis* **17**, 396–403 (1978).
15. K. Nordström, *Sven. Kem. Tidskr.* **76**, 86–119 (1964).
16. R. F. Simpson, *Vitis* **17**, 274–287 (1978).
17. D. D. Ramey and C. S. Ough, *J. Agric. Food Chem.* **28**, 928–934 (1980).
18. W. R. Sponholtz, *Dtsch. Lebensm.-Rundsch.* **75**, 277–279 (1979).
19. T. Shinohara, J. Shimizu, and Y. Shimizu, *Agric. Biol. Chem.* **43**, 2351–2358 (1979).
20. T. Shinohara and J. Shimizu, *Nippon Nogei Kagaku Kaishi* **55**, 679–687 (1981).
21. E. Parkkinen, *Cell. Mol. Biol.* **26**, 147–154 (1980).
22. E. Parkkinen and H. Suomalainen, *J. Inst. Brew.* **88**, 34–38 (1982).
23. E. Parkkinen and H. Suomalainen, *J. Inst. Brew.* **88**, 98–101 (1982).
24. M. Spaepen and H. Verachtert, *J. Inst. Brew.* **88**, 11–17 (1982).
25. K. Yoshioka and N. Hashimoto, *Agric. Biol. Chem.* **45**, 2183–2190 (1981).
26. P. A. Thurston, D. E. Quain, and R. S. Tubb, *J. Inst. Brew.* **88**, 90–94 (1982).
27. W. W. D. Wagner and G. W. W. Wagener, *S. Afr. J. Agric. Sci.* **11**, 469–476 (1968).
28. T. Shinohara, *Agric. Biol. Chem.* **40**, 2475–2477 (1976).
29. J. P. Snyman, *Vitis* **16**, 295–299 (1977).
30. C. Reinhard, *Allg. Dtsch. Weinfachztg.* **110**, 1004–1009 (1971).
31. C. S. Ough and M. A. Amerine, *Am. J. Enol. Vitic.* **18**, 157–164 (1967).
32. C. S. Ough, J. A. Cook, and L. A. Lider, *Am. J. Enol. Vitic.* **19**, 254–265 (1968).
33. A. T. R. Williams and W. Slaving, *J. Agric. Food Chem.* **25**, 756–759 (1977).
34. D. J. Casimir, J. C. Moyer, and L. D. Mattick, *J. Assoc. Off. Anal. Chem.* **59**, 269–272 (1976).
35. R. R. Nelson, T. E. Acree, C. Y. Lee, and R. M. Butts, *J. Assoc. Off. Anal. Chem.* **59**, 1387–1389 (1976).
36. R. R. Nelson, T. E. Acree, *Am. J. Enol. Vitic.* **29**, 83–86 (1978).
37. R. R. Nelson and T. E. Acree, C. Y. Lee, and R. M. Butts, *J. Food Sci.* **42**, 57–59 (1977).
38. J. W. R. Liu and J. F. Gallander, *J. Food Sci.* **50**, 280–282 (1985).
39. C. S. Ough and M. A. Amerine, *Bull.—Calif. Agric. Exp. Stn.* **827**, 1–36 (1966).
40. M. Onishi, E. A. Crowell, and J. F. Guymon, *Am. J. Enol. Vitic.* **29**, 54–59 (1978).
41. L. Nykanen, I. Nykanen, and H. Suomalainen, *J. Inst. Brew.* **83**, 32–34 (1977).
42. W. Postel and L. Adam, *Dtsch. Lebensm.-Rundsch.* **80**, 1–5 (1984).
43. T. Shinohara, Y. Shimazu, and M. Watanabe, *Agric. Biol. Chem.* **43**, 2569–2577 (1979).
44. A. D. Webb, *Proc. Bienn. Int. CODATA Conf., 5th, 1976* pp. 101–108 (1977).
45. T. Shinohara and M. Watanabe, *Agric. Biol. Chem.* **45**, 2903–2905 (1981).

46. P. Schreier, F. Drawert, and A. Junker, *Chem., Mikrobiol., Technol. Lebensm.* **5**, 35-52 (1977).
47. B. C. Rankine, *Vitis* **7**, 22-49 (1968).
48. C. Corison, C. S. Ough, H. W. Berg, and K. E. Nelson, *Am. J. Enol. Vitic.* **30**, 130-134 (1979).
49. M. C. Meilgaard, *Tech. Q. Master Brew. Assoc. Am.* **12**, 151-168 (1975).
50. S. Engan, *J. Inst. Brew.* **78**, 33-36 (1972).
51. A. A. Williams, *J. Inst. Brew.* **80**, 445-470 (1974).
52. J. F. Guymon, *Wines Vines* **35**(1), 28-29 (1974).
53. M. C. Meilgaard, *Tech. Q. Master Brew. Assoc. Am.* **12**, 107-117 (1975).
54. T. B. Selfridge and M. A. Amerine, *Am. J. Enol. Vitic.* **29**, 1-6 (1978).
55. *The Toxic Substance List.* U.S. Dep. of Health, Education and Welfare, Rockville, MD, 1973.
56. F. Grundschober, *Toxicologia* **8**, 387-390 (1977).
56a. *Official Methods of Analysis*, 14th ed. Association of Official Analytical Chemists, Arlington, VA, 1984, pp. 220-230.
57. V. L. Singleton and C. S. Ough, unpublished data (1969).
58. P. Ribéreau-Gayon, *Qual. Plant. Mater. Veg.* **11**, 249-255 (1964).
59. A. Rapp, H. Hastrich, and L. Engel, *Vitis* **15**, 29-36 (1976).
60. J. Marais and A. C. Houtman, *Am. J. Enol. Vitic.* **30**, 250-252 (1979).
61. J. C. Moyer and L. R. Mattick, *Am. J. Enol. Vitic.* **27**, 134-135 (1976).
62. B. Y. Ong and C. W. Nagel, *Am. J. Enol. Vitic.* **29**, 277-281 (1978).
63. F. M. Romeyer, J. J. Macheix, J. P. Goiffon, C. C. Reminiac, and J. C. Sapis, *J. Agric. Food Chem.* **31**, 346-349 (1983).
64. J. Schimizu and M. Watanabe, *Nippon Nogei Kagaku Kaishi* **52**, 289-291 (1978).
65. G. I. Kozub, Z. A. Mamakova, and S. A. Skorbanova, *Sadovod. Vinograd. Vinodel. Mold.* **36**(3), 30-32 (1981).

Six
NITROGEN COMPOUNDS

A large number of nitrogen-containing compounds, such as ammonia, amino acids, proteins, vitamins, amines, and nitrate, have been found in musts and wines (1, 2). Nitrogen compounds are of extreme importance in yeast growth and in some instances can represent a limiting factor. The presence of certain nitrogen compounds can be deleterious. Careless "blue fining" can leave residual hydrogen cyanide. Much of the history of a wine or a must can be deduced from a careful nitrogen analysis.

Ammonia is present in musts and to a lesser extent in wines. The yeast rapidly depletes the free ammonia as an easily assimilable nitrogen source. The ammonia is present primarily as ammonium ion (NH_4^+) at the wine pH. The concentration range of ammonia in California grape juice is about 24–309 mg/L, with an average of 123 mg/L; the range in wines is from a few milligrams per liter to about 50 mg/L, with an average of about 12 mg/L (3). Most of the residual ammonia found in commercial wines is due to the grape juice used to sweeten the table wines just before bottling. There are great differences in the ammonia content of the juice as a result of varietal differences and vineyard treatments (4).

Amino acids are present in various amounts in musts and wines; Table 35 lists some reported values (5). Proline is usually present in the highest concentration: The average value for 78 determinations of grapes was 742 mg/L and for 42 determinations of wines was 869 mg/L (6). In California Cabernet Sauvignon wine as much as 90% of the total nitrogen can be present as proline (7). Results of other studies show arginine generally higher, followed by α-aminobutyric acid, glutamine, proline, α-alanine, and threonine (8). The amounts of each amino acid also vary widely by variety (9–11), yeast strain (10), region (12), treatment (13), and age (14).

The amino acids are of significance in the growth of lactic acid bacteria (15). After growth of these bacteria there is usually a small decrease in amino acids. Some strains decompose arginine, glutamic acid, histidine, and tyrosine completely. Only one strain (of 28) of *Pediococcus cerevisiae* decarboxylates histidine to histamine (15). Arginine is a demonstrated nitrogen source for *Leuconostoc oenos* (16, 17).

Table 35. Range of the Amino Acid Concentrations (mg/L) Reported in Grapes and Wines

Amino acid	Must			Wine		
	Min.	Max.	Number of samples	Min.	Max.	Average
Alanine	10	632	72	6	504	67
α-Aminobutyric acid	9	420	—	2	90	—
Arginine	4	2360	71	0	2311	299
Asparagine	24	24	—	1	2	—
Aspartic acid	0	290	70	1	136	33
Cysteine	0	2	—	1	2	—
Cystine	0	5	65	0	64	5
Glutamic acid	0	1330	70	3	86	67
Glutamine	24	97	—	1	310	—
Glycine	2	29	68	1	109	24
Histidine	7	143	71	0	159	24
Hydroxyproline	—	—	—	1	4	—
Isoleucine	0	100	72	1	75	15
Leucine	9	88	72	3	86	26
Lysine	0	74	72	0	117	33
Methionine	0	119	67	0	37	8
Ornithine	—	—	—	1	10	—
Phenylalanine	0	122	72	1	152	27
Proline	0	4600	964	30	3558	812
Serine	5	822	70	1	90	29
Threonine	10	400	70	2	108	27
Tryptophane	0	65	40	1	81	20
Tyrosine	0	58	71	1	81	10
Valine	0	88	71	1	103	25

From References 5 and 35.

Heat treatment of musts is uncommon. It results in higher amounts of amino acids and peptides in musts. These may be important to microbial instability of the resulting wines (18).

Proteins are present in grapes, and some remain in wine and may cause clouding in white wines during storage. Actually the "protein haze" may be due to protein bound to plant phenolics (19). Various researchers reported different values of proteins in wine, from 15–20 (20) to 60–411 (21) to 1.0–62.6 mg/L (22). In California musts and wines the reported values range from about 30 to 275 mg/L (23).

Different researchers have identified a number of proteins (23-27), and some have studied the amino acids these proteins contain. The heterogeneity of the proteins in grapes can be demonstrated; by thin-layer isoelectric focusing, up to 25 different components may be detected (28). Even the peptide fraction is complicated. The protein sequences of 16 peptides in a white wine have been determined (29). Enzyme preparations, to hydrolyze proteins and enhance wine stability, have been tested (30). Bentonite fining or heat treatment reduces the protein content (31-34).

Vitamins are also present in wines in very small concentrations. Table 36 lists the concentration values reported for musts and wines (36). Few, if any, musts are deficient in vitamins to the point of preventing adequate yeast growth or fermentation (35).

The amines are present in minor amounts in wines. The list of the amines that have been reported in grapes, wine, or yeast includes methylamine, ethanolamine, ethylamine, n-propylamine, isopropylamine, n-butylamine, isobutylamine, n-amylamine, isoamylamine, n-hexylamine, β-alanine, β-aminobutyric acid, agmatine, cadaverine, putrescine, spermidine, spermine, histamine, tryptamine, serotonin, β-phenethylamine, and tyramine. Only a few studies have been made of the nucleotides of wine (e.g., Reference 37). This method uses a cation-exchange column (Dowex 50 × 4) and passage through columns of Sephadex® G10 and Sephadex G25 as well as separation on an anion exchanger (Dowex 1). The nucleotides ADP, AMP, ATP, CMP, UDP, UMP, and UTP were found from 0.6 to 1.4 mg/L. A combination of liquid-liquid extraction

Table 36. Concentrations of Some Vitamins Present in Musts and Wines

Component	Must (mg/L)		Wine (mg/L)	
	Range	Average	Range	Average
Thiamine (B_1)	0.1–1.2	0.33	2–58	8.7
Riboflavin (B_2)	0.0–1.5	0.02	0.008–0.245	0.155
Pyridoxine (B_6)	0.1–2.9	0.42	0.22–0.82	0.455
Cobalamine (B_{12})	0.0–0.00013	0.00005	0.0–0.00016	0.000065
Pantothenic acid	0.25–10.5	0.82	0.47–1.87	0.89
Nicotinic acid	0.3–8.8	3.26	0.99–2.19	3.73
Biotin	0.001–0.06	0.0026	0.0006–0.0046	0.0021
p-Aminobenzoic acid	0.015–0.092	0.047		
Choline	0.024–0.039	0.033		
Folic acid	0.0–0.05			

and GLC has also been used (38). The nucleotides may act as reinforcers of flavors and in flavor formation in sparkling wines.

Nitrates are present in small concentration. The ratio of N_2O_5 to ash can be used if the analyst suspects that a wine is made of watered must containing nitrate; without the addition of water, the value in wines should be below 7–8 (39). In wines the nitrite (as N_2O_3) content was 3–78 μg/L (average 30 μg/L) and 3–92 μg/L (average 59 μg/L) (40). The nitrates and nitrites constitute only about 0.3% of the total nitrogen.

A number of other compounds containing nitrogen may be present in wine: pyrimidines, purines, nucleic acids, urethane, peptides, glycoproteins, hexoseamines, and so on.

Many of the nitrogen-containing substances arise directly by plant metabolism; the others arise from yeast metabolism. Toxic nitrogen compounds that may arise or be present from the addition of materials to wine are cyanic acid, certain pesticides, and herbicides. Use of these materials is closely regulated. Among the general changes that occur during yeast growth is the fairly rapid depletion of most of the amino acids and ammonia from the juice. Proline is usually spared and, in fact, is often secreted from the yeast into the medium (6). This is due to the membrane transport system of the yeast (41). The other amino acids and ammonia repress the permease that is responsible for the proline transport. By the time these amino acids have been depleted enough to cease repression of the proline permease, the fermentation is nearly completely anaerobic. The proline oxidase, the first step of the catabolism of proline, requires cytochrome c, which consequently requires molecular oxygen for regeneration. Since the medium is anaerobic, little catabolism occurs. In fact, the yeasts at this point have synthesized sufficient proline for their use and in some cases are excreting it to the medium.

Some amino acids are precursors of higher alcohols: amino butyric of 1-propanol, valine of 2-methyl-1-propanol, leucine of 3-methyl-1-butanol (42, 43) and other compounds. Amino acids are synthesized from sugar and nitrogen sources as they are required by the yeast. Some of the intermediates are shown on Table 25. For further information see Chapter 3. They are surely involved in aroma formation in many other ways (44, 45). It is of interest that the amino acid content of musts of botrytized grapes is only 40–86% that of nonbotrytized grapes of the same variety (46).

In sparkling wine production, the concentrations of arginine (47) and lysine (48) appear to be higher in bottle-fermented wines than in tank-fermented wines.

Use of paper chromatography "fingerprints" of amino acids has not yet been useful in identifying wines (49). Fining with sodium bentonite results in exchange with proline (in the cation form) and is pH dependent (50). Use of amount of total nitrogen, proline, and ammonia (or of their ratios) as measures of authenticity of wines has been suggested in Italy (51–55). Variability of results appears to be too great for legal standards, but further research should be conducted, particularly between regions. Adulterated wines seem to have much lower concentrations of proline (56).

The amino acid composition of the must (with regard to kind and amount) has been discussed (57–59) in relation to wine quality with some disagreements.

After the conversion of sugar to ethanol and glycerol, the biggest compositional changes in the fermenting wine are the uptake of nitrogen into the yeast cells and the subsequent loss of total nitrogen from the wine after clarification. There are excretions of other amino acids and nitrogen compounds back into the wine by autolysis. However, under modern winery practices this amount is small. The treatment of must with ammonium salts is seldom practiced but can result in wines of high nitrogen content. Growing conditions, crop level, and other factors can cause very large differences in the nitrogen content of grapes and, hence, in the wine. Predictions of fermentation rates can be made based primarily on nitrogen content (60).

TOTAL NITROGEN

The total nitrogen content of wines is determined primarily by using the Kjeldahl method; for details see Reference 61.

The bulk of the nitrogen in must and wines is in the form of amino acids, peptides, or proteins. If the total nitrogen content of musts or grape juices is to be determined, it is advantageous to ferment them first. Once the digestion is complete, the ammonia is distilled over into a boric acid solution and titrated to a methyl red end point. The boric acid forms ammonium borate, which titrates as if it were free ammonia:

$$(NH_4)_2B_4O_7 + 2H^+ + 5H_2O \rightleftharpoons 2NH_4^+ + 4H_3BO_3$$

Table 37 (62–64) lists the amounts of total nitrogen found in some musts and wines. The procedure given below is a variation of that of Amerine (65).

Procedure

Prepare a sulfuric acid–salicylic acid reagent by dissolving 33 g of salicylic acid in 1 L of concentrated sulfuric acid.

Prepare a boiling mixture by mixing 1 g of copper sulfate pentahydrate, 2 g of ferrous sulfate heptahydrate, and 20 g of sodium sulfate.

Pipet 50 mL of the dry wine sample or 25 mL of the must or dessert wine sample into an 800-mL Kjeldahl flask. If time allows, ferment the must or dealcoholized (by boiling) dessert wines in the digestion flask to remove the sugar. This saves on digestion reagents and digestion time.

Evaporate to 10 mL. Add 40 mL of the sulfuric–salicylic acid reagent, 10 g

Table 37. Total Nitrogen Concentrations (mg/L) Reported for Musts and Wines

	Total nitrogen						
	Musts			Wines			
Country	Number of samples	Range	Average	Number of samples	Range	Average	Reference
Brazil	144	57–437	203	—	—	—	62
France	—	—	—	177	77–952	351	63
Germany	—	—	—	122	102–980	523	63
Italy	—	—	—	145	46–201	102	63
United States (Calfornia)	70	542–2385	985	129	78–700	296	64
Various	28	98–1130	390	164	70–781	350	62

of the boiling mixture, and 3 drops of selenium oxychloride, mix, and immediately heat with a small flame. If excessive foaming occurs, add a drop of a silicone antifoam agent. As soon as the initial charring of the carbon compounds has subsided, heat vigorously until the solution becomes clear and continue to heat for another 20 min. With excess sugar present, add additional sulfuric acid. If the sulfuric acid is dissipated, the salts become molten, ammonia escapes, and the results are low. After the completion of the heating period, allow the solution to cool to room temperature, add 300 mL of water, mix, slowly pour 100 mL of 12 N sodium hydroxide down the side of the flask, add several drops of a phenolphthalein solution, and mix. If the color does not turn pink, indicating that the solution is basic, add more sodium hydroxide solution until the color turns to pink. Attach the flask to a macrodistillation apparatus (Figure 11) and distill 150 mL into a 500-mL Erlenmeyer flask containing 30 mL of 4% boric acid solution and a few drops of a 0.2% methyl red indicator solution in 60% ethanol. Remove the flask after termination of the distillation and rinse off the condenser tip into the flask. Titrate the distillate to a red end point with 0.1 N hydrochloric acid.

Carry a blank consisting of 50 mL of water instead of wine through the whole procedure.

$$\text{Nitrogen (mg/L)} = \frac{(A - B)(N)(14)(1000)}{v}$$

where A = volume of hydrochloric acid used for sample titration (mL)
B = volume of hydrochloric acid used for blank titration (mL)
N = normality of hydrochloric acid
v = volume of wine sample (mL)

The automation of the removal of ammonia from the digestion of the total nitrogen has been suggested (66). Further work on automation was reported by Davidson et al. (67), and Maddix et al. (68) investigated fermented beverages. The AutoAnalyzer® was used, and the reaction was with phenol–nitrophenol reagent. Twenty samples per hour could be determined with a standard deviation range of 1–32%. The ammonia-specific ion electrode has been used for determining the ammonia from the Kjeldahl digestion without distillation (69–72). The procedure can be automated (73). The results are comparable with those obtained by distillation.

AMMONIA

The determination of ammonia by the use of a weak cation-exchange resin (74) has been recommended by the OIV (75). It is satisfactory but less sensitive and much more time-consuming than the ammonia-specific electrode (76).

Procedure

Make up a synthetic grape juice (20 g/100 mL of glucose, 10 g/L malic acid, and sufficient K_2HPO_4 to bring the pH to 3.3) or a synthetic wine (as above but substitute 10 mL/100 mL of ethanol for the glucose). Prepare a stock solution of ammonia (1000 mg/L NH_3, 6.706 g of $(NH_4)_2SO_4$ per liter). Pipet 1, 5, 10, 20, and 30 mL of the stock solution into volumetric flasks. Bring to volume with the synthetic juice or wine. Mix and pipet 25 mL into beakers containing stir bars. Place a beaker on the magnetic stirrer and add 10 mL of 6 N NaOH; while stirring at a constant rate, immerse the ammonia electrode; after equilibrium is reached (1 or 2 min) read the millivolt value and record. Repeat for each level of standard. Draw a standard curve for juice and wine using log-linear paper. The slope should be in the prescribed range for the electrode. If drift or unreasonable answers are obtained, check the membrane for leaks and recharge the electrode with filling solution. Take the juice or wine and treat as above. Read the millivolt response and from the proper standard curve get the value. The only interference is ethylamine, which is seldom present in enough concentration to matter.

A method of addition is also described and has some advantages. The formula for determining the unknown ammonia concentration is:

$$\text{Unknown mg/L} = \frac{\text{Concentration of the standard (mg/L)}}{10^{\Delta E/\text{slope}}(1 + V_x/V_s) - V_x/V_s}$$

where ΔE = changes in current (in mV)
slope = Nerstian slope of the electrode (in mV)
V_x = volume of unknown addition (in mL)
V_s = volume of standard added (in mL)

The concentration of the standard should be estimated to be near the concentration of the unknown for best results.

AMINO ACIDS

The early results of amino acid determinations were obtained by microbiological assay (77, 78) and by paper chromatography or thin-layer chromatography. Ion-exchange chromatography has superseded this technique, mainly through the work of Moore, et al., (79–83). Since their early work, the technique has been modified and improved many times. Ion-exchange chromatography has been applied to musts (84–86) and wines (87–89). The directions of the amino acid analyses being used should be carefully followed. By use of different columns, almost all the amino acids can be determined (90).

The analysis of amino acids by paper chromatography belongs to the classic application of the technique (91). The technique has been applied to the analysis of grape juices and wines (92–97). Thin-layer chromatography can also be used.

More recently, amino acids have been analyzed by gas chromatography as various derivatives such as the N-trifluoracetyl n-butyl esters (98) and trimethylsilyl derivatives (99, 100), or by fluorometric analysis using fluorescamine (101).

To determine the free amino acid content of musts the ion-exchange column chromatography method (102) can be used.

High-performance liquid chromatography (HPLC) is available in most laboratories. It can be converted to do amino acid analysis. The types of derivatization, column, elution solvent, and detector that are available are numerous. Ninhydrin detection with post-column derivatization and cation-exchange resin developed from earlier work (79–83) has become the standard method to compare to other newer methods. Dansyl and o-phthaldialdyhyde (OPA) derivatives give greater sensitivity. There has been extensive publication comparing these to the ninhydrin method. Some comparisons have been favorable and some not. A report (103) gives the OPA method for juice or wine using precolumn derivatization. Another recent report (104) suggests dansylation and fluorescent detection is practical for juices, wines, and vinegar. A variation of the post-column ninhydrin method (105) seems to work well with juice and wine. It is the classic method using modern techniques and equipment.

Procedure

An IBM 9533 HPLC with an attached 9505 Automatic Sampler was used with a sulfonated polystyrene, divinyl benzene cross-linked cation-exchange column to effect separation of the amino acids. Post-column detection was done utilizing ninhydrin, and a reaction coil was maintained at 100°C. Lithium buffers were used in the time sequence listed in the accompanying tabulation.

Time (min)	Buffer[a]	
0–20	100% A	
20–70	100% A to 65% A + 35% B	Linear gradient
70–100	65% A + 35% B to 100% B	Linear gradient
100–132	100% B to 92% B + 8% C	Linear gradient
132–137	92% B + 8% C	
137–165	100% A	Step to A

[a]Buffer A = 0.24 N lithium citrate, pH 2.75, with 1% propanol; buffer B = 0.64 N lithium citrate, pH 7.50; buffer C = 0.3 N lithium citrate, pH 11.0.

Buffer flow was maintained at 0.3 mL/min. Ninhydrin reagent flow was also maintained at 0.3 mL/min. A 550 interference filter was used in the detector.

Samples were prepared by dilution with at least two volumes of lithium buffer (pH 2.20) followed by filtering through a 0.45-μm filter. Sample volume was 90 μL.

The buffers, regenerant (buffer A), diluting solutions, and post column were all products of Pickering Laboratories, Inc., 1951 Colony Street, Suite S, Mountain View, CA 94043.

A reconstructed chromatogram using this system is shown in Figure 22.

An example of data gathered from a limited number of California grape samples is given in Table 38 (105a). These data are biased on the low side but give a reasonable idea of what the relative relationships between the various amino acids will be.

For details of a GLC procedure for free amino acids that gave good agreement with classic ion-exchange amino acid analyzer results, see Reference 100. Individual amino acids also can be determined by various special procedures (5). An ion-selective membrane electrode has been used for arginine (106, 107). Methods for proline (108–110) in musts and wines are available.

α-Amino Nitrogen

The results for α-amino nitrogen vary, depending on the method used; hence they are generally considered to be estimates. Since the ninhydrin reaction color intensity is not the same for each amino acid, corrections are necessary, particularly when the proline concentration in wines is high.

The formol titration (111) is not specific for amino acids and also measures other components with the result that a higher value is usually obtained than for

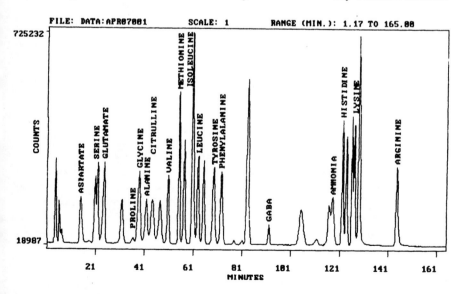

Figure 22. Typical standard chromatogram showing some of the key amino acid retention times.

Table 38. A Comparison of Amino Acids of 53 Paired Juices[a] and Wines (mg/L)

Amino acids	Juices			Wines		
	Low	High	Mean	Low	High	Mean
Aspartic acid	5.0	54.4	23.0	0.0	20.4	3.3
Threonine	9.6	117.2	41.9	0.0	9.9	0.8
Serine	11.3	198.0	48.4	0.0	156.3	12.3
Asparagine/glutamic acid	134.5	742.3	378.3	0.0	346.0	37.2
Glutamine	7.2	455.5	77.2	0.0	43.8	4.7
Proline	42.9	1782.2	649.1	0.0	1952.2	435.5
Glycine	0.0	11.5	3.0	0.0	69.2	6.3
Alanine	10.2	297.3	84.8	0.0	176.8	12.9
Citrulline	0.0	14.8	1.5	0.0	14.6	1.2
Valine	3.7	50.8	21.6	0.0	10.7	2.2
Methionine	0.0	12.5	3.6	0.0	11.6	1.5
Isoleucine	1.6	34.3	12.9	0.0	10.0	1.2
Leucine	1.8	34.0	15.3	0.0	26.4	2.9
Tyrosine	0.0	60.2	11.8	0.0	14.7	1.8
Phenylalanine	3.2	36.6	13.7	0.0	15.7	1.7
γ-Aminobutyric acid	1.2	163.4	67.3	0.0	156.0	16.3
Tryptophane	0.0	30.0	3.6	0.0	0.6	0.0
Histidine	2.8	100.1	21.3	0.0	20.6	2.9
Ornithine	0.4	18.8	5.3	0.0	19.0	6.1
Lysine	1.2	23.0	6.5	0.0	35.7	5.9
Arginine	6.6	649.2	179.8	0.0	200.7	9.7

From Reference 105a.
[a] Includes 6 Chardonnay, 10 Chenin blanc, 12 Petite Sirah, 18 Cabernet Sauvignon, and 7 White Riesling samples.

the α-amino acids by the ninhydrin method. The reaction is:

$$\text{R}-\underset{\underset{NH_2}{|}}{\text{CH}}-\text{COOH} \xrightarrow{\text{HCHO}} \text{R}-\underset{\underset{NH-CH_2OH}{|}}{\text{CH}}-\text{COOH} \xrightarrow{\text{HCHO}} \text{R}-\underset{\underset{HO-CH_2-N-CH_2OH}{|}}{\text{CH}}-\text{COOH}$$

The final product of the amino acid, after it reacts with 2 mol of formaldehyde, is a much stronger acid. However, this method is quick. In a modification of the method, the titration is applied after ion-exchange separation. Since ammonia would also react, it has to be removed prior to the determination.

A chemical method (112) using 2,4,6-trinitrobenzene sulfonic acid (TNBS) is useful for juice analysis, but it is not sensitive enough for wine. The method of Rosen (113) used previously for wine is adequate for both juice and wine.

Procedure

Extract 10 g of crushed berries with a 40% v/v ethanol solution or take 10 mL of wine. Place the prepared sample onto a 2 × 20-cm column loaded with Dowex

50W-X8 (H^+) resin. Wash with water (50 mL), elute with 3 N NH_4OH (150 mL), and then wash with 3 bed-volumes of water. Combine the eluates, other than the initial wash, in a beaker, hold at 45–50°C, and play a stream of clean air over the surface until dry. Take up the amino acids with 5 mL of 10% isopropanol. Take an aliquot containing 0.2–0.4 mol of amino acids and add 0.5 mL of cyanide buffer mixture (490 mg of sodium cyanide brought to 1 L with water of which 2.0 mL is added to a 998 mL of a buffer solution made up by combining 540 g of sodium trihydrate, 100 mL of glacial acetic acid, and 400 mL of water and brought to 1.5 L) and 0.5 mL of 3% ninhydrin in methyl cellusolve. Heat for 15 min in a boiling water bath in a capped tube. Add 5 mL of a 1:1 mixture of isopropanol and water. Cool to room temperature and immediately determine absorbance at 570 nm. Prepare a standard curve using leucine. Except for proline the amino acids all have a similar response factor.

Table 39 lists some data on the α-amino nitrogen values reported in wines, also indicating the method used for the determination.

Arginine

Next to proline, arginine is the most abundant amino acid in grapes. It is easily metabolized by yeast and is depleted during fermentation. Arginine in the juice is now being used as an indicator of the vine's nitrogen nutrition status (114). The procedure below is that of Gilboe and Williams (106) as adapted for musts (115). When the analyst is familiar with the procedure, six to eight samples can be run at the same time.

Procedure

Prepare 10% sodium hydroxide (10 g/100 mL of water), 0.02% 8-hydroxyquinoline (0.20 g/100 mL of ethanol, and then dilute to 1 L with distilled water), 1% sodium hypobromite [1 g of bromine (0.34 mL) diluted to 100 mL with 5% sodium hydroxide (use a pro-pipet, work in the hood; reagent is stable for 1 month if stored in a dark bottle in the refrigerator)], 40% urea (40 g of urea/100 mL of water), and 10 mg/L arginine standard (made by diluting a stock solution of 60.5 mg of arginine hydrochloride/100 mL of water 1 to 50 with distilled water).

Put all reagents in an ice bath. Prepare 5 test tubes (20–25-mm diameter) as follows:

Arginine (μg)	Arginine standard (mL of 10 μg/mL)	Distilled water (mL)	Total volume (mL)
0	0	5	5
10	1	4	5
20	2	3	5
30	3	2	5
40	4	1	5

Table 39. Total α-Amino Nitrogen Concentrations, Reported as Milligrams of Nitrogen per Liter in Musts and Wines

Region	Method	Musts			Wines		
		Number of samples	Range	Average	Number of samples	Range	Average
France	Formol	8	15–176	70	36[a]	10–67	34
	Formol	28	15–182	75	15[a]	62–168	105
	Formol				35[b]	41–131	80
	Formol				164	3–348	73
	Microbiological	6		141	9	81–207	110
	Paper chromatography				8[b]	91–463	228
Italy	Formol	8	13–36				
	Formol	2	100–149				
Soviet Union	Formol	8	298–436				
United States (Calfornia)	Chemical	139	24–624	89	67	7–452	66

[a] White wines only.
[b] Red wines only.

Using the blank, bring the colorimeter or spectrophotometer to zero. Keep test tubes containing standards or samples in an ice bath. To each test tube, add 1 mL of 0.02% 8-hydroxyquinoline plus 10 mL of 10% NaOH, mix, and return to ice bath for 2 min. Add rapidly 0.2 mL of 1% sodium hypobromite, shake, and within 15 sec add 1 mL of 40% urea, again mix; and within 1 min add 5 mL of cold distilled water and mix again. Read absorption at 500 nm within 5 min. Against the distilled water–reagent blank prepare a standard curve.

For musts, crush the berries and filter through cheesecloth, or centrifuge. Grapes contain 200–1500 μg/mL. Dilute an aliquot with distilled water so that 5 mL contains 5–40 μg of arginine. Develop color as above and calculate arginine content from the standard curve. For red wines with high color, pretreatment with charcoal is desirable.

Proline

Proline is the amino acid present in greatest amount in most wines and in most grape juices. Proline reacts with ninhydrin (116) in the presence of formic acid (117) to form a colored derivative. Ough (108) modified the procedure to permit the direct measurement in wine or grape juice.

Procedure

Prepare a proline stock solution by dissolving, in a 1-L volumetric flask, 0.575 g of proline in water. The concentration of this solution is 5 μmol/mL. From this stock solution, pipet 0-, 1-, 2-, 3-, 5-, 7-, and 10-mL aliquots into 100-mL volumetric flasks and dilute with water to volume. Take 0.5 mL of each diluted standard solution and add to separate 130 × 15-mm screw-cap test tubes. Add 0.25 mL of formic acid and 1 mL of a 3% ninhydrin solution in methyl cellosolve. Mix the tubes well, close them, and immerse in a boiling water bath for exactly 15 min. At the end of this period remove the tubes and place into a 20°C water bath, and add to each tube 5 mL of a 1:1 isopropanol–water mixture. Mix the tubes, then transfer the solutions into 1-cm cuvettes, and read the absorbance at 517 nm against a water blank carried through the procedure. The loss in color in time is about 2%/hr. Plot the absorbance values against concentration. For most wines, pipet 1 mL of the sample into a 50-mL volumetric flask and dilute to volume with water. Take a 0.5-mL aliquot of the diluted sample and carry it through the procedure as given above. If a sample of wine or grape has an absorbance value exceeding that for 0.5 μmol, dilute the sample 1:1 with 1:1 isopropanol–water and repeat the determination.

$$\text{Proline (mg/L)} = \frac{(A)(0.115)(1000)(F)}{v}$$

where A = amount present in the sample aliquot as read from the calibration curve (μmol)

F = dilution factor
v = volume of sample aliquot (mL)

Because of the interference of colored pigments, Wallrauch (109) recommends shaking the proline–ninhydrin complex with butyl acetate and evaluating the color of the organic phase. It has also been reported (110) that more reproducible results are obtained by replacing the 1:1 isopropanol–water mixture with n-butanol.

BIOGENIC AMINES

A number of amines are present in wines. They include 1,3-diaminopropane, putracine, cadaverine, spermine, spermidine, histamine, tyramine, β-phenethylamine and other nonvolatile amines. The volatile amines have been summarized (2). Of the biogenic amines found in wines, histamine and tyramine have been most studied:

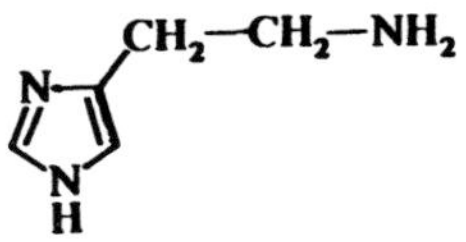

Histamine

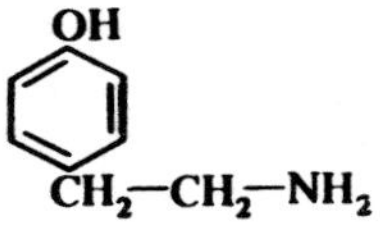

Tyramine

Both of these amines are found in relatively small amounts in wines (Table 40) (118–129). Tyramine is a pressor amine (130), causing increases in blood pressure. Histamine has an opposite effect, causing lowered blood pressure (131).

Table 40. Histamine and Tyramine Amounts (mg/L) Reported in Wines

Region	Histamine			Tyramine		
	Number of samples	Range	Average	Number of samples	Range	Average
Canada	58		3.32	63		2.37
Europe	13	0.04–15.6	4.9			
Finland	3	0–0.5	0.2	3	0.1–1.0	0.4
France	72	0.05–30.0	4.8			
Germany	56	0–16	1.57			
Hungary	35	0.19–3.10	1.06	35	0.29–1.38	0.73
Italy	148	0–7.5	0.89	205	0–12.2	1.08
USSR		0.6–4.2	2.0			
South Africa	184	0–49.1	3.1	156	0–6.4	0.40
United States	269	0.3–15.5	2.42			

From References 118–129.

Unless a person has been taking medically prescribed monoamine oxidase inhibitors, the small amount of these biogenic amines in wine has little physiological effect. However, if a person is under this specific medication, as little as 6 mg of tyramine can cause increased blood pressure (132) and 25 mg would be considered dangerous (133). Schneyder (134) indicates no observed symptoms in a normal adult not under medication. The red wines generally have greater amounts of histamine and tyramine than do white wines.

The main source of these amines in wines appears to be the decarboxylation of the precursor amino acid (135) (histidine or tyrosine, respectively, for histamine and tyramine). The usual biological source of the decarboxylation is *Lactobacillus* sp. (136). Other malolactic bacteria, such as *Leuconostoc gracile*, also are reported to form histamine (137). Lafon-Lafourcade (137) showed that treatment of wines with 100 mg/L histidine inhibited the formation of histamine. She also found that hetero- and homofermentive rods and heterofermentive cocci produced histamine in about equal amounts but that homofermentive cocci produced less under cell conditions. Treatment of wines with bentonite significantly reduces the histamine level (138–140).

Recent work has verified (129, 141) earlier work (118) that there is no general correlation between wines that have undergone malolactic fermentation and the amounts of histamine found. Since only a few strains have the ability to decarboxylate histidine, more or less random patterns of relationship result.

Mack (125) has given an automated method using an amino acid analyzer. Detection was only 0.3 mg/L in the sample with a standard deviation of ± 0.27 mg/L. Newer automated methods using fluorescent detectors and high-performance liquid chromatographs improve both accuracy and speed. A fairly rapid and accurate method that allows both tyramine and histamine to be determined has been developed (129). The tyramine was partially oxidized with perchloric acid after extraction and prior to OPA derivatization. The OPA derivatives were then separated by reverse-phase HPLC and detected fluorometrically. A more elaborate method (142) allows measurement of 20 amines in wine using dansyl derivatives. Poly(vinylpyrrolidone) has been used (143) to eliminate interferences and to measure 20 amines and some amino acids in a less complicated HPLC procedure.

Sen (120) reviewed both a spectrofluorometric and a gas chromatographic method for tyramine analysis. Some minor modifications will be required for wine analysis, separations, and semiquantification.

PROTEINS

Protein content can be estimated from the total nitrogen measurement, but this value is high in wine and is completely wrong when applied to grape juice.

The proteins can be precipitated by the addition of phosphomolybdic acid or by trichloroacetic acid (TCA), and the isolated protein can be determined by one or two color reactions, namely the biuret test or the Folin–Ciocalteu test, or its amount can be calculated by determining the total nitrogen content of the isolated protein and multiplying it by 6.25. The Folin–Ciocalteu test is about 100 times more sensitive than the biuret test, but the color varies among different proteins. It has been noted that unless peptides and polyphenols are removed, they will be precipitated and react in the biuret or Folin–Ciocalteu reactions to produce color. The proteins then can be isolated by TCA precipitation after removal of amino acids, and the peptides can be isolated by dialyzation; the precipitate must be washed several times with 5% TCA. To avoid interference of polyphenols, the protein content may be estimated from the sum of the amino acid analysis of all the amino acids in the TCA precipitate (144). The vertical polyacrylamide gel electrophoresis procedure given by Ornstein-Davis has been used for wine protein (145). Somers and Ziemelis (19, 146) recommended (a) separation of the protein fraction by gel filtration (Sephadex G-25) and (b) qualitative interpretation of the fraction by ultraviolet. It must be admitted that accurate data on protein content are difficult to obtain (147–150).

An improved and more rapid protein measurement procedure (151), of a previous method (146), is useful.

Procedure

Prepare a series of standard bovine albumin solutions (0–200 mg/L) in 3% acetic acid. Load a 15-cm × 1.3-cm-i.d. column with Sephadex G-25 (superfine) gel. Protect the top and bottom with filter paper circles. Using glass or silicone rubber only, connect the outlet of the column to the flow-through cell with a UV detector (280 nm) to monitor the flow. Apply 1.0 mL of each of the standards to the top of the gel column and adjust the flow rate of the 3% acetic acid while eluting solvent through the column at 3–4 mL/min. Use a peristaltic pump to achieve uniform flow. With the detector set to read full scale at 0.2 absorbance, record the peak heights or area of the standards.

Prepare a standard curve. For juice or wine, place 1.0 mL of membrane-filtered material on the column and proceed as above. The results are reported as milligrams per liter of bovine albumin equivalents.

Other methods using ion-exchange or reverse-phase separation will be useful in the future for rapid protein measurement of the more specific heat-unstable proteins.

NITRATE

The methods for determining nitrate in musts or wine were inadequate until the development, by Rebelein (152), of a method involving the reduction of the nitrate to nitrite by metallic cadmium and subsequent colorimetric determination.

Most of the data shown in Table 41 (153–161) has been measured using this method or a variation of it (152).

A comparison of four different methods (162), including the original Rebelein method, indicated that the use of a cadmium column or *E. coli* to reduce the nitrate was advantageous in increasing reproducibility. The *E. coli* method used *p*-cresol for the color development and is safer in that the use of cadmium is eliminated.

Nitrate has been determined by an ion-selective electrode (163).

A spectrophotometric method for nitrates and nitrites is based on the reaction of nitrite with 2,3-dimethyl-1-phenylpyrazolones-5. The color is measured at 345 or 600 nm. The nitrate is reduced to nitrite by passing through a column filled with metallic cadmium. The amounts of nitrite present were below sensitivity of the method. The Rebelein procedure is given here.

Procedure

Take 5 g of cadmium acetate (CdAc · $2H_2O$), add 1 mL of acetic acid, and bring to 100 mL with water. Add together 50 mL of acetic acid and 200 mL of water to give a 20% acetic acid solution. Prepare one Greiss reagent by dissolving 1.5 g of sulfanilic acid with 50 mL of acetic acid and dilute to 250 mL with water. Prepare the second Greiss reagent by adding 75 g of naphthylamine in 50 mL of acetic acid and dilute to 250 mL with water.

Into a 50-mL volumetric flask place 5.0 mL of wine (or juice), 5.0 mL of water, and 2.0 mL of concentrated ammonium hydroxide. Add 500 mg ± 25 mg of zinc dust. Immediately add 1 mL of 5% cadmium acetate, then mix so that the reaction will go rapidly. It is desirable to inject the cadmium acetate vigorously into the center of the flask. Without again mixing, allow the flask to stand for 5 min (to allow the cadmium sponge to form). Bring the flask to volume with water, shake vigorously, and filter.

To 10 mL of the filtrate add 10 mL of each of the Greiss reagents (prepared just previously). To another 10 mL of the filtrate add 10 mL of the 20% acetic acid solution. After 15 min determine the absorbance at 530 nm using a 1-cm cuvette. The filtrate treated with 20% acetic acid is used as a blank. Determine the N_2O_5 concentration from the standard curve.

To prepare a standard curve, take 0.374 g of KNO_3 dried at 105°C, dissolve, and make to 100 mL with water. Dilute 15 mL of this to 100 mL with water. Take 25 mL of this solution and dilute to 250 mL with water. (30 mg/L as N_2O_5.) Take 20, 40, 60, 80, and 100 mL of this solution and dilute to 100 mL. This gives standard solutions of 6, 12, 18, 24, and 30 mg/L. Take 5 mL of each and treat as in the preceding two paragraphs. Substitute standards for wine (or juice). Draw a standard curve plotting absorbance versus N_2O_5 mg/L.

The concentrated ammonium hydroxide may be replaced by a mixture of 5 mL of 28% ammonium chloride and 2.5 mL of 40% sodium hydroxide (153).

Table 41. Range and Concentration of Nitrate in Juices and Wines

			Nitrate, NO_3 (mg/L)			
Country	Wine type	Number of samples	Range	Mean	References
Austria	Table	59	3.88–48.0	18.4	153
Germany	Juice	233	0–111.8	9.6	154
	Juice	75	1.4–65.1	10.1	155
	Table	200		14.6	
	Dessert	26	0–75.7	11.8	
	Red table	39		6.6	
	Wine	500		3.0	156
	White juice and wine	23	0.8–29.6	7.7	157
	Red juice and wine	17	5.4–23.2	8.8	
Italy	Table	73	0.1–4.7	2.6	158
	White table	32	0.0–9.8	3.3	40
	Red table	60	4.0–4.9	5.0	
	Table	102	1.9–38.2	8.7	39
	Specialty	31	0.9–15.1	11.5	
	Table	44	1.4–5.5	3.3	
Spain	White table	11	5.6–17.1	11.5	159
	Red table	14	7.7–18.1	11.8	160
United States	White table	122	2.1–53.7	16.4	161
	Red table	75	0.9–41.4	8.5	

REFERENCES

1. D. Tercelj, *Ann. Technol. Agric.* **14**, 307-319 (1965).
2. C. S. Ough, *Proc. ALKO Symp. Flavour Res. Alcohol. Beverages, 1984* Vol. 3, pp. 199-225 (1984).
3. C. S. Ough, *Am. J. Enol. Vitic.* **20**, 213-220 (1969).
4. C. S. Ough and A. Kriel, *S. Afr. J. Enol. Vitic.* **6**, 7-11 (1985).
5. C. S. Ough and O. Bustos, *Wines Vines* **50**(4), 50-58 (1969).
6. C. S. Ough, *Vitis* **7**, 321-331 (1968).
7. C. S. Ough and V. L. Singleton, *Am. J. Enol. Vitic.* **19**, 129-138 (1968).
8. G. Lotti and G. Anelli, *Riv. Sci. Tecnol. Aliment. Nutr. Um.* **1**, 25-32 (1971).
9. A. Salzedo, P. Vian, and C. Mattarei, *Vini Ital.* **18**, 415-421 (1976).
10. R. DiStefano, *Vini Ital.* **19**, 89-93 (1977).
11. W. M. Kliewer, *J. Food Sci.* **34**, 274-278 (1969); **35**, 17-21 (1970).
12. C. Nicolosi Asmundo and C. M. Lanza, *Riv. Sci. Tecnol. Aliment. Nutr. Um.* **6**, 185-186 (1976).
13. C. Bertolini and L. Paronetto, *Vignevini* **3**(5), 13-18 (1976).
14. G. Cerutti, C. Mazzolini, and R. Ziliotto, *Riv. Sci. Tecnol. Aliment. Nutr. Um.* **5**, 309-311 (1975).
15. H. G. Weiller and F. Radler, *Z. Lebensm.-Unters. -Forsch.* **161**, 259-266 (1976).
16. U. Kuensch, A. Temperli, and K. Mayer, *Am. J. Enol. Vitic.* **25**, 191-193 (1974).
17. M. Feuillat, M. Gouilloux-Benatier, and V. Gerbaux, *Sci. Aliments* **5**, 103-122 (1985).
18. C. Poux, M. Caillet, and R. Joubert, *Ind. Aliment. Agric.* **91**, 695-699 (1974).
19. T. C. Somers and G. Ziemalis, *Am. J. Enol. Vitic.* **24**, 47-50 (1973).
20. R. D. Begunova and A. E. Linetskaya, *Vinodel. Vinograd. SSSR* **27**(5), 31-34 (1967).
21. W. Diemair and G. Maier, *Z. Lebensm.-Unters. -Forsch.* **118**, 148-152 (1962).
22. D. Motoc and M. Bulancea, *Ind. Aliment. (Bucharest)* **20**, 557-559 (1969).
23. R. Moretti and H. W. Berg, *Am. J. Enol. Vitic.* **16**, 69-78 (1965).
24. F. C. Bayly and H. W. Berg, *Am. J. Enol. Vitic.* **18**, 18-32 (1967).
25. D. Zakov, B. Mesrob, T. Iwanov, Z. Nikova, and S. Valtscheska, *Mitt., Rebe Wein, Obstbau Fruechteverwert.* **19**, 437-447 (1969).
26. I. Molnar, *Szolesz. Boraszat* **1**, 303-335 (1975).
27. L. Usseglio-Tomasset and R. DiStefano, *Riv. Vitic. Enol.* **30**, 452-469 (1977).
28. F. Drawert and W. Muller, *Z. Lebensm.-Unters. -Forsch.* **153**, 204-212 (1973).
29. K. Yokotsuka, Y. Umehara, T. Aihara, and T. Kushida, *Hakko Kogaku Zasshi* **53**, 620-625 (1975).
30. T. S. Nanitashvili, *Appl. Biochem. Microbiol.* **9**, 85-87 (1973); see also *Prikl. Biokhim. Mikrobiol.* **9**, 102-105 (1973).
31. L. Erchzheggi and A. Asvany, *Bull. O.I.V.* **40**(442), 1379-1390 (1967).
32. V. I. Nilov, T. P. Chazova, and V. I. Safonov, *Appl. Biochem. Microbiol.* **5**, 64-68 (1969).
33. J. Jurubita, J. Gritsch, and C. Stanescu, *Bull. O.I.V.* **40**(442), 1390-1399 (1967).
34. C. Tarantola, *Bull. O.I.V.* **40**(437-438), 773-787 (1967).
35. C. S. Ough, unpublished data.

36. M. A. Amerine, J. W. Berg, R. E. Kunkee, C. S. Ough, V. L. Singleton, and A. D. Webb, *The Technology of Wine Making*, 4th ed. Avi Publishing Co., Westport, CT, 1980.
37. M. Feuillat and J. V. Morfaux, *Connaiss. Vigne Vin* **10**, 33-49 (1976).
38. M. Castegnaro, B. Pignatelli, and E. A. Walker, *Analyst (London)* **99**, 156-162 (1974).
39. G. Lotti and P. V. Baldacci, *Riv. Vitic. Enol.* **23**, 262-272 (1970).
40. V. Coppola, *Riv. Vitic. Enol.* **30**, 248-257 (1977).
41. B. Duteurtre, C. Bourgeois, and B. Chollot, *J. Inst. Brew.* **77**, 28-35 (1971).
42. M. A. Amerine and M. A. Joslyn, *Table Wines, The Technology of Their Production*, 2nd ed. Univ. of California Press, Berkeley and Los Angeles, 1970.
43. P. Bidau, *Bull. O.I.V.* **48**(536), 842-867 (1975).
44. M. Giaccio and R. Angelucci, *Quad. Merceol.* **16**, 141-149 (1977).
45. F. Drawert, A. Rapp, and H. Ullemeyer, *Vitis* **6**, 177-197 (1967).
46. A. Rapp and K. H. Reuther, *Vitis* **10**, 51-58 (1971).
47. F. De Francisco and G. Margheri, *Vini Ital.* **15**, 257-261 (1973).
48. K. G. Bergner and H. Wagner, *Mitt., Rebe Wein, Obstbau Fruechteverwert.* **15A**, 181-198 (1965).
49. A. Daghetta, P. Resmini, S. Saracchi, and G. Volonterio, *Riv. Vitic. Enol.* **23**, 207-225 (1970).
50. O. Colagrande, F. Griselli, and A. A. Del Re, *Connaiss. Vigne Vin* **7**, 93-106 (1973).
51. G. Pallotti, B. Bencivenga, G. Brighena, and A. Palmioli, *Riv. Soc. Ital. Sci. Aliment.* **5**, 303-305 (1976); see also *ibid.* **4**, 331-353 (1975).
52. G. Pallotti, B. Bencivenga and C. Botre, *Rass. Chim.* **27**, 297-305 (1975).
53. G. Modi and M. Guerrini, *Boll. Chim. Unione Lab. Prov. Ital.* **27**, 297-305 (1975).
54. O. Colagrande, V. Mazzoleni, and A. Del Re, *Connaiss. Vigne Vin* **10**, 23-32 (1976); see also *Ann. Fac. Agrar. (Univ. Cattol. Sacro Cuore) Milan* **15**, 11-21 (1975).
55. B. Mincione, S. Spagna Musso, and V. Coppola, *Riv. Vitic. Enol.* **29**, 492-515 (1976).
56. A. Minguzzi and A. Amati, *Riv. Sci. Tecnol. Aliment. Nutr. Um.* **3**, 371-372 (1973).
57. O. Juhasz, P. Kozma, and D. Polyak, *Acta Agron. Acad. Sci. Hung.* **33**, 3-17 (1984).
58. O. Juhasz and D. Torley, *Acta Aliment. Acad. Sci. Hung.* **14**, 101-112 (1985).
59. A. A. Bell, C. S. Ough, and W. M. Kliewer, *Am. J. Enol. Vitic.* **30**, 124-129 (1979).
60. C. S. Ough and R. E. Kunkee, *Appl. Microbiol.* **16**, 572-576 (1968).
61. C. L. Ogg, in *Treatise on Analytical Chemistry*, Part II, Sect. B, Vol. 11, E. W. D. Huffman and J. Mitchell, Jr., Eds. Wiley, New York, 1965, pp. 457-489.
62. P. Fenocchio and G. M. Pezzi, *Pesqui. Agropecu. Bras., Ser. Cigron.* **1**, 121-124 (1976).
63. M. A. Amerine, *Adv. Food Res.* **5**, 353-510 (1962).
64. C. S. Ough, unpublished data (1972).
65. M. A. Amerine, *Laboratory Procedures for Enologists*. Associated Students Bookstore, Davis, CA, 1970.
66. A. Bouat and C. Crouzet, *Ann. Agron.* **16**, 107-118 (1965).
67. J. Davidson, J. Mathieson, and A. W. Boyne, *Analyst (London)* **95**, 181-193 (1970).
68. C. Maddix, R. L. Norton, and M. J. Nicolson, *Analyst (London)* **95**, 738-742 (1970).
69. A. R. Deschreider and R. Meaux, *Analusis* **2**, 442-445 (1973).
70. G. K. Buckee, *J. Inst. Brew.* **80**, 291-294 (1974).

71. C. G. Barraso, J. L. Hidalgo, and J. A. Perez-Bustamante, *Afinidad* **40**(383), 41-46 (1983).
72. F. M. Paillar, *Ann. Falsif. Expert. Chim. Toxicol.* **75**, 431-439 (1982).
73. A. R. Descheider and R. Meaux, *Rev. Ferment. Ind. Aliment.* **28**, 238-244 (1974).
74. V. Dimotaki-Kourakau, *Ann. Falsif. Expert. Chim.* **53**, 337-348 (1960).
75. *Recueil des Méthodes Internationales d'Analyse des Vins*, 5th ed. Office International de la Vigne et du Vin, Paris, 1978.
76. D. J. McWilliam and C. S. Ough, *Am. J. Enol. Vitic.* **25**, 67-72 (1974).
77. R. J. Block, *Amino-Acid Handbook; Methods and Results of Protein Analysis.* Thomas, Springfield, IL, 1956.
78. F. Kavanaugh, in *Amino Acids*, G. D. Shockman, Ed. Academic Press, New York, 1963, pp. 567-673.
79. S. Moore and W. H. Stein, *J. Biol. Chem.* **176**, 337-388 (1948).
80. S. Moore and W. H. Stein, *J. Biol. Chem.* **211**, 663-681 (1954).
81. S. Moore and W. H. Stein, *J. Biol. Chem.* **211**, 893-906 (1954).
82. S. Moore and W. H. Stein, *J. Biol. Chem.* **211**, 907-913 (1954).
83. D. H. Spackman, W. H. Stein, and S. Moore, *Anal. Chem.* **30**, 1190-1206 (1958).
84. F. Drawert, *Vitis* **4**, 49-56 (1963).
85. B. S. Luh and H. N. Daoud, *Fruechtsaft-Ind.* **13**, 50-59 (1968).
86. W. Diemair, J. Koch, and E. Sajak, *Z. Lebensm.-Unters. -Forsch.* **116**, 209-215 (1961).
87. A. K. Rodopulo and A. F. Pisarnitskii, *Vinodel. Vinograd. SSSR* **28**(1), 5-7 (1968).
88. K. Mayer, G. Pause, U. Velsch, U. Kunsch, and A. Temperli, *Mitt., Rebe Wein, Obstbau Fruechteverwert.* **23**, 331-340 (1973).
89. A. K. Rodopulo, I. A. Egorov, A. F. Pisarnitskii, A. A. Martakov, and T. N. Levchenko, *Prikl. Biokhim. Mikrobiol.* **5**, 186-188 (1969).
90. A. Seppi, *Riv. Soc. Ital. Sci. Aliment.* **75**, 109-112 (1975).
91. R. Cosden, A. H. Gordon, and S. P. Martin, *Biochem. J.* **38**, 224-232 (1944).
92. K. Henning and S. M. Flintje, *Wein-Wiss.* **8**, 121-125, 129-140 (1954).
93. P. A. Dimotaki, *Chem. Chron. A* **20**(10), 3 (1955); *Am. J. Enol. (Engl. Transl.)* **9**, 79-85 (1957).
94. A. R. Nassar and W. M. Kliewer, *Proc. Am. Soc. Hortic. Sci.* **89**, 281-294 (1966).
95. V. Lepadatu and I. Tanase, *Ann. Technol. Agric.* **16**, 321-331 (1967).
96. K. J. Bergner, *Bull. O.I.V.* **41**(446), 460-467 (1968).
97. W. M. Kliewer, *Am. J. Enol. Vitic.* **19**, 166-174 (1968).
98. C. W. Gehrke and D. L. Stalling, *Sep. Sci.* **2**, 101-138 (1967).
99. K. Ruhlmann and G. Michael, *Bull Soc. Chim. Biol.* **47**, 1467-1475 (1965).
100. P. Fantozzi and G. Montedoro, *Am. J. Enol. Vitic.* **25**, 151-156 (1974); see also *Sci. Tecnol. Aliment.* **3**, 53-54 (1973).
101. A. M. Felix and G. Terkelsen, *Arch. Biochem. Biophys.* **157**, 177-182 (1973).
102. R. M. Kluba, L. R. Mattick, and L. R. Hackler, *Am. J. Enol. Vitic.* **29**, 102-111 (1978).
103. E. M. Sanders and C. S. Ough, *Am. J. Enol. Vitic.* **36**, 43-45 (1985).
104. P. Martin, C. Polo, M. Cabezudo, and M. V. Dabrio, *J. Liq. Chromatogr.* **7**, 539-558 (1984).
105. A. Henshall, M. J. Pickering, and D. Soto, *Chromatogr. Rev.* **9**, 8-10 (1983).

105a. C. S. Ough. Unpublished data (1986).
106. D. D. Gilboe and J. N. Williams, Jr., *Proc. Soc. Exp. Biol. Med.* **91**, 535-536 (1956).
107. T. A. Neubecker and G. A. Rechnitz, *Anal. Lett.* **5**, 653-659 (1972).
108. C. S. Ough, *J. Food Sci.* **34**, 228-230 (1969).
109. S. Wallrauch, *Fluess. Obst* **43**, 430, 435-437 (1976).
110. G. Cavasino and L. Farsaci, *Riv. Sci. Tecnol. Aliment. Nutr. Um.* **5**, 113-114 (1975).
111. S. P. L. Sorensen, *Biochem. Z.* **7**, 45-101 (1908).
112. E. A. Crowell, C. S. Ough, and A. Bakalinsky, *Am. J. Enol. Vitic.* **36**, 175-177 (1985).
113. H. Rosen, *Arch. Biochem. Biophys.* **67**, 10-15 (1957).
114. W. M. Kliewer and J. A. Cook, *Am. J. Enol. Vitic.* **25**, 111-118 (1974).
115. W. M. Kliewer, private communication (1970).
116. F. P. Chinard, *J. Biol. Chem.* **199**, 91-95 (1952).
117. A. D. Lashkhi and T. P. Tsiskarishivili, *Vinodel. Vinograd. SSSR* **27**(1), 19-21 (1967).
118. C. S. Ough, *J. Agric. Food Chem.* **19**, 241-244 (1971).
119. L. Jakob, *Weinwirtshaft* **114**, 126-127 (1973).
120. N. P. Sen, *J. Food Sci.* **34**, 22-26 (1969).
121. R. E. Zappavigna, E. Brambati, and G. Cerutti, *Riv. Vitic. Enol.* **27**, 285-294 (1974).
122. R. E. Zappavigna and G. Cerutti, *Lebensm.-Wiss. Technol.* **6**, 151-152 (1973).
123. G. Cerutti and L. Remondi, *Riv. Vitic. Enol.* **25**, 66-78 (1972).
124. E. Paputti and H. Soumalainen, *Mitt., Rebe Wein, Obstbau Fruechteverwert.* **19**, 184-192 (1969).
125. D. Mack, *Z. Lebensm.-Unters. -Forsch.* **152**, 321-323 (1973).
126. M. Kallay, G. Bajoczy, J. Nedelkovits, and M. Bodyne, *Borgazdasag* **32**, 27-31 (1984).
127. J. A. Zee, R. E. Simard, L. L'Heureux, and J. Tremblay, *Am. J. Enol. Vitic.* **34**, 6-9 (1983).
128. M. A. Koreisha, L. A. Tkachuk, and L. A. Furtune, *Sadovod. Vinograd. Vinodel. Mold.* **39**(11), 42-43 (1984).
129. J. D. Cilliers and C. J. Van Wyk, *S. Afr. J. Enol. Vitic.* **6**, 35-40 (1985).
130. G. Barger and C. S. Walpole, *J. Physiol. (London)* **38**, 343-352 (1901).
131. F. Franzen and K. Eyesill, *Biologically Active Amines Found in Man*. Pergamon Press, New York, 1969.
132. S. L. Rice, R. R. Eitenmiller, and P. E. Koehler, *J. Milk Food Technol.* **39**, 353-358 (1976).
133. T. A. Weaton and I. Steward, *Phytochemistry* **8**, 85-92 (1969).
134. J. Schneyder, *Bull. O.I.V.* **46**(511), 821-831 (1973).
135. V. A. Lagerborg and W. E. Chapper, *J. Bacteriol.* **63**, 393-397 (1952).
136. K. Mayer, G. Pause, and U. Vetsch, *Mitt., Rebe Wein, Obstbau Fruechteverwert.* **21**, 278-288 (1971).
137. S. Lafon-Lafourcade, *Connaiss. Vigne Vin* **9**, 103-115 (1975); see also *Rev. Fr. Oenol.* **15**(63), 33-38 (1976).
138. L. Jakob, *Weinberg Keller* **15**, 550-560 (1968).
139. G. Cerutti and G. Colombo, *Riv. Vitic. Enol.* **25**, 451-458 (1972).
140. K. Mayer and G. Pause, *Schwiez. Z. Obst -Weinbau* **121**, 203-208 (1985).
141. K. Mayer and G. Pause, *Schweiz. Z. Obst -Weinbau* **118**, 723-727 (1982).
142. P. Lehtonen, *Z. Lebensm.-Unters. -Forsch.* **183**, 177-181 (1986).

143. K. Mayer and G. Pause, *Lebensm.-Wiss. Technol.* **17**, 177-179 (1984).
144. K. Yokotsuka, M. Yoshii, T. Aihara, and T. Kushida, *Hakko Kogaku Zasshi* **55**, 510-515 (1977).
145. T. P. Chazova and V. I. Nilov, *Appl. Biochem. Microbiol.* **6**, 14-17 (1970); see also *Prikl. Biokhim. Mikrobiol.* **6**, 18-22 (1970).
146. T. C. Somers and G. Ziemelis, *J. Sci. Food Agric.* **23**, 441-453 (1972).
147. S. Ferenczy, *Bull. O.I.V.* **39**(429), 1313-1336 (1966).
148. R. Cordonnier, *Bull. O.I.V.* **39**(430), 1475-1489 (1966).
149. Z. Kichovsky and N. Mekhoulza, *Bull. O.I.V.* **40**(439), 926-933 (1967).
150. C. Tarantola, *Bull. O.I.V.* **44**(479), 47-54 (1971).
151. T. C. Somers and G. Ziemelis, *Aust. Grapegrower Winemaker* **18**(208), 66-67 (1981).
152. H. Rebelein, *Dtsch. Lebensm.-Rundsch.* **63**, 233-239 (1967).
153. J. Schneyder and G. Vlcek, *Mitt., Rebe Wein, Obstbau Fruechteverwert.* **18A**, 92-97 (1968).
154. C. Junge, *Dtsch. Lebensm.-Rundsch.* **66**, 421-424 (1970).
155. K. Wucherpfenning and K. Otto, *Mitt. Klosterneuburg* **36**, 254-258 (1986).
156. H. Rebelein, *Bull. O.I.V.* **41**(445), 344-354 (1968).
157. C. Reinhard, *Weinwirtsch., Tech.* **8**, 211-214 (1984).
158. G. Beneventi, G. Sala, E. Gavioli, and G. Barbieri, *Riv. Vitic. Enol.* **36**, 153-172 (1983).
159. C. Di Leo, *Riv. Vitic. Enol.* **35**, 118-121 (1982).
160. A. Gonzalez, P. C. Baluja-S., and L. Reija-B, *Quim. Anal.* **30**, 289-294 (1976).
161. C. S. Ough and E. A. Crowell, *Am. J. Enol. Vitic.* **31**, 344-346 (1980).
162. K. D. Millies, J. Meissner, B. Schmidt, and K. Schork, *Mitt. Klosterneuburg* **36**, 264-268 (1986).
163. P. Sudraud and E. Flores, *Connaiss. Vigne Vin* **1**, 57-63 (1969).

Seven
PHENOLIC COMPOUNDS

The phenolic compounds in wine range from relatively simple compounds produced by the grape vine to complex tannin-type substances extracted from the wood of the barrels during aging. Phenols may also be derived from special flavoring or other agents added to certain wine e.g., vermouth. The phenolic compounds are important for several reasons: In addition to furnishing the color for wine, imparting an astringent taste, and possibly causing pungent odors, they are a reservoir for oxygen reduction and a source of browning substrate. The complexity of terminology and the large number of compounds in this group have led to some confusion in analytical work. For example, Drawert (1) found 58 different phenolic compounds present in Tokai-type wine. For help in clarifying terminology and furnishing specific information on wine phenols, see References 2–6.

The anthocyanins are the only significant pigments in red grapes. The major ones found in the *Vitis* species are listed in Table 42. These compounds are found only as the glucosides (at position 3) or as the diglucosides (at positions 3 and 5) in the case of non-*vinifera Vitis* species, or of hybrids between *V. vinifera* and other species. In addition, certain of the glycoside residues may be acylated with *p*-coumaric, caffeic, or acetic acid, most likely on the 6-position of the glucose. Two reports (7, 8) gave data on the composition of anthocyanins in *V. vinifera* red wines. Malvidin glycoside was the main form, with the acylated malvidin glycoside being next in predominance.

Other flavonoid phenols occurring in wine (Table 43) (9–11) are anthocyanogens, catechins, flavonols, and flavanones. The first two groups of compounds are found polymerized and make up the bulk of what are called *condensed tannins* in wine. Neither of these form glucosides in the wine. The flavonols found in wine are usually present as glucosides. Bachmann (12) analyzed 90 species of *Vitaceae* and found the main flavonol-*O*-glycoside to be quercitin-3-glycoside. The corresponding glucuronides were sometimes found for quercetin and kaempferol but not for myricetin. They are also called *anthoxanthin pigments* because of their light yellow color. All the previously named phenolic compounds are mostly found in the grape skins, but the flavanones usually occur mainly in the seeds.

Table 42. Anthocyanins Occurring in Wines

Specific name	R_3	R_4	R_5
Cyanidin	OH	OH	
Peonidin	OCH_3	OH	
Delphinidin	OH	OH	OH
Petunidin	OCH_3	OH	OH
Malvidin	OCH_3	OH	OCH_3

Derivatives	Structure
Monoglucoside	R_1 = glucose (bound at the glucose 1-position)
Diglucoside	R_1 and R_2 = glucose (bound at the glucose 1-position)

Some nonflavonoid phenols occurring in wine are listed in Table 44. The benzaldehyde and cinnamaldehyde derivatives come from wood. The benzoic and cinnamic acid derivatives are produced from the grape's metabolic process. Recently, Singleton et al. (13) found *cis-* and *trans-p-*coumaroyl-(+)-tartaric (coutaric) acid and *trans-*caffeoyl-(+)-tartaric (caftaric) acid. They did not detect the ferulic acid derivatives in white wine. Japanese workers (14) found that coutaric and caftaric acids were in greater amounts in Koshu wines than those from vinifera varieties and that the amounts present contributed to the bitterness. Singleton et al. (15) have shown that caftaric acid is a significant component in the prevention of browning in grapes. The component that results from the enzymic oxidation is (2S)-glutathionylcaftaric acid (16), a colorless compound. Both coutaric and cafteric acids show (17) little change during ripening. Table 45 (18) gives a summary of the ranges and average values for caftaric and coutaric acids in grapes.

A number of less well understood and more complex tannins are found in wines that are kept in barrels or have oak chips or oak chip extracts added to them. These are mainly gallo- or ellagitannins and some lignin derivatives.

The pathways (19, 20) for the formation of the nonflavonoid grape phenols

Table 43. Some Flavonoid Phenols Found in Wine

General name	Basic structure[a]	Specific common name	Additional structure $-OH(-OCH_3)$	Molecular weight[b]	Melting point[c] (°C)
Anthocyanogens (Leucocyanidins) Flavan-3,4-diols		Cyanidinol	3', 4'	306.28	
		Delphinidinol	3', 4', 5'	322.28	
		Malvidinol	4'(3', 5')	350.28	
		Petunidinol	4', 5'(3')	336.28	
Catechins		(+)-Catechin (2,3H-trans)	3', 4'	290.28	174
		(−)-Epicatechin (2,3H-cis)	3', 4'	290.28	236
		(+)-Gallocatechin (2,3H-trans)	3', 4', 5'	306.28	
		(−)-Epigallocatechin (2,3H-cis)	3', 4', 5'	306.28	245

Flavonols	Kaempferol	4'	286.24	277
	Quercetin	3', 4'	302.24	314
	Quercitrin (-3-rhamnoside)	3', 4'	448.37	168
	Myricitrin (-3-rhamnoside)	3', 4', 5'	463.37	198
	Rutin (-3-rhamnoglucoside)	3', 4'	610.51	215[d]
Flavanones	Naringenin	4'	272.25	251
	Naringin (-7-rhamnoglucoside)	4'	580.53	82
	Hesperitin	3', (4')	302.27	251[d]
	Hesperidin (-7-rhamnoglucoside)	3', (4')	610.55	260

From References 9–11.

[a] The numbering system for all these ring structures is the same as that given for the anthocyanogens.
[b] Molecular weight data for flavonols are for the sugar-free molecules except for rutin. For the flavonones the molecular weights are for the sugar-free molecules.
[c] Melting point data for flavonols are for the sugar-free molecules except for rutin. The melting point of naringin with 8H$_2$O is given.
[d] Decomposes.

Table 44. Some Nonflavonoid Phenols Found in Wine

General type	General structure[a]	Common name	Additional structure −OH	Additional structure −OCH₃	Molecular weight	Melting point (°C)
Benzaldehyde	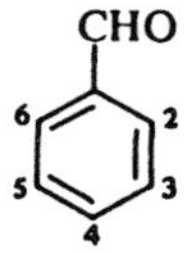 CHO-C₆H₄	Vanillin	4	3	152.16	80
		Syringaldehyde	4	3, 5	182.18	113
Cinnamaldehyde	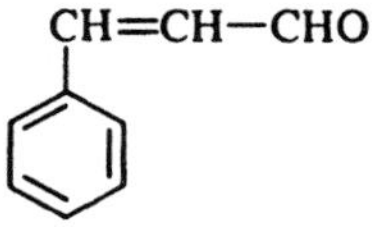 CH=CH−CHO-C₆H₅	Coniferaldehyde	4	3,	178.19	84
		Sinapaldehyde	4	3, 5	208.22	
Benzoic acid	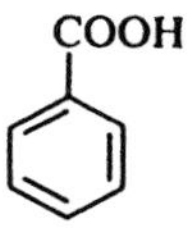 COOH-C₆H₅	Salicylic acid	2		138.12	159
		p-Hydroxybenzoic acid	4		138.12	215
		Vanillic acid	4	3	168.16	214
		Gentesic acid	2, 5		154.12	205
		Syringic acid	4	3, 5	198.19	204
		Gallic acid	3, 4, 5		170.12	253[b]
		Protocatechuic acid	3, 4		154.12	201[b]
Cinnamic acid	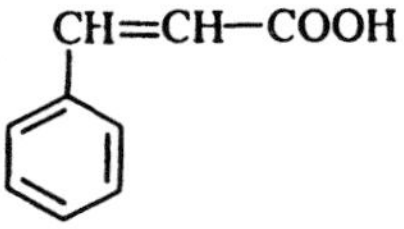 CH=CH−COOH-C₆H₅	p-Coumaric acid	4		164.17	215[b]
		Ferulic acid	4	3	194.19	171
		Caffeic acid	3, 4		180.17	225[b]

[a] The parent compound is represented by the general structure and probably is not present in grapes or wine in significant amounts.
[b] Decomposes.

(shikimic acid through eugenol) and the flavonoid grape phenols are shown in the scheme on p. 201.

Coumarin probably arises from the ring closure of the corresponding cinnamic acid precursor. For example, cinnamic acid forms the glycoside, then undergoes a trans → cis isomerization; the glucose is hydrolyzed off, and the ring closes to form coumarin:

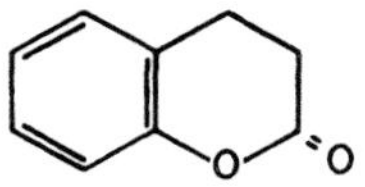

The precursors p-coumaric acid, caffeic acid, and ferulic acid give umbelliferone, esculetin, and scopoletin, respectively. Other grape phenolic compounds

NONFLAVONOID

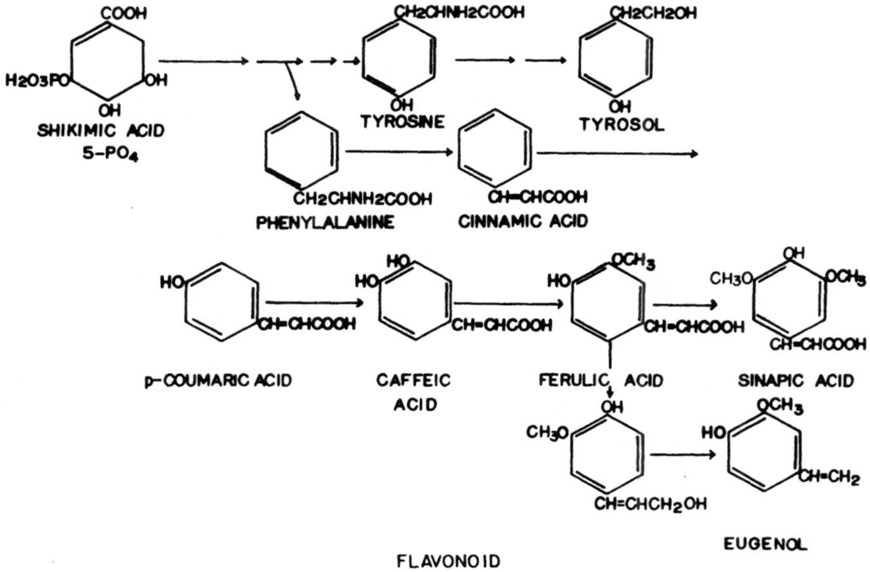
COOH
H2O3PO
OH
OH
SHIKIMIC ACID
5-PO4
CH2CHNH2COOH
OH
TYROSINE
CH2CH2OH
OH
TYROSOL
CH2CHNH2COOH
PHENYLALANINE
CH=CHCOOH
CINNAMIC ACID
HO
CH=CHCOOH
p-COUMARIC ACID
HO
HO
CH=CHCOOH
CAFFEIC
ACID
OCH3
HO
CH=CHCOOH
FERULIC ACID
OH
CH3O
OCH3
CH=CHCOOH
SINAPIC ACID
OH
CH3O
CH=CHCH2OH
OCH3
HO
CH=CH2
EUGENOL

FLAVONOID

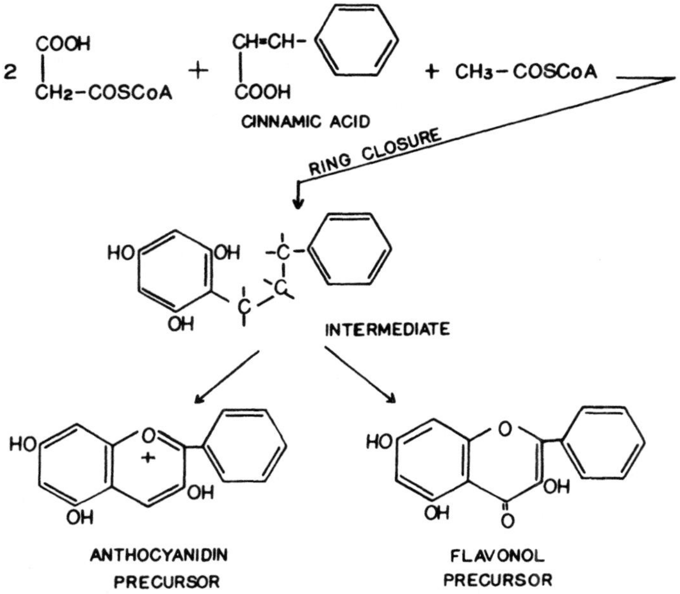
COOH
2
CH2-COSCoA
+
CH=CH-
COOH
CINNAMIC ACID
+ CH3-COSCoA
RING CLOSURE
HO
OH
OH
INTERMEDIATE
HO
O
+
OH
OH
ANTHOCYANIDIN
PRECURSOR
HO
O
OH
OH
O
FLAVONOL
PRECURSOR

Table 45. Caftaric and Coutaric Acid Content of Berries of Various Vitis sp.[a]

Type	Number of samples	Caftaric acid (mg/L)				Coutaric acid (mg/L)			
		Trans		Cis		Trans		Cis	
		Range	Average	Range	Average	Range	Average	Range	Average
Vitis vinifera									
White	23	16–295	127	0–4	2.4	tr–47	14	tr–5	3
Red	23	57–430	163	1–6	2.8	5–39	18	tr–8	3.4
Vitis labruscana	6	104–322	202	1–5	2.7	10–41	23	tr–4	2
Hybrids	7	14–107	57	tr–2	—	2–66	16	tr–10	2.6
Red juice var.	4	87–238	166	2–3	2.3	28–44	33	4–7	5
Vitis sp.[b]	30	0–1337	292	0–16	—	0–326	—	4–16	—

From Reference 18.
[a] tr, trace.
[b] A cross section of 27 different species of *Vitis*.

originate from these pathways, along with individual oxidations, reductions, hydroxylations, methylations, and certain rearrangements.

There is no legal limitation on phenols in wines. The toxicities of some of the phenolic compounds of wine are given in Table 46 (21).

The sensory aspects of phenols were summarized by Singleton and Noble (6). The main components contributing flavor to red wines are the flavonoids, anthocyanins, flavonols, catechins, and anthocyanogenic tannins. The authors estimate these to be present at 5–10 times threshold levels. Volatile phenols individually are present in less than threshold levels but, by additive or synergistic effects, may contribute to the odor of wines. The acid phenols and tyrosol are present at near threshold levels and probably contribute to harshness of wines.

TOTAL PHENOLS

A number of publications give data on the amount of total phenols present in wines. Table 47 summarizes some data on the total phenol content (22–24); the

Table 46. Minimum LD_{50} Doses for Rats (Unless Otherwise Noted)

Substance	Toxicity (mg/kg body weight)
Gallic acid	5000 subcutaneously
Anisole	3700 oral
Eugenol	2680 oral
Vanillin	1580 oral
Methyl salicylate	887 oral
Rutin (mice)	950 intraveneously
Coumarin	680 oral
Quercetin	160 oral

From Reference 21.

Table 47. Total Phenol Concentrations for Some Wine Types

Type	Total phenol (mg/L)	
	Range	Average
White table	40–1300	225
Red table	190–3800	1800
White dessert	100–1100	350
Red dessert	400–3300	900

From References 22–24.

latest trends in wine-making practices, such as quick separation of juice from skins for white wines and the earlier pressing of red wines for quicker maturity suggest that these values are slightly higher than those found at present. White wines average closer to 250 mg/L and red table wines about 1400 mg/L now, at least in California wines (6).

The reported (2) maximum average total phenolic contents for red grapes and for white grapes are 5500 and 4000 mg/kg, respectively. If stems were also completely extracted they would increase phenol content of a wine by 2000 mg/L. The authors give the distributions of the total phenol in the grapes between the skins, pulp, juice, and seeds as 33.3, 0.7, 3.4, and 62.6% for reds and 23.2, 0.9, 4.5, and 71.4% for whites. If the grapes are seedless, the total phenol is reduced accordingly.

Several factors affect the amount of total phenol found in a wine. Skin and seed contact time, ethanol concentration, fermentation temperature, agitation of juice and skins, intensity of pressing, grape variety, and their total phenol content all effect the final total phenol content of the wine.

There are several methods for estimating the total phenol content of wines: precipitation with heavy metals, precipitation by the addition of organic compounds, oxidation under controlled conditions, and formation of colored products with various chemicals. Of these only two have survived: the Neubauer-Löwenthal method (25) and the Folin–Denis method (26). The first was, for a while, the official method of the AOAC but was replaced by the Folin–Denis method. More recently, the Folin–Denis reagent was replaced by the Folin–Ciocalteu reagent.

The main differences in the two reagents are the use of lithium sulfate, longer heating time in the reagent preparation, and the presence of hydrochloric acid in the Folin–Ciocalteu reagent. The Folin–Ciocalteu reagent contains a higher percentage of molybdate in the complex and is more easily reduced. The lithium prevents precipitation problems, which bothered most workers using the Folin–Denis reagent. Lithium substitutes for sodium and its salts are more soluble. The sulfate ion may also improve the solubility of the salt complexes.

The response of the test depends on the phenol present. The number of —OH groups or potentially oxidizable groups control the amount of color formed. The phenol group must be in the phenolate form

for the heteropoly molybdo- and tungstophosphate anions to cause the oxidation. The molybdophosphates are stronger oxidizing agents than the corresponding tungstophosphates. Singleton (3) suggests that the mixture of the two heterpoly molecules is superior because the oxidizing ability of one is stronger, but the other will accept one-electron transfers easily at the required pH. The reduced heteropoly molecules are blue, whereas the unreduced molecules are yellow. These yellow molecules are slowly decomposed at the higher pH required to maintain the phenol in the phenolate form. There is competition for the oxidizing substance: (a) natural decomposition under the alkali condition; (b) reduction by the phenolate ion to the more stable, blue heteropoly molecule. The latter is a one-electron change of some of the molybdenum molecules from a valence of 6+ to 5+. The phenolate molecule is changed to the quinoid structure.

The procedure given below is that of Singleton and Rossi (27).

Procedure

Folin-Ciocalteu Reagent. In a 2-L, round-bottom flask dissolve 100 g of sodium tungstate ($Na_2WO_4 \cdot 2H_2O$) and 25 g of sodium molybdate ($Na_2MoO_4 \cdot 2H_2O$) in 700 mL of water. Add 50 mL of 85% phosphoric acid and 100 mL of concentrated hydrochloric acid. Connect a reflux condenser to the flask, drop in a few glass beads, and reflux for 10 hr. The refluxing does not necessarily have to be continuous. Rinse down the condenser with 50 mL of water and remove. Add 150 g of lithium sulfate monohydrate and a few drops of bromine and boil for 15 min in a hood. The final color should be yellow with no trace of blue or green. Cool the solution, transfer to a 1-L volumetric flask, dilute to volume with water, and filter and store in an amber bottle.

Sodium Carbonate Solution. Dissolve 200 g of anhydrous sodium carbonate in 800 mL of water by boiling. Cool to room temperature, seed with a few crystals of sodium carbonate, filter after 24 hr, and bring to 1 L.

Phenol Stock Solution. In a 100-mL volumetric flask, dissolve 0.500 g of dry gallic acid in 10 mL of ethanol and dilute to volume with water.

For the preparation of the calibration curve, pipet 0-, 1-, 2-, 3-, 5-, and 10-mL aliquots of the phenol stock solution into 100-mL volumetric flasks, and dilute each to volume with water. The phenol concentrations of these solutions (expressed as gallic acid equivalents, GAE) are 0, 50, 100, 150, 250, and 500 mg/L. From each solution, pipet 1 mL into separate 100-mL volumetric flasks; to each flask add 60 mL of water, mix, add 5 mL of the Folin-Ciocalteu reagent, then mix well; after 30 sec and before 8 min, add 15 mL of the 20% sodium carbonate solution, mix, and bring to volume with water. Let the solutions stay for 2 hr at 20°C, then determine the absorbance of each solution at 765 nm against the blank in 10-mm cells and plot absorbance against concentration.

In the case of white wines, pipet a 1-mL aliquot into a 100-mL volumetric flask and proceed as above, starting with the addition of 60 mL of water. In the case of red wines, dilute a 10-mL sample to 100 mL and use a 1-mL aliquot of the diluted sample for the determination.

Sugar (fructose) interferes with the determination (28). For corrections see Table 48. No correction is necessary for dry or nearly dry wines.

Table 48. Correction for Total Phenol (Folin–Ciocalteu) for Sugar Interference

Sugar concentration (g/100 mL)	Divide total phenol result by factor of:
1.0– 2.5	1.03
2.5–10.0	1.06
10.0–20.0	1.10

From Reference 28.

There has been criticism of the method (29). SO_2 was identified as a component causing high readings. It is true that when comparing wines, both white and red, made with and without SO_2, wines with average SO_2 values seem to have GAE values that are 30–40 mg/L higher (30). However, the statement (29) that the measure of total phenolic in white wines by the Folin–Ciocalteu method could be magnified as much as fivefold seems exaggerated.

Orthofer et al. (31) conclude that only free SO_2 can have a significant influence on the Folin–Ciocalteu method for phenols. The percentage error is higher for white wines than for red ones because of the usually higher SO_2 and lower phenol values in the whites. They found that pretreatment with acetaldehyde in excess would control the free SO_2 effects and that the SO_2 adduct effect could be readily calculated and used as a correction factor to subtract from the total phenol value. The SO_2 is already bound to a great part in most wines, and the values do not change greatly; however, it is a safety measure to add the acetaldehyde (500 mg/L). The calculation for the SO_2 adduct correction is:

$$\text{Total } SO_2 \text{ (mg/L)} \times 0.122 = \text{GAE correction}$$

A spectral method has been suggested (32). It has significant problems (33).

The Folin–Ciocalteu method lends itself well to automation. Figure 23 is a schematic representation of one layout described recently (34).

GRAPE PIGMENTS

The grape pigments or anthocyanins are present in red grapes only. The amounts vary with variety. The range reported (35) is 42–5933 mg of pigment monoglucosides per kilogram of fruit. For young wines it is 200–500 mg/L. The separation of these pigments by paper chromatography has proved useful in distinguishing some varieties. Table 49 lists the distribution of the main anthocyanins from Cabernet Sauvignon grapes by high-pressure liquid chromatography (36). In Europe, in particular, the identification of the presence of

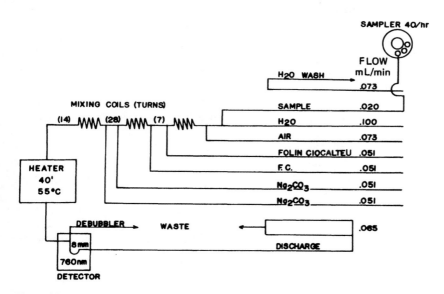

Figure 23. Flow diagram and pump manifold for automated analysis of total phenolic content.

Table 49. Percentage Distribution of Anthocyanins in Six Red Grape Varieties

	Distribution of anthocyanins (%)	
Compound	Range[a]	Mean
Delphinidin-3-monoglucoside	4.3–10.0	7.1
Cyanidin-3-monoglucoside	0.5–2.8	1.5
Petunidin-3-monoglucoside	5.5–11.4	7.1
Peonidin-3-monoglucoside	2.7–13.0	7.75
Malvidin-3-monoglucoside acetate	33.8–62.5	45.3
Delphinidin-3-monoglucoside acetate	tr–25	0.7
Cyanidin-3-monoglucoside acetate	tr–0.3	0.1
Petunidin-3-monoglucoside acetate	tr–2.2	1.0
Peonidin-3-monoglucoside acetate	0.15–2.9	1.2
Malvidin-3-monoglucoside acetate	1.5–20.5	7.9
Delphinidin-3-monoglucoside p-coumarate	0.4–1.7	0.7
Cyanidin-3-monoglucoside p-coumarate	0.2–2.4	1.4
Petunidin-3-monoglucoside p-coumarate	0.2–0.6	0.5
Peopnidin-3-monoglucoside p-coumarate	0.5–2.7	1.5
Malvidin-3-monoglucoside p-coumarate	6.4–22.6	15.4

From References 7 and 36.

[a] tr, trace.

malvidin diglucoside at levels of 5 mg/L or more is considered proof of adulteration of *V. vinifera* wine with that of other *Vitis* species. The pigments tend to condense into forms not yet fully explained. Free anthocyanins quickly disappear as the wines age. The newly formed polymers have red color-absorptive characteristics, not unlike those of the original anthocyanins. The anthocyanins can be decolorized by the addition of sulfurous acid or hydroxide ion. Further treatment with OH^- will give the colored anhydro base, which can range from brown to green, blue, or purple, depending on the pigment makeup (Figure 24).

In some wineries the separation of grape pigment by thin-layer paper chromatography was a routine tool for variety verification. The literature on the technique is extensive, but the major differences involve the extraction procedures and the solvent systems (37, 38).

With the improvement of HPLC equipment, columns, and techniques, anthocyanins can be quantified much more easily and accurately (8, 36). Research (8) comparing the pH shift method to the HPLC found reasonable agreement if the wine was young.

Anthocyanins

The most difficult problem in measuring the anthocyanin content of grape juice or wine by spectrophotometric means is the separation of the condensed or polymeric fractions. In most separation techniques, changes that invalidate the conversion of the direct color measurement into a meaningful quantitative value occur in both fractions. An example of the variable results that can be obtained by recognized methods was demonstrated by Ribéreau-Gayon and Nedeltchev

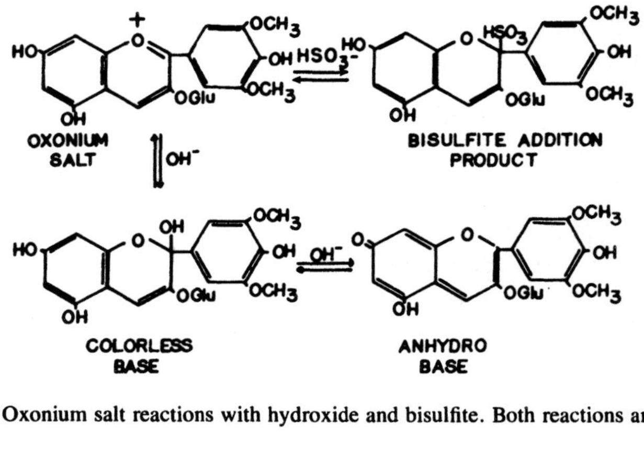

Figure 24. Oxonium salt reactions with hydroxide and bisulfite. Both reactions are reversible.

(39). Two of the methods (pH shift and bisulfite bleaching) have been standardized (40) and are given below. The changes in pH cause corresponding changes in the absorbance of the unpolymerized pigments but cause only slight absorbancy changes in the polymerized pigments. Similarly, sulfur dioxide additions cause changes in absorbance in the unpolymerized pigments but not in the condensed or polymerized pigments.

Procedure

pH Shift. Place 1 mL of wine into each of two test tubes. Add 1 mL of 0.1% hydrochloric acid solution in 95% ethanol to each tube. To one tube add 10 mL of 2% hydrochloric acid and to the second tube add 10 mL of a pH 3.5 buffer, prepared by mixing 303.5 mL of 0.2 M disodium hydrogen phosphate solution and 696.5 mL of 0.1 M citric acid solution. Determine the absorbance at 520 nm, using a water blank and a 10-mm light-path cuvette. Calculate the difference in absorbance at the two values of pH. For the preparation of calibration curves, prepare purified pigment extracts, as described in Reference 38, using larger amounts of grapes. Dissolve the dried powder in a 0.1% solution of hydrochloric acid in 95% ethanol and, by proper dilution with the same HCl–ethanol, prepare standard solutions with pigment concentrations between 37.5 and 375 mg/L. For the purpose of calculation the pigment is considered to be cyanidin.

Prepare a tartaric acid solution in 1:9 ethanol in the following way. Dissolve 0.5 g of tartaric acid in 100 mL of 1:9 ethanol. Divide the solution into two parts with the respective volumes of 33.3 and 66.7 mL; titrate the first with 2 N sodium hydroxide to pH 8.6, then recombine the two solutions. To two 1-mL aliquots of each pigment standard solution, add 1 mL of the tartaric acid solution; place them into separate test tubes and carry out the determination as described above. Plot the results as absorbance against pigment concentration expressed as cyanidin.

Bisulfite Bleach. Mix 1 mL of wine, 1 mL of 0.1% hydrochloric acid solution in 95% ethanol, and 20 mL of 2% hydrochloric acid, then pipet 10-mL aliquots of the mixture into each of two test tubes. Add 4 mL of 15% sodium bisulfite solution to one tube and 4 mL of water to the second. After 20 min, measure the absorbance of each mixture at 520 nm against water and calculate the difference in absorbance of the two determinations.

Prepare a standard curve by preparing pigment standard solutions as in the procedure for pH shift and by analyzing 1-mL aliquots of the standard solutions as above.

The molar absorbance values for grape pigments provide an accurate measure of the individual pigment values. The extinction coefficients for a number of diglucosides and malvidin-3-monoglucoside at pH 1.0 at 520 nm are as follows (41):

Pigment	Molar absorbance value, e
Malvidin-3-glucoside	28,000
Malvidin-3,5-diglucoside	37,700
Peonidin-3,5-diglucoside	36,654
Petunidin-3,5-diglucoside	33,040
Cyanidin-3,5-diglucoside	30,175

To express extracted and separated pigments, we write

$$C(\text{g/L}) = \frac{A}{e}(\text{MW})(F)$$

where C = weight of pigment (g/L)
A = absorbance
e = molar absorbance value
MW = molecular weight of pigment
F = dilution factor

For extracted pigment from *V. vinifera* grapes or wine, the use of the monoglucoside e value and a molecular weight of 529 in the formula is suggested. For hybrids or Concord type, use the malvidin-3,5-diglucoside e value and a molecular weight of 691.

Diglucosides

The detection of malvidin diglucoside is used to determine whether a red wine has been made even partially from a hybrid or a non-*vinifera* variety.

Hadorn et al. (42) found less than 5 mg/L of malvidin diglucoside in a number of commercial wines of *V. vinifera*, whereas blends of *vinifera* wine with only 3% of hybrid wine usually exceeded 5 mg/L. They suggested a limit of 5 mg/L of malvidin diglucoside for wine.

The methods given below are essentially as given by Bieber (43). See also Reference 44. These methods are based on the reaction of Dorier and Verelle (45), which diazotizes the malvidin diglucoside in the presence of alcoholic ammonia. The resulting compound has a specific fluorescence that can be measured.

Procedure

Qualitative Determination. In the case of red wines containing less than 40 mg/L of free sulfur dioxide, place 1 mL of wine into a test tube, add 0.05 mL of 1 N hydrochloric acid and 1 mL of 1% sodium nitrite solution, and mix for 2 min. Add 10 mL of 5% ammonia in 96% ethanol and mix for 5 min. Filter the solution and examine under ultraviolet light at 365 nm. If fluorescence is detectable, the presence of malvidin diglucoside is suspected. In the case of rosé wines with less than 40 mg/L of free sulfur dioxide, place 5 mL of wine into a test tube, add 0.05 mL of 1 N hydrochloric acid and 1 mL of 1% sodium nitrite solution, and mix for 2 min. Add 6 mL of a 5% solution of ammonia in 96% ethanol and treat as above.

In the case of wines containing more than 40 mg/L of free sulfur dioxide, place 10 mL of wine in a test tube, add 1.5 mL of a 10% acetaldehyde solution in 1:1 ethanol-water, mix, and let stand for 15-20 min. In the case of red wines, add 1.15 mL of this solution into the test tube; in the case of rosé wines, use 5.75 mL and proceed as above.

Quantitative Determination. Pipet 1 mL of the red or rosé wine in a test tube; add 0.05 mL of 1 N hydrochloric acid, 1 mL of 1% sodium nitrite solution, and mix for 2 min. Add 10 mL of the ammonia-alcohol solution specified in the preceding procedure, then let stand for 5 min. Filter, place the filtrate into a 1-cm cuvette, excite with a fluorometer at 365 nm, and measure the emitted fluorescence at 490 nm. If the sample reads in excess of 35, dilute the wine with a wine free of malvidin diglucoside and consider the dilution factor in the calculation. Wine free of malvidin diglucoside will give a reading of about 5 or 6. A reading of unity is equivalent to a malvidin diglucoside concentration of 0.426 mg/L. Calculate the concentration in the wine sample as follows:

$$\text{Malvidin diglucoside (mg/L)} = (A - 6)(0.426)(F)$$

where A = fluorescence reading of the sample
F = dilution factor

Standardize the instrument to read 100 with a quinine sulfate solution prepared in the following way. Into a 100-mL volumetric flask, add 10 mg of quinine sulfate and dilute to volume with 0.1 N sulfuric acid. Pipet a 2-mL aliquot of this solution into a 1-L volumetric flask, dilute to volume with 0.1 N sulfuric acid, and use this solution for standardization.

There are a number of other ways to determine the presence of the diglucosides. Many of the paper chromatographic methods separate these slower moving bands (46).

Deibner et al. (47) separated the diglucosides from the other pigments on cellulose TLC using a two-dimensional technique and using 2:100 acetic acid-water and 6:0.1:94 *tert*-amyl alcohol-acetic acid-water as the solvent systems. As little as 5 mg/L of malvidin diglucoside was detected by UV examination of the plates. A one-dimensional TLC method (48) is suitable for determination in both juice and wine. The plate is prepared with Kieselgel G, and a 3:1:3:3 mixture of isoamyl alcohol-*n*-hexane-acetic acid-water is used as the developing solvent. The R_f for malvidin-2,3-diglucoside is 0.14–0.15. The chromatogram is sprayed with a diazotizing reagent, and the presence or absence of the diglucose is determined by the presence of fluorescence. Column chromatography can also be used. The 3,5-diglucosides of the five major anthocyanidins can be separated and isolated by passing the grape skin extracts through an insoluble polyvinylpyrrolidone column (49).

FLAVONOIDS AND OTHER PHENOLS

The possible use of formaldehyde to precipitate the flavonoid phenolic compounds has been proposed for wine (3). The method has been further developed (50). Formaldehyde will react with the 6- or 8-position on the 5,7-dihydroxy flavonoids forming a methylol derivative that will attach to another 6- or 8-position on another flavonoid, and so on (Figure 25). These condensed molecules can be removed by filtration. The residual nonflavonoid phenolic tannin is analyzed by the Folin–Ciocalteu method.

Procedure

Pipet 10 mL of wine (filtered through a 0.45-μm filter) into a 25-mL tube or flask. Mix in 5 mL 1:4 HCl and 5 mL of formaldehyde (13 mL of 37% formaldehyde diluted to 100 mL with distilled H_2O). Displace air in the container with nitrogen, then stopper. Leave at room temperature for 24 hr. Filter through a 0.45-μm membrane filter and determine total phenol content by using the Folin–Ciocalteu method. The difference between the value for the untreated wine and that of the treated one is the flavonoid pigments.

For white wines or young red wines with less flavonoids, the addition of 1 mL (for white wines) and 3 mL (for red wines) of a solution of 10 g/L of phoroglucinol helps with the precipitation (3).

The method has also been applied to measure the amount of nonflavonoid phenols extracted from wood (51). Burkhardt (52) tested three methods for sepa-

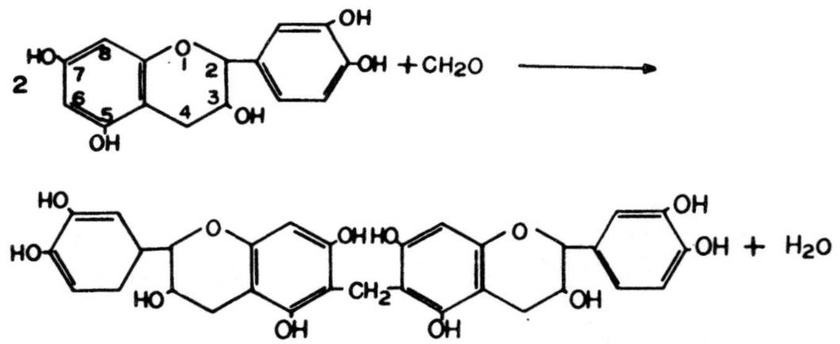

Figure 25. Reaction of flavonoid phenols with formaldehyde.

ration of the nonflavonoid phenols from flavonoid phenols. She found that the cinchonine sulfate method (53), the polyvinylpyrrolidine and trichloroacetic acid method (54), and the methyl cellulose and ammonium sulfate method (55) gave equally good separation of the two fractions.

HYDROLYZABLE TANNIN PHENOLS

Peri and Pompei (53) have suggested precipitation of the hydrolyzable tannin phenols by the addition of cinchonine sulfate according to the method of Brugirard and Tavernier (56). The precipitate would contain (a) the hydrolyzable tannins, polyesters of sugars, or related polyhydric alcohols with gallic or ellagic acid and (b) the condensed tannins, which are polymers and copolymers of catechins and leucoanthocyanins. The hydrolyzable tannins would normally come from wood extract. The rest of the phenols are in the supernatant liquid.

Procedure

pH 7.9 Buffer Solution. In a 500-mL volumetric flask dissolve 1.36 g of potassium dihydrogen phosphate, 8.35 g of disodium hydrogen phosphate dodecahydrate, and 12.5 g of sodium hydrogen carbonate in water, and dilute to volume.

Cinchonine Sulfate Solution. In a 100-mL volumetric flask dissolve 1.5 g of cinchonine base in 2 mL of 1:3 sulfuric acid–water, and dilute to volume with water.

Ethanolic Hydrochloric Acid Solution. Into a 100-mL volumetric flask, add 10 mL of concentrated hydrochloric acid, and dilute with 95% ethanol to volume.

Place 50 mL of white wine into a 100-mL centrifuge tube, then neutralize to pH 7.0 with 1 N sodium hydroxide solution. Add 25 mL of the pH 7.9 buffer solution and 12.5 mL of the cinchonine sulfate solution. Mix and let stand for 20 min at room temperature. Centrifuge to remove precipitate. Remove the clear supernatant into a 200-mL volumetric flask, wash the precipitate twice with 10 mL of a 10% sodium sulfate solution, and combine wash with the supernatant. Acidify the condensed wash and supernatant to pH 3.5 with 1 N hydrochloric acid and bring to volume with water. Dissolve the precipitate in 50 mL of the ethanolic hydrochloric acid solution. Determine the total phenol content of both solutions and report as gallic acid. The solution made from the precipitate will give the amount of hydrolyzable tannin phenols, whereas the determination from the combined supernatant and wash will give the amount of other phenolic compounds.

Catechins

Catechins are primarily associated with skins and seeds. The following data indicate that they make up a large portion of the total phenols. Rebelein (57, 58) gives the following catechin concentrations for a number of wines:

	Range (mg/L)	Average (mg/L)
White wines	20–330	74
Red wines	450–7900	2150

Herrmann and Berger (59) suggest normal levels of 100–200 mg/L for white wine and 1000 mg/L for red. To convert these figures to gallic acid equivalents, multiply the values by the ratio of 3 equivalents of catechin to 2 equivalents of gallic acid, then multiply by the molecular weight ratios of the two compounds ($3/2 \times 170/290$). Under moldy growing conditions with increased temperatures, the catechins in grapes are partially decomposed, as follows (59):

CATECHIN → PHLOROGLUCINOL (R=OH) or 1,3,5 HYDROXYBENZOIC ACID (R=COOH) + PROTOCATECHUIC ACID

Flavonoids with a reactive phloroglucinol moiety can be reacted with vanillin or other similar aldehydes, resulting in a colored compound that can be quantitatively determined colorimetrically (57, 58, 60). Flavonols and flavanones (4-keto compounds) do not react (61). The main reactants are the flavan-3-ols and flavan-3,4-diols. Condensed tannins also react but to a lesser degree (43, 62). Pompei and Peri (63) suggested increasing the ethanol content to improve the stability of the colored compound. This reaction (Figure 26) is used to estimate the flavonoid content of wines.

The following procedure is the Pompei–Peri variation (63) of Rebelein's method (57, 58). Singleton and Esau (2) question whether the method is appropriate for red wines.

Procedure

Standard Catechin Solution. In a 100-mL volumetric flask, dissolve 0.250 g of (+)-catechin in 96% ethanol and dilute to volume with this solvent. In a second 100-mL volumetric flask, dilute 10 mL of the first solution to volume with 96% ethanol. This solution has a catechin concentration of 0.25 mg/mL.

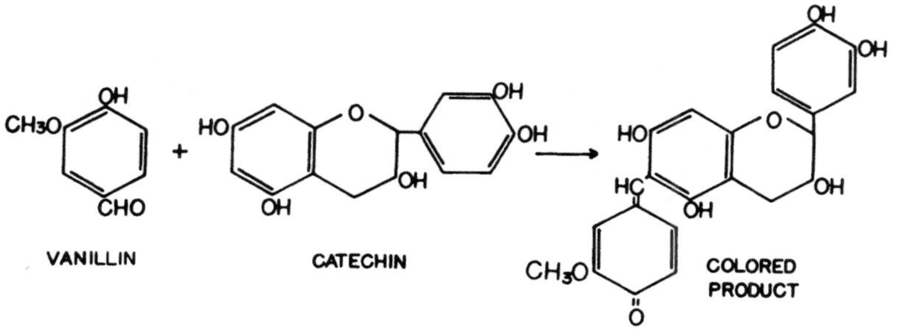

Figure 26. Reaction of vanillin with catechin.

To individual 25-mL volumetric flasks, add 0-, 1-, 3-, 5-, and 10-mL aliquots of the standard catechin solution, 10 mL of 11.5 N hydrochloric acid, and 5 mL of ethanolic vanillin 1% solution, then dilute each to volume with 96% ethanol. After mixing and allowing to stand for 20 min, measure the absorbance at 500 nm using a 1-cm cuvette. After 30 min, the color values change appreciably. Use a water solution blank with white wine; note that for red wine the blank should contain wine but not vanillin. Plot a calibration curve.

In the case of white wines, use a 1-5-mL sample aliquot for the determination, and for red wines carry out a 1:10 dilution and use an aliquot of the diluted sample for the measurement. Pipet the aliquot into a 25-mL volumetric flask and carry out the determination as above.

The importance of catechins in white grapes has been stressed (64), and the methods of determination have been reviewed.

Leucoanthocyanins

Leucoanthocyanins are found in lowest levels in German white grapes at about optimum maturity (65). Before this point, or if the grapes increase further in maturity, higher values were noted. In 554 wine samples the leucoanthocyanin averaged 12.8 mg/L of musts for nine varieties; the range was 2.3–116 mg/L. The intensity and harshness of the press determine the amount for white juices and wines.

Bate-Smith (66) utilized strong mineral acid treatment of flavan-3,4-diols to produce anthocyanins. However, Swain and Hillis (61) reported only semiquantitative results, mainly because insoluble phlobaphene was produced. Ribéreau-Gayon and Stonestreet (67) measured the total leucoanthocyanins directly, determining the increase in anthocyanin color after acid hydrolysis, and compared this to the unheated sample. The difference in absorbance at 550 nm could be referred to prepared anthocyanin standard curves. Nylon or polyvinylpyrrolidone (PVP) are good adsorbents for leucoanthocyanins (68).

The following procedure is that of Pompei et al. (69).

Procedure

Ferrous Sulfate Solution. In a 1-L volumetric flask dissolve 0.150 g of ferrous sulfate in 500 mL of hydrochloric acid and dilute to volume with *n*-butanol.

Regenerated Nylon. Rinse the nylon with 5% sodium carbonate solution and dissolve in hot formic acid. Pour the solution slowly into a large volume of hot water with high-speed stirring (20,000 rpm) to prevent clumping. Allow the nylon to flocculate for 24 hr, then homogenize the mass with a high-speed blender at 40,000 rpm. After drying, it is ready for use.

Dilute a sample of the white wine 1:10 or 1:20 with water, pipet a 20-mL aliquot of the diluted sample into a 20 × 180-mm test tube, and add 1 g of the regenerated nylon. Agitate the tube with a Vortex-type mixer for 30 sec, then filter the solid on a Whatman No. 1 paper or equivalent and rinse with several portions of water. Transfer the solid to a 20 × 180-mm tube and add 20 mL of the ferrous sulfate solution. Heat the tube in a boiling water bath for 50 min, then cool to room temperature in a cold water bath. Transfer the liquid by decantation to a 25-mL volumetric flask, wash the residue with a few milliliters of the ferrous sulfate solution, combine the washing with the solution in the volumetric flask, and dilute to volume with the ferrous sulfate solution. Determine the absorbance of this solution at 550 nm in a 1-cm cuvette. Zero the instrument with a blank containing water instead of wine, which had been carried through the whole procedure. Express the total anthocyanogens as cyanidin chloride:

$$\text{Anthocyanogens, as cyanidin chloride (mg/L)} = \frac{(A)(1.217)(F)(1000)}{(20)}$$

where A = absorbance measured at 550 nm
F = dilution factor

FLAVONOLS AND FLAVANONES

The flavonols and the flavanones are present in lesser amounts. Flavonols are reported (35) to range 40–310 mg/kg (as rutin) in grapes and are present in extremely small amounts in white wine. Red wine can have 20–100 mg/L. Rutin and the two flavanones, hesperitin and naringin, have been reported (70) at 3–9 mg/L in the skins of white grapes. Measurements have been achieved by TLC by adsorption onto PVPP and elution, silylation, and gas chromatographic separation and quantification.

NONFLAVONOID PHENOLS

Nonflavonoid phenols originate both from the grapes and from oak barrels, as well as from mold, bacteria, and yeast activity. The nonflavonoid phenols are located primarily in the juice (50) and were not found to differ greatly for white or red grapes (183–322 mg/L as gallic acid). The ethyl phenols and vinyl phenols originate by bacterial action on coumaric acid phenols (50). Also *Hansenula anomala* and four species of *Saccharomyces* were found to decarboxylate ferulic and *p*-coumaric acids (71). The volatile phenols in Table 50 were determined from a sherry by Tressl et al. (20), and the nonvolatile phenols were found by Valuiko et al. (72); a great many of the recognized wine nonflavonoid phenols are incorporated in this list. Values (73) for three young red wines are also in the table. Estrella et al. (74) reported values for sherries, from various solera stages, to exceed the values in the table 10–1000 fold.

The simpler phenols can be determined by gas chromatography of the trimethylsilyl derivatives (75). Christensen and Caputi (76) described the following procedure for wine analysis.

Procedure

Prepare a series of standard solutions of the nonflavonoid phenols (vanillic acid, etc.) in 1:9 ethanol with concentrations of 2, 5, 7.5, and 10 mg/L each. Similarly for gallic acid, the flavan-3-ols, and quercetin, prepare standards with concentrations of 10, 25, 50, 75, and 100 mg/L. Take a 100-mL aliquot of the standard or wine and extract with 20 mL of ethyl acetate. Pipet a 3-mL aliquot of the extract into a 4-mL reaction vial and evaporate to dryness under nitrogen. Treat the residue with 0.1 mL of *O,N*-bis(trimethylsilyl)acetamide (BSA); allow 20 min for the reaction. Inject an aliquot of the reaction mixture into the gas chromatograph operated under the following conditions:

Column dimensions	10 ft × $\frac{1}{8}$-in. o.d.
Column packing	3% SE-30 on acid-washed and DMCS-treated Chromosorb W, 100/120 mesh
Temperature	
Injection port	325°C
Column	Programmed from 100°C to 136°C at 6°C/min, then at 15°C/min up to 325°C
Detector	325°C

The following HPLC method can be used to separate and determine lower-molecular-weight phenols (73) such as vanillic or syringic acids.

Procedure

Add 20 mL of wine to 20 mL of water and concentrate in a rotary evaporator at 35°C to 20 mL. Adjust pH to 7.0 with 2 *N* NaOH, then extract three times with 20, 10, and 10 mL of ethyl acetate. Centrifuge to separate emulsion, combine the extracts, and evaporate in a rotary evaporator to dryness. Redissolve the neutral phenols with 2 mL of methyl alcohol for injection into HPLC. The aqueous

Table 50. Nonflavinoid Phenols in Several Wines (in mg/L)

Volatile[a]		Nonvolatile[b]	
Phenol	<0.01	Salicylic acid	16.0
o-Cresol	<0.01	Vanillic acid	6.0 (0.3, 0.7, 0.6)[c]
m-Cresol	0.01	Gentisic acid	4.5
p-Cresol	0.01	Syringic acid	5.5
2-Ethyl phenol	0.05	p-Coumaric acid	4.0 (0.4, 0.8, 1.3)
4-Ethyl phenol	0.35	Gallic acid (red wine)	35.0 (5.8, 3.9, 2.7)
4-Vinyl phenol	0.02	Gallic acid (white wine)	2.0
2-Hydroxybenzaldehyde	0.015	Ferulic acid	3.5 (0.1, 0.1, 0.1)
4-Hydroxybenzaldehyde	0.03	Caffeic acid	5.0 (0.9, 0.5, 0.3)
4-Methyl guiacol	<0.01		
4-Ethyl guiacol	0.08		
4-Vinyl guiacol	0.05		
Vanillin	0.04		
Eugenol	0.01		
Eugenol, methyl ester	<0.01		
Ethyl syringinol	0.04		
Tyrosol	4.90 (3.7, 3.1, 1.7)[c]		
Tyrosol acetate	0.37		
Tryptophol	— (0.9, 1.2, 1.1)		
Acetovanillin	0.15		
Syringaldehyde	0.035		
Isopropyl syringinol	<0.01		

[a] From Reference 20.
[b] From Reference 72.
[c] The values in parentheses are those of Reference 73.

phase contains the acid phenols. Reduce the pH of the aqueous phase to 2.0 with 2 N HCl and treat as above (in this paragraph) from extraction through injection.

For chromatography use a 20-μL loop, 30-cm × 4-mm-i.d. Micropack C_{18} column, or equivalent, at 25°C. Detect at 280 nm. Elution solvents are methanol (A) and distilled water adjusted to pH 2.5 with perchloric acid (B). Run a linear gradient from 2% A to 32% A for 35 min, then from 32% A to 98% A for 30 min. Set flow rate at 1.5 mL/min.

Extraction efficiencies for the neutral and acid phenols vary from 32 to 98%. The reproducibility error was calculated to be less than ±25% for most components.

The separation of this group of phenols by paper chromatography has been demonstrated (77). Miskov and Bourzeix (78) give details for thin-layer separation with two solvent systems and quantification with densitometer. Gel filtration (molecular exclusion) chromatography has been used (79) to separate the phenols of the grape into various groups. Electrophoresis has also been used for the separation (80). Ribéreau-Gayon and Glories (81) precipitated the phenolic compounds as lead salts, then fractionated them on Sephadex G-25 gel. The isolated tannin molecules were separated with respect to size by passing through various pore size membrane filters. Drawert et al. (1), first adsorbing the phenols on PVPP, and then eluting with aqueous dimethylformide, concentrating, and making the trimethylsilyl derivatives, could separate and identify by GLC about 45 nonflavonoid phenols and related compounds. In addition, 15 flavonoid phenols were separated and identified.

REFERENCES

1. F. Drawert, V. Lessing, and G. Leupold, *Chem., Mikrobiol., Technol. Lebensm.* **5**, 65–70 (1977).
2. V. L. Singleton and P. Esau, *Phenolic Substances in Grapes and Wine, and Their Significance*, Adv. Food Res., Suppl. 1. Academic Press, New York, 1969.
3. V. L. Singleton, *Adv. Chem. Ser.* **137**, 184–211 (1974).
4. V. L. Singleton, *Adv. Chem. Ser.* **137**, 254–277 (1974).
5. P. Ribéreau-Gayon, *Adv. Chem. Ser.* **137**, 50–87 (1974).
6. V. L. Singleton and A. C. Noble, *ACS Symp. Ser.* **26**, 47–70 (1976).
7. J. P. Roggero, B. Ragonnet, and S. Coen, *Vignes Vins* **327**, 38–42 (1984).
8. E. LaNotte and D. Antonacci, *Riv. Vitic. Enol.* **38**, 367–398 (1985).
9. R. C. Weast, *Handbook of Chemistry and Physics*, 52nd ed. Chem. Rubber Publ. Co., Cleveland, OH, 1971.
10. *Dictionary of Organic Compounds*, 5 vols. Eyre & Spottiswoode, London, 1965.
11. P. G. Stecher, *The Merck Index of Chemicals and Drugs*, 7th ed. Merck, Rahway, NJ, 1960.
12. O. Bachmann, *Vitis* **17**, 234–257 (1978).
13. V. L. Singleton, C. F. Timberlake, and A. G. H. Lea, *J. Sci. Food Agric.* **29**, 403–410 (1978).

14. C. Yoneyama and T. Kushida, *Yamanashi Daigaku Hakko Kenkyusho Kenyu Hokoku* **15**, 9-13 (1980).
15. V. L. Singleton, J. Zaya, E. Trousdale, and M. Salgues, *Vitis* **23**, 113-120 (1984).
16. V. F. Cheynier, E. K. Trousedale, V. L. Singleton, M. L. Salgues, and R. Wylde, *J. Agric. Food Chem.* **34**, 217-221 (1986).
17. V. L. Singleton, J. Zaya, and E. Trousedale, *Vitis* **25**, 107-117 (1986).
18. V. L. Singleton, J. Zaya, and E. Trousedale, *Phytochemistry* **25**, 2127-2133 (1986).
19. J. Bonner and J. E. Varner, *Plant Biochemistry*. Academic Press, New York, 1965.
20. R. Tressl, R. Renner, and M. Apetz, *Z. Lebensm.-Unters. -Forsch.* **162**, 115-122 (1976).
21. V. L. Singleton and F. H. Kratzer, *J. Agric. Food Chem.* **17**, 497-512 (1969).
22. M. A. Joslyn and M. A. Amerine, *Dessert, Appetizer and Related Flavored Wines, The Technology of Their Production*. University of California, Division of Agricultural Sciences, Berkeley, 1964.
23. V. L. Singleton and C. S. Ough, *J. Food Sci.* **27**, 189-196 (1962).
24. M. A. Amerine, *Adv. Food Res.* **5**, 353-510 (1954).
25. J. Löwenthal, *Z. Anal. Chem.* **16**, 33-48 (1877).
26. O. Folin and W. Denis, *J. Biol. Chem.* **12**, 239-243 (1912).
27. V. L. Singleton and J. A. Rossi, Jr., *Am. J. Enol. Vitic.* **16**, 144-158 (1965).
28. E. Donko and E. Phiniotis, *Szolesz. Boraszat.* **1**, 357-366 (1975).
29. T. C. Somers and G. Ziemelis, *J. Sci. Food Agric.* **31**, 600-610 (1980).
30. C. S. Ough, *Am. J. Enol. Vitic.* **36**, 18-22 (1985).
31. R. Orthofer, D. Goldfield, and V. L. Singleton, personal communication (1987).
32. T. C. Somers and G. Ziemelis, *J. Sci. Food Agric* **36**, 1275-1284 (1985).
33. J. Bakker, N. W. Preston, and C. F. Timberlake, *Am. J. Enol. Vitic.* **37**, 121-126 (1986).
34. K. Slinkard and V. L. Singleton, *Am. J. Enol. Vitic.* **28**, 49-55 (1977).
35. M. Bourzeix, *Rev. Fr. Oenol.* **14**(49), 15-25 (1973).
36. L. W. Wulf and C. W. Nagel, *Am. J. Enol. Vitic.* **29**, 42-49 (1978).
37. D. W. Anderson, E. A. Julian, R. E. Kepner, and A. D. Webb, *Phytochemistry* **9**, 1569-1578 (1970).
38. R. A. Fong, R. E. Kepner, and A. D. Webb, *Am. J. Enol. Vitic.* **22**, 150-155 (1971).
39. P. Ribéreau-Gayon and D. Nedeltchev, *Ann. Technol. Agric.* **14**, 321-330 (1965).
40. P. Ribéreau-Gayon and E. Stonestreet, *Bull. Soc. Chim. Fr.* pp. 2649-2652 (1965).
41. G. K. Niketic-Aleksic and G. Hrazdina, *Lebensm.-Wiss. Technol.* **5**, 163-165 (1972).
42. H. Hadorn, K. Zurcher, and V. Ragnarson, *Mitt. Geb. Lebensmittelunters. Hyg.* **58**, 1-30 (1967).
43. H. Bieber, *Dtsch. Lebensm.-Rundsch.* **63**, 44-46 (1967).
44. M. H. B. Biegas de Barros, *Estud., Notas Relat. Vinho Verdes (Porto.)* **(7)**, 59-69.
45. P. Dorier and L. P. Verelle, *Ann. Falsif. Expert. Chim.* **59**, 1-10 (1966).
46. P. Ribéreau-Gayon, *Ind. Aliment. Agric.* **80**, 1079-1084 (1963).
47. L. Deibner, M. Bourzeix, and M. Cabibel-Hugues, *Ann. Technol. Agric.* **13**, 359-378 (1964).
48. H. Schmidt-Hebbel, W. Michelson, L. Masson, and H. Stelzer, *Z. Lebensm.-Unters. -Forsch.* **137**, 169-171 (1968).
49. G. Hrazdina, *J. Agric. Food Chem.* **18**, 243-245 (1970).

50. T. E. Kramling and V. L. Singleton, *Am. J. Enol. Vitic.* **20**, 86-92 (1969).
51. V. L. Singleton, A. R. Sullivan, and C. Kramer, *Am. J. Enol. Vitic.* **22**, 161-166 (1971).
52. R. Burkhardt, *Mitteilungsbl. GDCh-Fachgruppe Lebensmittelchem. Gerichtl. Chem.* **30**, 206-213 (1976).
53. C. Peri and C. Pompei, *Am. J. Enol. Vitic.* **22**, 55-58 (1971).
54. S. Mitjovila, M. Schiavan, and R. Derache, *Ann. Technol. Agric.* **20**, 335-346 (1971).
55. G. Montedoro and P. Fantozzi, *Lebensm.-Wiss. Technol.* **7**, 155-161 (1974).
56. A. Brugirard and J. Tavernier, *Ann. Technol. Agric.* **3**, 311-343 (1952).
57. H. Rebelein, *Dtsch. Lebensm.-Rundsch.* **61**, 182-183 (1965).
58. H. Rebelein, *Dtsch. Lebensm.-Rundsch.* **61**, 239-240 (1965).
59. K. Herrmann and W. G. Berger, *Weinberg Keller* **19**, 559-568 (1972).
60. R. Burkhardt, *Weinberg Keller* **10**, 274-285 (1963).
61. T. Swain and W. E. Hillis, *J. Sci. Food Agric.* **10**, 63-68 (1959).
62. J. L. Goldstein and T. Swain, *Nature (London)* **198**, 587-588 (1963).
63. C. Pompei and C. Peri, *Vitis* **9**, 312-316 (1971).
64. U. Pallotta and C. Canterelli, *Vignevini* **6**(4), 19-46 (1979).
65. K. Wucherpfenning, S. Hadikhadem, R. Hensl, and W. Goergen, *Weinberg Keller* **19**, 449-466 (1972).
66. E. C. Bate-Smith, *Biochem. J.* **58**, 122-125 (1954).
67. P. Ribéreau-Gayon and E. Stonestreet, *Chim. Anal.* **48**, 188-196 (1966).
68. V. L. Singleton, *Wines Vines* **48**(3), 23-26 (1967).
69. C. Pompei, C. Peri, and G. Montedoro, *Ann. Technol. Agric.* **20**, 21-34 (1971).
70. F. Drawert, *Bull. O.I.V.* **43**(467), 19-27 (1970).
71. G. Albagnac, *Ann. Technol. Agric.* **24**, 133-141 (1975).
72. C. G. Valuiko, N. V. Stankova, and N. M. Pavlenko, *Appl. Biokhim. Mikrobiol.* **10**, 832-836 (1976).
73. M. H. Salagoity-Auguste and A. Bertrand, *J. Sci. Food Agric.* **35**, 1241-1247 (1985); see also *Ann. Falsif. Expert. Chim. Toxicol.* **74**, 17-28 (1981).
74. M. I. Estrella, M. T. Hernandez, and A. Olano, *Food Chem.* **20**, 137-152 (1986).
75. T. Furuya, *J. Chromatogr.* **19**, 607-610 (1965).
76. E. N. Christensen and A. Caputi, Jr., *Am. J. Enol. Vitic.* **19**, 238-245 (1968).
77. K. Hennig and R. Burkhardt, *Weinberg Keller* **5**, 542-552, 593-600 (1958); see also *Am. J. Anol. Vitic.* **11**, 64-79 (1960).
78. O. Miskov and M. Bourzeix, *Ind. Aliment. Agric.* **87**, 1515-1518 (1970).
79. T. C. Somers, *J. Sci. Food Agric.* **18**, 193-196 (1967).
80. R. Burkhardt, *Naehrung* **12**, 615-621 (1968).
81. P. Ribéreau-Gayon and Y. Glories, *C. R. Hebd. Seances Acad. Sci. Ser. D.* **273**, 2369-2371 (1971).

Eight
CHEMICAL ADDITIVES

At present the only antiseptic agents that may be used in winery practice are sulfur dioxide (or salts that yield sulfur dioxide in acid solution) and sorbic acid (or sorbates), and their amounts are strictly controlled. Additives that were used in the past as preservatives but that are now prohibited include sodium azide, diethyl pyrocarbonate, 5-nitro furylacrylic acid, cyanide (from the use of potassium ferrocyanide to remove copper and iron), salicylic, benzoic, monochlor- and monobromacetic acids, and the parabens (p-hydroxybenzylcarboxylic acid esters). The inadvertent introduction of excess amounts of pesticides or fungicides into musts from contaminated grapes and subsequently into wines is rare. In some cases most of the pesticide or fungicide is lost during fermentation. However, since the producer is responsible for the purity of the product, enologists must be aware of the possibility of accidental contamination and take appropriate measures.

SULFUR DIOXIDE

The practice of employing sulfur dioxide (SO_2) as an antiseptic agent in wines is of very ancient origin. The gas was originally obtained by burning sulfur. It is difficult to regulate the amount of SO_2 formed by this method or the amount absorbed in the wine. The more common sources of SO_2 nowadays are potassium metabisulfite, cylinders of compressed SO_2, or solutions of the gas in water.

The usual amount of SO_2 added to sound crushed grapes is 0–50 mg/L. This is accomplished by in-stream gas injection at the must pump. In smaller operations the distribution is accomplished by addition of potassium metabisulfite at the crusher or by water solutions added after the crusher or after juice separation. Table 51 (1) gives theoretical SO_2 yields and solubilities. For unsound fruit SO_2, up to 200 mg/L may be required to prevent oxidation, unwanted yeast, and bacterial growth. Certain yeasts are very susceptible to SO_2 inhibition (*Pichia* sp., *Kloeckera*). Others (*Zygosaccharomyces* sp.) are much

Table 51. Common Chemicals with SO_2 Available for Use

Chemical	Formula	Yield (g/100 g of SO_2)	Solubility (g/L)
Sulfur dioxide	SO_2	100	110 (20°C)
Sodium sulfite (anhydrous)	Na_2SO_3	50.82	280 (40°C)
Sodium sulfite heptahydrate	$Na_2SO_3 \cdot 7H_2O$	25.41	240 (25°C)
Sodium metabisulfite	$Na_2S_2O_5$	67.39	540 (20°C)
Potassium metabisulfite	$K_2S_2O_5$	57.60	250 (0°C)

From Reference 1.

more resistant. Wine bacteria are well inhibited by SO_2. As little as 100 mg/L of total SO_2 in some red wines is sufficient to inhibit growth of malolactic bacteria.

During fermentation, some of the SO_2 is oxidized to sulfate. One logical route is oxidation caused by peroxide formed enzymatically:

$$H_2O_2 + SO_3^{2-} \longrightarrow H_2O + SO_4^{2-}$$

Addition of 35 mg/L of SO_2 to most musts was sufficient to completely stop oxygen uptake by polyphenoloxidase enzymes (2).

As a rule SO_2 is maintained in wines between 5 and 40 mg/L as free SO_2. Free SO_2 is the oxide not bound to acetaldehyde or some other aldehyde or organic compound.

When dissolved in water, sulfur dioxide exists as bisulfite (HSO_3^-), sulfite (SO_3^{2-}), or SO_2. Sulfurous acid (H_2SO_3) per se does not exist according to Schroeter (3).

There is an equilibrium between these various inorganic forms of sulfur dioxide, depending on the amounts present, the pH, and the temperature. In general the following reactions can be presumed:

$$SO_2(g) \rightleftharpoons SO_2(aq)$$

$$SO_2(aq) + H_2O \rightleftharpoons H^+ + HSO_3^-$$

$$HSO_3^- + HSO_3^- \rightleftharpoons S_2O_5^{2-} + H_2O$$

$$HSO_3^- \rightleftharpoons SO_3^{2-} + H^+$$

Temperature shifts these equilibria in a complex fashion as sulfur dioxide is lost in the form of gaseous sulfur dioxide at higher temperatures. Increasing the acidity also shifts the equilibrium to the left. The influence of the pH can be approximately calculated from the following equation:

$$\log \frac{[\text{acid salt}]}{[\text{acid}]} = \text{pH} - \text{p}K_1$$

The pK for this equilibrium is about 1.8. Thus for a pH of 2.8, about 10% of the acid is present in the un-ionized state while 90% exists in the form of the HSO_3^- ion. At a pH of 3.8, only 1% would be in the un-ionized state while 99% exists as HSO_3^-. This is also significant in indicating the relatively greater antiseptic ability of sulfur dioxide at low, as compared to high, pH values. Figure 27 indicates the amounts of each form over the pH range of wine. There are insignificant amounts of the SO_3^{2-} form at wine pH values.

In musts and wines the bisulfite ion reacts with acetaldehyde to form acetaldehyde hydroxysulfonate (also called the *bisulfite complex*). It also reacts with aldose sugars (such as glucose), with glyoxylic, pyruvic, α-ketoglutaric, and galacturonic acids, with some unsaturated compounds, and with phenolic compounds (such as caffeic and *p*-coumaric acids). These reactions are reversible. The reaction for carbonyls is:

$$R-\underset{\underset{\text{Carbonyl}}{}}{C\overset{O}{\underset{X}{\diagdown}}} + \underset{\text{Bisulfite}}{HSO_3^-} \longrightarrow \underset{\alpha\text{-Hydroxysulfonate}}{\overset{R}{\underset{X}{\diagup}}C\overset{OH}{\underset{SO_3^-}{\diagdown}}}$$

X = H or R_1
R and R_1 are generally saturated carbon chains

Amines, particularly tertiary amines, also form this complex (4). Moldy fruit has more binding sites for SO_2 (5). Medinger (6) reviewed the bisulfite reaction

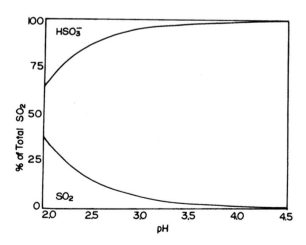

Figure 27. Percentage of bisulfite and free SO_2 of the total SO_2 present at differing pH values.

substrates for SO₂ in wine (Table 52). All the bisulfite reaction products are in equilibrium with the main pool of SO_2. The anthocyanins in red wine also bind SO_2, but rather loosely. This causes much difficulty and disagreement in some published free SO_2 methods. For example, it is extremely difficult to determine free SO_2 in red wine by using the Ripper method.

The antimicrobial activity is much greater (ca. 500-fold) for the SO_2 in the un-ionized form (7). The mechanism for the reaction that causes the death of the organism is not known (8). Some of the reactions that can take place are as follows:

1. The breaking of disulfide bonds.

$$R_1-S-S-R_2 + SO_2 \longrightarrow R_1-SH + R_2S-SO_3^-$$
  Disulfide molecule   Breakdown products

2. The reaction with thiamine.

Thiamine + SO_2 ⟶ Breakdown + Products

Table 52. Bisulfite Reaction Substrates for SO_2 in Wine

Substrate	% Bound[a]	Concentration in wine (mg/L)
Acetaldehyde	99	20–60
Pyruvate	66	10–40
α-Ketoglutaric acid	47	30–80
1-Xylose	27	—
Galacturonic acid	2.5	—
Glucose	0.12	—

From Reference 6.
[a] With 50 mg/L of free SO_2 added at pH 3–4.

3. The deamination of cytosine to form uracil.

$$\text{CYTOSINE} + SO_2 \longrightarrow \text{URACIL}$$

Sulfur dioxide has an unpleasant odor if present in excess. The free SO_2 reacts vigorously with the receptors in the nose, causing sneezing, pain, and choking sensations. The amount of sensory response depends a great deal on the pH of the wine. Wine with a high pH has very little SO_2 in the gaseous form, whereas wine with a low pH has a much higher percentage in the active form. A taste sensation, likewise unpleasant, also accompanies excess SO_2. In levels usually present in consumed wine (60–130 mg/L with less than 15 mg/L of free SO_2), the adverse sensory effects noted are very slight. Wines without normal amounts of SO_2 seem to lack aroma. A normal amount may well be necessary in a white table wine to attain maximum quality.

A joint committee of the Food and Agriculture Organization (FAO) and the World Health Organization (9) recommended human acceptable daily intake (ADI) be no more than 0.35 mg of SO_2 per kilogram of body weight, but it later (10) raised the ADI to 0.7 mg/kg; well below toxic levels for normal individuals. The Select Committee of the U.S. Food and Drug Administration (FDA) on GRAS (acronym for "generally regarded as safe") substances (11) reports that there is no evidence that SO_2 is hazardous to the public at present levels of use. The FDA surveyed oral feeding studies and noted, in general, no significant effects except when the percentage of SO_2 in the diet exceeded 0.1–0.215%. Jaulmes (12) reported the oral LD_{50} for rats using dilute (5%) solutions at 2 g per kilogram of body weight. The aldehyde–SO_2 complex was found to have an oral LD_{50} of 1.5 g/kg for rats. This can be compared to the oral LD_{50} of 1.9 g of acetaldehyde per kilogram of body weight. Shapiro (13) concedes that no proof exists that the ingestion of SO_2 or its salts, at normally accepted levels, causes any toxic or genetic problems in humans.

A reexamination of the GRAS (generally regarded as safe) status of sulfiting agents (14) in 1985 still found no toxicological reasons for further limiting the use of SO_2, but it did note the problems resulting by exposure to SO_2 by certain asthmatics. Bound SO_2 as well as free SO_2 can cause reactions in certain asthmatics at low levels of concentration.

Table 53 gives the amounts of SO_2 legally allowed in musts, fruit juices, and wines. The European Common Market values (effective September 1977)

Table 53. Legal Limits for Free and Total SO_2

Country; wine or juice	Total, maximum (mg/L)	Free, maximum (mg/L)
Algeria	450	
Argentina		
White table (ordinary)	300	
White table (special)	250	
Red table (ordinary)	300	
Red table (fine)	200	
Special wines	300	
Grape concentrates	500	
Australia[a]	300	
Austria		
White and rosé table	200	50
Red table	160	50
Late harvest	300	50
Auselese	350	60
Ice wine	475	
Belgium	250	50
Brazil	350	50
Bulgaria	200	40
Canada	350	70
Chile	200	50
Common Market		
White dry	135,[b] 155	
Red dry	95,[b] 120	
White sweet	275	
Red sweet	225	
Late harvest	300	
Certain sweets	400	
France		
White and rosé dry table	150	
Red dry table	125	
White sweet table	175	
Red sweet table	150	
Other	300	
Germany (West)		
White and rosé dry table	210,[b] 260	
Red dry table	160,[b] 210	
Spatlese	300	
Auslese	350	
Other	400	
Great Britain	450	
Greece	450	100
Hungary		
White dry table	300	60
Red dry table	200	40

Table 53. (*Continued*)

Country; wine or juice	Total, maximum (mg/L)	Free, maximum (mg/L)
Israel	350	
Italy		
White and rosé dry table	200	
Red dry table	160	
Sweet red table	175	
Fruit juice	10	
Japan		
Wines	350[c]	
Natural fruit juices	150[c]	
Netherlands		
Fruit wines	200	20
Sherry	200	15
Liqueur wine	450	45
Sparkling wine	235	
New Zealand		
Dry table	200	
Medium sweet table	300	
Sweet table	400	
Portugal		
Ordinary wines	300	70
Sweet wines	400	100
Romania	450	100
South Africa		
Table wines	200	
Late harvest	300	
Soviet Union		
Ordinary wines	200	20
Special wines	400	40
Fruit juices	125	
Spain		
White sweet wines	260	10
White dry wines	210	5
Red dry wines	160	3
Red sweet wines	210	
Switzerland		
Table wines	250	35
Special wines	400	50
Fruit juices	80	
United States (California)	350	
Yugoslavia	400	40

[a] Queensland allows 400 mg/L total.
[b] For superior quality wines.
[c] As mg/kg.

reflect the trend toward more rational levels. An example of this is the report of Gounelle et al. (15). At that time, 48 French dry white wines averaged 252 mg/L for total SO_2 and 39 mg/L for free SO_2. Ranges were 128–589 and 10.2–124 mg/L for total and free SO_2, respectively. The total SO_2 average exceeded the allowable maximum established only a few years later.

A free SO_2 concentration of 20–30 mg/L at bottling is usually sufficient, but less should be used if possible. To obtain a given amount of free SO_2 by theoretical calculation is difficult because the ratio of free acid to total acid varies from wine to wine. The fixing of SO_2 by aldehydes, sugars, and other compounds makes it difficult to measure the free SO_2 accurately, since it is subject to rapid change. For this reason the present California regulations omit any limitation on free SO_2. Where small amounts of SO_2 are added to the wine over a period of time, there will be a progressive increase in fixed acidity, which may be undesirable in low-alcohol–high-acid wines. For this reason it has been recommended that SO_2 be added to these wines as the bisulfite or metabisulfite rather than as liquid SO_2 or the aqueous solution of SO_2 (16).

Analysis of total SO_2 in one of several countries was summarized (17, 18). Recent data for some countries shows significant decline in SO_2 use.

Free Sulfur Dioxide

The Ripper determination of SO_2 is based on the oxidation–reduction reaction:

$$SO_2 + I_3^- + H_2O \longrightarrow SO_3 + 3I^- + 2H^+$$

In the determination of free (i.e., not bound) SO_2, the wine is first acidified to reduce the oxidation of polyphenols by iodine, then titrated with iodine to a starch end point. The method is not exact, since wines that contain no SO_2 still consume some iodine because of the presence of certain nonsulfite iodine-reducing substances. Red wines, because of their higher content of tannin and coloring matter, may consume an appreciable quantity of iodine. The errors involved in this method were reviewed (4).

Procedure

Pipet 50 mL of the wine sample into a 250-mL Erlenmeyer flask. Add 5 mL of a starch indicator solution and 5 mL of 1:3 sulfuric acid. Rapidly titrate the free sulfurous acid using 0.02 N iodine solution. The end point is the first darkening of the solution to a bluish color, which persists for 1–2 min. The temperature of the solution should not exceed 20°C. For red wines place a strong source of yellow light so that the light is transmitted through the solution from the side, to make the end point more readily distinguishable.

$$\text{Free SO}_2 \text{ (mg/L)} = \frac{(V)(N)(32)(1000)}{v}$$

where V = volume of iodine solution used for the titration (mL)
N = normality of the iodine solution
v = volume of the wine sample (mL)

Burroughs (19) suggests that the anthocyanin–SO_2 complex is largely responsible for the larger titration of free SO_2 values in red wines than in whites. This complex dissociates very readily in strong acid solution, releasing titratable free SO_2. He offers a method for measurement of free SO_2 based on color changes in the wine at 520 nm due to the addition of excess bisulfite; these changes then are compared to the original color.

Schneyder and Vlcek (20) discuss the use of iodate rather than iodine as the titrant. An excess of iodide is added to the wine and is then titrated with iodate:

$$8I^- + IO_3^- + 6H^+ \longrightarrow 3I_3^- + 3H_2O$$

$$I_3^- + SO_2 + H_2O \longrightarrow SO_3 + 3I^- + 2H^+$$

As soon as excess iodine is present, the starch end point is noted. The advantage is the stability of the iodate solution as compared to iodine solutions. The method compares favorably to the Ripper determination.

Procedure

Prepare a potassium iodide–starch solution by dissolving 10 g of KI and 2.5 g of soluble starch with distilled water in a 1-L flask and bring to volume. Dissolve 0.11135 g of potassium iodate with 200 mL of 2 N sulfuric acid and bring to volume in a 1-L flask with distilled water.

Add 10 mL of wine and 2.0 mL of iodide–starch solution to a 250-mL flask and titrate with the 0.0005203 M potassium iodate solution.

$$\text{Free SO}_2 \text{ (mg/L)} = 10 \times V$$

where V = volume of 0.0005203 M KIO_3 used

If any other molarity or volumes are used, then

$$\text{Free SO}_2 \text{(mg/L)} = \frac{(V)(N)(32)(1000)}{v}$$

where V = KIO$_3$ solution used (mL)
N = normality of KIO$_3$ solution (1 M = 6 N)
v = volume of wine used

A method (aeration–oxidation) for free SO$_2$ has been described (21, 22). Acidified wine is subjected at room temperature to a stream of air. The SO$_2$ stripped off is then oxidized in a hydrogen peroxide solution. The H$^+$ formed is titrated. The reaction gives a net gain of 2H$^+$ for each SO$_2$ stripped out by the gas:

$$SO_2 + H_2O_2 \longrightarrow SO_3 + H_2O$$

$$H_2O + SO_3 \longrightarrow 2H^+ + SO_4^{2-}$$

Burroughs and Sparks (23) modified the procedure slightly. They found that very little of the bound SO$_2$ dissociated during the 15-min period of the test. The method was compared to the Ripper and distillation methods (24). Figure 28 is a diagram of the Lieb–Zacherl apparatus. Details of the method are given below.

Procedure

Prepare standardized 0.01 N sodium hydroxide. Restandardize weekly. Keep protected with a soda lime tube. Dilute 1 mL of 30% hydrogen peroxide solution to 100 mL with water. Make up daily and store in refrigerator. Add 0.100 g of methyl red and 0.050 g of methylene blue to a volumetric flask and make to 100 mL with 50 vol% ethanol. Dissolve 280 mL of phosphoric acid (90%) in distilled water in a 1-L flask and bring to volume.

Pipet 10 mL of 0.3% hydrogen peroxide into the 50-mL pear-shaped flask (D in Figure 28). Add 3 drops of indicator to turn the solution purple. Adjust color to olive green by addition of a few drops of 0.01 N sodium hydroxide. Connect vacuum line to adapter C. Pipet 20 mL of wine and 10 mL of 25 vol% phosphoric acid into the round-bottom flask G. The flask should be submerged in an ice bath to prevent dissociation of bound SO$_2$. Start water in the condenser B; this restrains volatile organic acids from distilling over. Turn on vacuum and draw air through the system at 1000–1500 mL/min for 10 min. Remove the flask (D) and titrate the acid formed with the 0.01 N sodium hydroxide back to the olive-green end point. It is advisable to run a blank and, if necessary, make this correction by subtracting milliliters of sodium hydroxide required for the blank from the sample. The calculations are:

$$\text{Free SO}_2 \text{ (mg/L)} = \frac{(N)(V)(32)(1000)}{v}$$

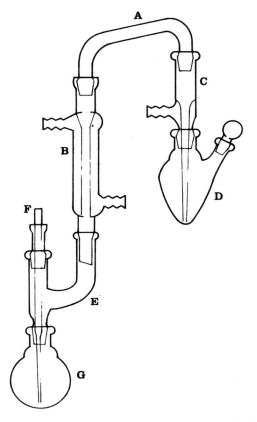

Figure 28. Semimicro apparatus for SO determination by aeration–oxidation: A, connecting adapter, 14/20, K-275050; B, condenser, Liebig 110 mm, 14/20 50 mL, K-282210; C, vacuum adapter, 14/20, K-276750; D, 50-ml pear-shaped flask, 14/20; E, Claissen adapter, 14/20, K-273750; F, Pasteur pipet sealed with 'O' ring; G, 50- or 100-ml round-bottom flask, 14/20 (catalog numbers from Kontes of Berkeley, CA).

where N = normality of NaOH
V = NaOH (mL)
v = wine (mL)

Total Sulfur Dioxide

For the determination of the total SO_2 content, it is necessary to hydrolyze the acetaldehyde-α-hydroxysulfonate and other bisulfite complexes. This can be done by using a strong alkali, then acidifying and titrating directly. This is the Ripper procedure.

Procedure

Pipet 20 mL of the wine sample and 25 mL of 1 N sodium hydroxide solution into a 250-mL Erlenmeyer flask. Mix, stopper the flask, and let stand for 10 min to permit hydrolysis of the acetaldehyde-sulfurous acid. Add 5 mL of a starch indicator solution and 10 mL of 1:3 sulfuric acid. Titrate rapidly with 0.02 N iodine solution to a bluish darkening of the solution, which persists for 30 sec. Calculate the total SO_2 content as given above for free SO_2.

Alternatively, add 50 mL of 0.02 N iodine solution after the addition of 1:3 sulfuric acid and then back-titrate with standardized sodium thiosulfate solution. Add the starch indicator solution only after most of the excess iodine has been used up. The end point in this case is the disappearance of the blue color.

$$\text{Total SO}_2(\text{mg/L}) = \frac{(V_1 N_1 - V_2 N_2)(32)(1000)}{v}$$

where V_1 = volume of iodine solution added (mL)
N_1 = normality of the iodine solution
V_2 = volume of the thiosulfate solution used for back-titration (mL)
N_2 = normality of the thiosulfate solution
v = volume of wine sample (mL)

The total SO_2 content of pink or light-colored wines can be determined by this procedure by placing the Erlenmeyer flask over a strong yellow light; however, the results are not too reliable. Darker wines obscure the end point, and this can cause overtitration by inexperienced analysis.

The disadvantage of this procedure is that iodine reacts with sugars, aldehydes, and other substances, and the amount varies from wine to wine depending on their composition, the temperature, and other conditions.

The aeration–oxidation method (24), given below, takes slightly longer than the method above but gives more accurate and reproducible results.

Procedure

Follow the procedure given on p. 231 for free SO_2, but heat the sample in the flask (Figure 28, G) to a gentle boil. Be sure that sufficient cooling water is flowing in the condenser B to cause good reflux; this prevents the distillation of volatile acids. The calculation is identical to that for free SO_2.

The official method of analysis of the AOAC for total SO_2 is the Monier–Williams method (24a). Recently (25), it was shown that the recovery was greater than 90% for sulfite added to certain foods but less than 85% when added to others. The method is similar to the oxidation–reduction method, with the exception that the extended distillation is not required for wine. If offers the option of the gravimetric determination of barium sulfate after the addition of barium, followed by drying and weighing the precipitate. For wine, this is not practical or necessary.

A total SO_2 analytical method (26) combines SO_2 with $HgCl_2$ to form $(HgCl_2SO_3)^{2-}$; then by acid cleavage a reaction with pararosaniline and formaldehyde occurs to give the colored compound:

$$HO_3S-CH_2-NH-C_6H_4-C(=C_6H_4=NH-CH_2-SO_3H)(C_6H_4-NH-CH_2-SO_3H)$$

The absorbance at 570 nm is measured. The method, which was determined to be as accurate as the official German method, could be automated.

The direct potentiometric method should theoretically be the preferred method. However, at present, despite some efforts, no satisfactory SO_2 electrode is available.

Continuous flow technique for SO_2 has been available for a number of years (27). Grekas and Calokerinos (28) passed N_2 through hot, acidified wine and measured the SO_2 by emission cavity analysis. Sensitivity was 4 mg/L. Another method (29) used a permeable membrane to separate the SO_2 into a stream of p-rosanaline where the color was developed and determined. Repeatability was 1% at 10 mg/L concentration of free SO_2. Similar results were obtained by List et al. (30).

Other methods include the recently developed enzymatic method (31, 32) which is specific for SO_2.

$$SO_3^{2-} + O_2 + H_2O \xrightarrow{\text{Sulfite oxidase}} SO_4^{2-} + H_2O_2$$

$$H_2O_2 + NaOH + H^+ \xrightarrow{\text{NADH peroxidase}} 2H_2O + NAD^+$$

The interference is from the reaction

$$SO_3^{2-} + H_2O_2 \longrightarrow SO_4^{2-} + H_2O$$

In order to successfully use the enzymatic reactions and measure NADH disappearance, the reactions must be carefully buffered in the alkaline region to restrict the chemical oxidation of sulfite to sulfate. The results compare favorably to others (33). Ascorbic acid interferes and must be removed by pretreatment with ascorbic oxidase. The enzymes are available commercially.

Headspace sampling techniques and subsequent gas chromatographic analysis have been proposed (34, 35). HPLC-ion chromatography was used (36) and agreed well with the Monier-William AOAC procedure. However, when tried with grape juice, variations did occur, with recovery ranging from 94 to 118%. Barnett and Davis (37) and Davis et al. (38) calculated the free SO_2 from Henry's Gas Law and the Henderson-Hasselbalch equation. They used headspace analysis, using a Hall detector in the sulfur mode, to quantify the molecular SO_2 in the gas above the solution. See p. 348.

SORBIC ACID AND SORBATES

The use of sorbic acid or sorbates as an antiseptic agent to inhibit the growth of yeast has been approved only recently. Sorbic acid is essentially inactive against bacteria. It must be used in sufficient amounts to inhibit growth fully, or it is ineffective. Alcohol has a synergistic action with sorbic acid in growth inhibition. Under usual bottling conditions, 100–200 mg/L of sorbic acid is required. The fewer yeast cells present, the more effective the treatment. It is relatively ineffective as a preservative of grape juice at allowable levels. Its only practical use in the winery is at bottling. From 50 to 500 mg/L, sorbic acid acts only to inhibit growth but will not inhibit fermentation. From 500 to 1500 mg/L, cells are quickly killed even on short exposure (39). Since enzyme activity is not slowed by sorbic acid it is assumed that activity is against the functions of the cell membrane. Like SO_2, it is most active in the un-ionized form or at lower wine pH values.

$$HOOC-CH=CH-CH=CH-CH_3$$
trans, trans,-2, 4-Hexadienoic acid (sorbic acid)

$$\rightleftharpoons H^+ + {}^-OOC-CH=CH-CH=CH-CH_3$$

It is generally used as the potassium salt, since the acid is only sparingly soluble in wine.

Studies (40) have shown that sorbic acid will react with SO_2 to form

$$CH_3-\underset{\underset{HSO_3}{|}}{CH}-CH_2-CH=CH-COOH$$

by addition to the C-4 double bond. It was suggested this reaction proceeds in a 1:1 ratio. Thus the amount of free SO_2 would be reduced, allowing for bacterial growth.

The sensory threshold of sorbic acid is difficult to establish. The material itself has no odor. In wine an off-odor develops slowly by chemical or bacterial action. In addition, some judges are extremely sensitive to the odor while others are not. Attempts to establish threshold values have led to a number of differing values. The most disagreeable odor produced is that caused by bacterial action. The chemical pathway is as follows (41):

$$CH_3-CH=CH-CH=CH-COOH \xrightarrow[H^+]{C_2H_5OH} CH_3-CH=CH-CH=CH-\overset{\overset{O}{\|}}{C}-OC_2H_5 + H$$
Sorbic acid $\qquad\qquad\qquad\qquad\qquad\qquad$ Ethyl sorbate

$\downarrow$ Lactic acid bacteria

$$CH_3-CH=CH-CH=CH-CH_2OH \xrightarrow{C_2H_5OH} CH_3-CH=CH-CH=CH-CH_2OC_2H_5 + $$
Sorbyl alcohol $\qquad\qquad\qquad\qquad\qquad\qquad$ Ethyl sorbyl ether

$\downarrow$ Rearrangement H^+

$$CH_2=CH-CH=CH-\underset{OH}{CH}-CH_3 \xrightarrow{C_2H_5OH} CH_2=CH-CH=CH-\underset{OC_2H_5}{CH}-CH_3 + H_2O$$
3,5-Hexandien-2-ol $\qquad\qquad\qquad\qquad\qquad\qquad$ 2-Ethoxyhexa-3,5-diene

The alcohols formed do not have objectionable odor. The 2-ethoxyhexa-3,5-diene is the most odorous and probably the main offender formed. Other products such as the ethyl sorbyl ether probably contribute off-odors also.

The amounts of sorbic acid that can be added legally to wine vary from country to country. Table 54 summarizes the legal limits. Much larger amounts are required for beverages or foods that do not contain alcohol or that are less acid.

Sorbic acid was fed to rats at 0, 1.5, and 10% of their diet for 2 years (42). No carcinogenic effects were found. At 10% of the diet feeding level, some liver and kidney enlargement occurred, but no toxic or physiological symptoms were seen. The no-effect level was established at 1.5% of the diet. The oral LD_{50} for rats for potassium sorbate is 4.0–7.16 g/kg of body weight; for the acid, investigators found 7.36–10.5 g/kg (43). This, along with other published results, indicates the relative harmlessness of this additive.

About 1300 German export wines were analyzed for sorbic acid over a 4-yr period (44): 31% contained up to 100 mg/L, and 5% had over that amount.

There are two AOAC methods for sorbic acid analysis for wine. The first is a chemical method involving the oxidation of sorbic acid to a dialdehyde, which

Table 54. Amounts of Sorbic Acid Allowed for Treatment of Wines

Country	Maximum amount (mg/L)	Country	Maximum amount (mg/L)
Algeria	200	Luxemburg	200
Argentina	250	Mexico	No limit
Australia	200	Monaco	200
Belgium	200	Netherlands	
Brazil	200	Grape wine	200
Bulgaria	1000	Fruit wine	300
Canada	500	New Zealand	200
Chile	200	Norway	1000
Colombia	No limit	Peru	No limit
Denmark	2000	Poland	2000
Germany (East)	250	Portugal	200
Germany (West)	200	Romania	200
Great Britain	200	South Africa	200
Finland	1000	Soviet Union	300
France	200	Spain	200
Greece	200	Sweden	2000
Hungary	400	Thailand	200
Israel	300	Uruguay	350
Italy	150	Venezuela	No limit
Japan	200	United States	1000

^a As mg/kg.

is condensed with thiobarbituric acid to give a colored compound that can be determined spectrophotometrically. The reaction is as follows (45, 46):

$$CH_3-CH=CH-CH=CH-COOH \xrightarrow{[O]} CH_3-\overset{O}{\underset{H}{C}}-\overset{O}{\underset{H}{C}}-\overset{O}{\underset{H}{C}}-\overset{O}{\underset{H}{C}}-C\overset{O}{\underset{OH}{}} \xrightarrow{\text{Autooxidation}}$$

$$CH_3-\overset{O}{\underset{H}{C}}-\overset{}{\underset{H}{C}}-CH_2-C\overset{O}{\underset{H}{}} + CO_2 \xrightarrow{[O]} CH_3-C\overset{OH}{\underset{O}{}} + \overset{O}{\underset{H}{C}}-CH_2-C\overset{O}{\underset{H}{}}$$

Malonyl aldehyde

$$\underset{\text{barbituric acid}}{\overset{NH-C\overset{O}{\diagup}}{\underset{NH-C\overset{}{\diagdown}}{C}}\overset{}{\underset{O}{\diagdown}}CH_2} + \overset{O}{\underset{H}{C}}-CH_2-C\overset{O}{\underset{H}{}} \longrightarrow S=\overset{NH-C\overset{O}{\diagup}}{\underset{NH-C\overset{}{\diagdown}}{C}}C=CH-CH_2-CH=\overset{C-NH}{\underset{C-NH}{C}}C=S$$

$$+ 2H_2O$$

The thiobarbituric acid method can be set up as a continuous method using automatic analyzers (47). The sorbic acid is volatilized in a stream of nitrogen and condensed into a water solution, and the appropriate chemicals are added. The only pretreatment involved for wine is dealcoholization and returning to original volume. The procedure is given as follows.

Procedure

Prepare reagent I by adding 750 g of magnesium sulfate heptahydrate to 0.5 N sulfuric acid and bring to 1 L in volumetric flask with 0.5 N sulfuric acid. For reagent II, put 370 mg of $K_2Cr_2O_7$ in a 1-L volumetric flask, dissolve, and bring to volume with 0.2 N sulfuric acid. Prepare reagent III: Add 1250 mg of 2-thiobarbituric acid in 150 mL of water and 25 mL of 1 N sodium hydroxide; then add 28 mL of hydrochloric acid and make up to 250 mL with water in a volumetric flask (solution stable for 2 weeks).

Standard solutions are prepared. Dissolve 500 mg of sorbic acid in 0.01 N sodium hydroxide and make up to 1L with water. Pipet 5, 10, 20, and 40 mL of the stock solution in five separate 100-mL volumetric flasks and make up to volume with distilled water. This gives standards of 25, 50, 100, and 200 mg/L. Store in refrigerator (stable for 2 weeks). Generate a standard curve for comparison to the samples. The wine is dealcoholized, then returned to its original volume prior to placing on the sampling tray.

The automatic procedure diagram appears in Figure 29 with details.

The chemical method for sorbic acid determination in wine given here is that developed by Caputi et al. (48), with follow-up work (49, 50). This is also an official AOAC chemical method. For accurate results, careful attention must be paid to details.

Procedure

Prepare a 2 N sulfuric acid solution (14.2 mL of sulfuric acid to 250 mL with water). Dilute 15 mL of the 2 N sulfuric acid solutions to 100 mL with water to make a 0.3 N sulfuric acid solution. Weight out 0.147 g of potassium dichromate and bring to 100 mL with water. Dissolve 0.25 g thiobarbituric acid in 5 mL of 0.5 N sodium hydroxide in a 50-mL flask, swirl, and add 20 mL of hot water, neutralize with 3 mL of 1 N hydrochloric acid, and bring to volume with water. Make up a stock solution of sorbic acid. Add 0.134 g of potassium sorbate to a 1-L volumetric flask and bring to volume with water. Take 0-, 5-, 10-, and 15-mL portions of stock solution and add each to a 500-mL volumetric flask and dilute to volume with water. Pipet 2 mL of each of the standards into 15-mL test tubes. Add 1.0 mL of the 0.3 N sulfuric acid and 1.0 mL of potassium dichromate, then heat in boiling water bath for exactly 5 min. Immerse tubes in an ice bath,

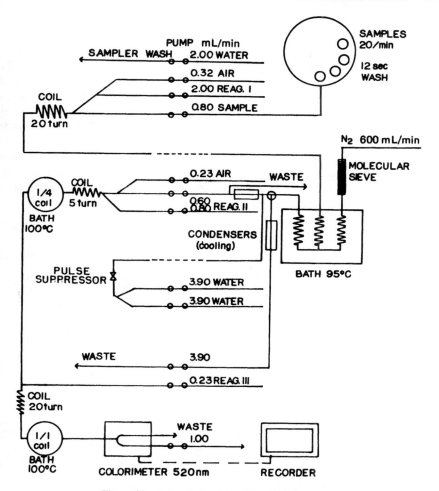

Figure 29. Automated sorbic acid analytical system.

add 2 mL of thiobarbituric acid, and place again in the boiling water bath. Heat for 10 min, then cool in an ice bath and determine absorbance at 532 nm with a 1-cm cell. Make a standard curve (plot 0, 1, 2, 3 mg/L sorbic acid vs. absorption values).

Pipet 2 mL of wine sample into a Cash still. Rinse with 2 or 3 mL of water. Steam distill 190 mL into a 200-mL volumetric flask. Bring to volume with water. Take 2 mL of this distillate and proceed as above. Read sorbic acid from the standard curve and multiply dilution factor by 100 to get concentration in the wine.

The UV method of sorbic acid determination is AOAC approved, as is the standard OIV method. It is simple and direct and involves the measurement of a distilled portion of the wine in a spectrophotometer at 260 nm. Mandrow et al. (51) have measured interfering substances that can arise unless the wine is neutralized and evaporated to dryness prior to determination. These compounds interfere at the 1–10 mg/L levels, and unless the measurements entail critical amounts, the presence of these substances need not be considered for routine analysis. For instances when benzoic acid is also present in the sample, several solutions have been offered to determine both substances by UV absorption (52–54). The method as given in the AOAC for wine is suggested.

Procedure

Prepare a 0.1 N hydrochloric acid solution. Weigh out 1.340 g of potassium sorbate, dissolve, and dilute with water to 1 L as a standard stock solution. Transfer 0, 10, 20, 30, and 40 mL of this stock solution to 100-mL volumetric flasks and dilute with water to volume. Transfer 2 mL of each solution into separate 200-mL volumetric flasks, add 0.5 mL of 0.1 N hydrochloric acid, and bring to volume with water. Measure the absorbance at 260 nm in a 1-cm cell. Plot absorbance against concentration.

Pipet 2 mL of wine sample into a Cash still. Rinse with 2 or 3 mL of water. Steam distill 1900 mL into a 200-mL volumetric flask containing 0.5 mL of 0.1 N hydrochloric acid. Bring to volume with water and read absorbance at 260 nm in a 1-cm cell. Compare to standard curve.

Bertrand and Sarre (55) have reported on a GLC method for sorbic acid. They suggested either 5% diethylene glycol succinate plus 1% phosphoric acid or 1% diethylene glycol adipate plus 1% phosphoric acid, both on Gas Chrome Q, 80/100 mesh, as liquid phases.

Procedure

Dissolve 1 g of undecanoic acid, for an internal standard, in 1 L of 95% ethanol. Load a 4-m × $\frac{1}{4}$-in.-i.d. stainless steel column with 5% diethylene glycol succinate plus 1% phosphoric acid, or 1% diethylene glycol adipate plus 1% phosphoric acid, as above. Preheat the column isothermally at 200°C. Add 20 mL of wine, 2 mL of internal standard, and 1 mL of concentrated sulfuric acid to the extraction funnel. Then add 10 mL of ethyl ether and shake funnel for 5 min. Inject 2 mL of the ether extract into the chromatograph. Run isothermally at 175°C. Make a standard curve for comparison.

Ito et al. (56) developed an extraction–GLC method capable of quantitating both sorbic and benzoic acids to 5 mg/L. They used columns and solvents similar to those used in the previous procedure.

Another version (57) uses a specially packed silica column to separate sorbic (and benzoic) acid(s) from other materials. They are then eluted and gas chromatographed using n-butyl benzoate as an internal standard. Sorbic and benzoic acids can also be determined by HPLC to 1 mg/L levels (58).

SALICYLIC ACID

Very small amounts of salicyclic acid have been reported to occur in some non-*Vitis vinifera* grapes. Its use as an antiseptic agent in wines was rather common in the latter part of the last century. The quantities used for antiseptic purposes varied from 100 to 300 mg/L. Today, state and federal laws generally prohibit its use, and its determination is seldom required. However, sweet table wines of low sulfur dioxide content that have not been bottle-pasteurized may be suspected of containing salycylic, benzoic, sorbic, or monochloracetic acid or combinations of these. One of the primary objections to these acids as food preservatives is the difficulty of their detection, since they are practically odorless. If used by careless persons in excessive quantities, they easily escape sensory detection. One of the advantages of sulfur dioxide is that it is easily detected when present in excessive amounts.

An AOAC method (24a, pp. 389–390) is a very sensitive test. Bertrand and Sarre (59) found that salicylic acid could be extracted and chromatographed, along with sorbic acid and benzoic acid. Their procedure is modified in that the ether extract is concentrated to about 0.5–1.0 mL before injection of 1 μL into a gas–liquid chromatograph. Oleic acid (0.5 g dissolved in 1 L of 60 vol% ethanol) is used as the internal standard. Oven temperature is 170°C.

BENZOIC ACID

The use of benzoic acid in wines has largely disappeared. The poor sensory character of pre-Prohibition period benzoic acid was partially responsible for this, but its rather inferior antiseptic properties were also a factor. Although state and federal laws generally permit its use in small amounts (less than 1000 mg/L), they usually insist that a statement of its presence be placed on the label. The unfavorable publicity associated with such a statement on a wine label has discouraged its use in this country. Attempts to use benzoic acid in connection with sulfur dioxide have generally given disappointing results.

Quantitative AOAC methods for benzoic acids are given (24a, pp. 378–379). Gas chromatographic methods for benzoic acid were presented earlier (p. 241). Revuelta et al. (60) gave a GLC method in which monochloracetic, benzoic, salicyclic, *p*-hydroxybenzoic, and *p*-chlorobenzoic acids are extracted, methylated, and gas chromatographed. The detection ranges are quite low (1–10 mg/L).

Procedure

Place 100 mL of wine in an extraction flask with 100 mL of ethyl ether and 20 mL of 10 vol% sulfuric acid. Mix thoroughly for 5 min, then draw off the ether layer and dry it with anhydrous sodium sulfate. Chromatograph 2 μL on a 1% SE-30 in Chromosorb W-AW (80/100 mesh) 1-m × 3-mm-i.d. column at 155°C. Detect by flame ionization. The monochloracetic acid and ethyl ester *p*-hydroxybenzoic are separable.

Concentrate the rest of the ether extract to dryness. Add 10 mL of methanol and 0.5 mL of concentrated hydrochloric acid. Reflux for 4 hr at 60°C. Extract the methyl esters by a chloroform layer to chromatograph the methyl esters of benzoic acid, salicylic acid, *p*-chlorobenzoic acid, and *p*-hydroxybenzoic acid.

p-HYDROXYBENZOIC ESTERS

The *n*-alkyl esters of 4-hydroxybenzoic acid have been studied extensively (61). These esters increase in effectiveness as bactericides as the chain length of the *n*-alkyl unit increases, but the solubility decreases with increasing chain length. These compounds are *not* permitted for use in wine. They also have the sensory property of causing slight numbness in the mouth. Chan et al. (62) showed that they are effective against yeast and bacteria in wine. If used in combinations of hexyl, heptyl, and octyl esters, they are effective at concentrations below saturation. These esters can be determined on dealcoholized wine by basic hydrolysis of the esters, followed by extraction with ethyl ether and chromatographing.

Procedure

Take a 100-mL sample of wine containing the hexyl, heptyl, or octyl esters and decrease by one-third volume by boiling over low heat. Transfer the residue to a 125-mL Erlenmeyer flask containing 5 mL of 50% w/w sodium hydroxide mix and place in an ice bath for 30 min. Transfer to an extraction funnel and extract three times with 10 mL of ethyl ether. The extracts of the hydrolyzed esters (add 5 g of NaCl if an emulsion forms) are combined, then dried by filtering through anhydrous sodium sulfate. Reduce volume to 2 mL and inject 1–2 μL onto gas chromatograph. Use a glass column 6 ft long, 2 mm i.d., packed with 5% Carbowax 20M on Gas Chrome Q, 60/80 mesh. Program temperature at 2°C/min, from 100°C to 125°C. An alcohol that is normally not present in the wine and that is not a hydrolysis product of the esters is chosen for the internal standard. The internal standard can be prepared by adding 0.5 g of the alcohol to a liter of water (if not soluble, use sufficient methanol to dissolve). Add 1–2 mL of the internal standard solution after the hydrolysis step (depending on the alcohol used). Compare to a standard curve.

HALOGENATED ACIDS

Monochloro-, monoiodo-, and monobromoacetic acids have considerable antiseptic activity toward yeasts and, to lesser extent, toward other microorganisms. They were suggested as a fermentation inhibitor for use by the food industries during the 1930s and were used in a number of cases. Monochloroacetic acid has been used in the United States, and monobromoacetic was used in Europe; in France it was legalized for a time during World War II as a substitute for sulfur dioxide, which was scarce. Since the war both the U.S. Food and Drug Administration and the California State Department of Public Health have forbidden the use of these acids in wines. The OIV (63) has published a qualitative method that detects 1.5–2 mg/L of monohalogen derivatives of acetic acid. It also has a method to differentiate monochloro- from monobromacetic acid.

The OIV method for the determination of monochloroacetic acid is based on its extraction from the wine with ethyl ether, destruction of the organic acid with nitric acid, and determination of the chloride ion using the standard Volhard procedure of precipitation as silver chloride and determination of the excess silver using thiocyanate.

Recently (64, 65) the three halogen compounds of acetic acid and their ethyl esters have been determined by GC/EC to μg/L levels. The suspect wines are extracted with ethyl ether. The extract is put through an adsorption column and then eluted with ether/ethanol (80/20). The eluate is evaporated to a low volume on a rotary evaporation and made acid and treated with ethanol. The ethyl esters formed are separated using GC headspace analysis and determined by EC measurment. A more elaborate scheme (66) was used for making pentafluorobenzylbromide derivatives of the acids. Sensitivity was 20 μg/L for the monochloracetic acid and 1 μg/L for the iodine and bromine analogues. Bandion and Valenta (67) investigated the detection of "Steril," a mixture of ethyl bromoacetate, benzyl alcohol, benzyl bromoacetate, and dibenzyl ether used to preserve wine in Austria. They found that the dibenzyl ether could be detected and was not a normal component of wine. The level of sensitivity of the test was 1 µg/L (20 µg/L is required for effective preservation).

5-NITROFURYLACRYLIC ACID

5-Nitrofurylacrylic acid (NFA) has been suggested as a yeast inhibitor in Europe and the Soviet Union (68–70). It is effective against yeast at a concentration of about 5 mg/L. The compound

$$O_2N-\underset{O}{\underline{}}-CH=CH-COOH$$

also has a strong reducing effect. It has exhibited an LD_{50} of 310 mg/kg for mice and 800 mg/kg for rats. It seems not to have any adverse sensory attributes. It is *not* legal for use in wines in this country.

Masquelier (71) discusses the toxicity of NFA. His conclusions were that it would never be approved by the European Economic Community because of its relatively high toxicity. The required effective levels of 5–10 mg/L exceed the normal safety factors for both Europe and United States. In addition, the presence of nitrate groups always suggests the possibility for nitrosamine formation.

Several reports (72, 73) show NFA to vary between 0 and 8.7 mg/L and average 1–2 mg/L for Czechoslovakian wine with residual sugar.

Junge (74) offers a thin-layer method for NFA detection. A further improvement has been offered (75) through the use of (a) *n*-butanol extraction and (b) phenol removed by PVPP adsorption. An HPLC method (76), which uses direct injection of the wine, allows for rapid screening. It has a 1-mg/L sensitivity using a 10-µL injection.

DIETHYL DICARBONATE

Because of trace amounts of ethyl carbamate formed by the reactions of dicarbonate and ammonia, the FDA banned its use. A rather complete report on this compound and its uses in grape juice and wines has been compiled (77).

DIMETHYL DICARBONATE

Dimethyl dicarbonate is the methyl analogue of diethyl dicarbonate. It is not yet approved for use in wine. It does not produce any known carcinogenic side-products. It reacts in the same manner as diethyl dicarbonate, and is just as effective.

The purity of dimethyl dicarbonate can be determined (78) by reacting it with a known, but excess, amount of morpholine. The excess is then back-titrated with methanolic HCl to pH 4.0. The purity of dimethyl dicarbonate should be in excess of 99%. Breakdown in storage is rapid and autocatalytic. Ough et al. (79) give efficiency data. Stafford and Ough (80) give details of the method of analysis.

Procedure

Add 50 mg of diethyl carbonate to 100 mL of 95% ethanol for an internal standard solution. Extract 100 mL wine with 1 mL internal standard solution added with 20 mL carbon disulfide. Centrifuge lower carbon disulfide layer and inject 5 μL of the clear solution into the gas chromatograph. Prepare a stock solution of ethyl methyl carbonate by adding 50 mg to 100 mL of 95% ethanol. Prepare a standard curve by adding 2, 5, 10, and 15 mL of these stock solutions to a 1-L volumetric flask; bring to volume with 12% v/v ethanol solution. Take 100-mL aliquots and proceed as above.

Prepare a 10-ft × $\frac{1}{8}$-in. stainless steel column packed with 10% Carbowax 20 M on 100/120 mesh Chromosorb G. The temperatures are as follows: injector, 150°C; column, 70°C; and detector, 200°C. A flame ionization detector is used. Inject 5 μL of solution. Make a standard curve plotting the peak height ratios of ethyl methyl carbonate–diethyl carbonate to the concentrations of ethyl methyl carbonate, and compare the peak height ratios for the unknowns to the standard curve.

To calculate the dimethyl dicarbonate added to a wine from the residual ethyl methyl carbonate and the ethanol, the following formulas are suggested:

$$\text{Dimethyl dicarbonate (mg/L)} = \frac{(\text{EMC})(100)}{(0.39)(\text{ET})}$$

where EMC = ethyl methyl carbonate residue (mg/L)
       ET = ethanol in wine (% v/v)

The factor 0.39 represents the average value % of ethyl methyl carbonate produced per % v/v ethanol from addition of 100 mg/L of dimethyl dicarbonate to both red (0.37) and to white (0.41) wines.

Capillary fused silica columns, Carbowax 20M or equivalent, can readily be substituted for packed columns. Methylene chloride can be substituted for carbon disulfide. Ethyl methyl carbonate can be synthesized by reacting methyl chloroformate with ethanol while neutralizing the HCl formed. It can be purified by careful distillation.

SODIUM AZIDE

The use of sodium azide (NaN_3) is forbidden in almost all countries. The compound is extremely active, and even small errors in its use could lead to fatal consequences. It should not be used. Only 40–50 mg of sodium azide per kilogram of body weight is sufficient to kill. As little as 0.01 mg/kg causes a drop in blood pressure.

Cence and Gallerani (81) describe five methods for analyses from wine. The favored ferric chloride method leads to false-positive results occasionally. An HPLC method was developed (82).

Procedure

To 250 mL of wine add 10 mL of 10% NaOH and distill off 100 mL in a rotary evaporator. Add 20 mL of 10% H_2SO_4 and connect immediately to a distillation apparatus. Distill over about 7 mL into a receiving volumetric flask (10 mL) containing 2 mL of 0.5 N NaOH. The distilling tube should dip into the base in the receiving flask. Adjust the volume to 10 mL, mix, and pipet 1 mL into a test tube. Add 2 drops of bromothymol blue and 0.5 mL of acetonitrile. Neutralize to a yellow color by the careful addition of 0.5 N HCl, then add 0.1 mL of 1 M acetate buffer (pH 4.7) and 0.05 mL of 3,5-dinitrobenzoyl chloride solution (2 g in 10 mL acetonitrile). Wait 3 min, then filter and inject 10 μL into HPLC. The derivative has a short-lived stability. Use a 10-cm × 5-mm-i.d. Hypersil 5 ODS or similar column. Use the mobile phase of 40% acetonitrile in water at 1.2 mL/min. Detect at 240 nm. Calculate the results on peak area or height. Relate to external standard.

DIETHYLENE GLYCOL

Diethylene glycol is a practically odorless substance with a sweetish taste. It has been illegally used in some instance in Europe to add body to a wine and improve the mouth feel. It is a very toxic substance and should never be used in food or beverage.

Because of its solubility in both polar and semipolar solvents, it is difficult to extract from wine. A number of GC, HPLC, and other methods have been suggested (83–88); at best, these methods are barely sensitive at the levels desired for detection in wines. Most are good to 10 mg/L, and a couple claim 1 mg/L. A procedure used in our laboratory is given below.

Procedure

Take 750 mL of wine, then vacuum (30 mm Hg) and evaporate (35°C to a volume of 350–400 mL). Add 100 g of $(NH_4)_2SO_4$, dissolve, and extract 3 × 150 mL ethyl acetate. Dry the extract with 10 g $MgSO_4$, then filter and evaporate to 6–7 mL with N_2. Take a separate sample, spiked with 10 mg/L of diethylene glycol, then extract and handle in the same manner.

Use a Supelcomax 10 60-m × 0.25-mm-i.d. (or Carbowax 20M) column at 180°C for the oven, 200°C for the injector, and 250°C for the detector. Use a split ration of 60:1. Put column flow at 1 mL/min. Inject 2 µL. Diethylene glycol will elute at about 10.9 min. Measure the area of both spiked and unspiked samples. Use the method of addition (p. 338) to calculate the amounts.

$$\text{DEG}(\text{mg/L}) = \frac{A_u \,(10 \text{ mg/L})}{A_s - A_u}$$

where A_u = area for the unknown DEG sample
A_s = area for the spiked sample

FERROCYANIDE AND CYANIDE

Excesses of iron and copper represent one of the important problems of white wine production. In Germany, France, and Greece, use of potassium ferrocyanide has been established as a legal method for removing these excess metals; however, in the United States this method is not accepted. The American point of view has been that, whereas potassium ferrocyanide itself is not highly toxic, if it is allowed to remain in contact with the wine for a period of time, it may break down to form the toxic and odorous hydrogen cyanide. This could result only from a highly improbable accident; nevertheless, where use of this substance is legal, it can be employed only under strict laboratory supervision. In this country a proprietary product, Cufex, is employed. Although there is less possibility that excess residues or ferrocyanide will remain in the wine with this procedure, it is possible that some Prussian blue could find its way into the wine because of poor filtration techniques (16). In both cases if the preliminary tests have been carefully followed there should never be any excess ferrocyanide.

The OIV (63) gives qualitative procedures for the detection of Prussian blue and cyanide ion and quantitative procedures for the determination of total and free cyanide (89).

The principle of the Hubach (89) procedure outlined here is the formation of a Prussian blue color on a white filter paper that has been impregnated previously with ferrous sulfate and sodium hydroxide solutions. For a modification see References 90 and 91, where a limit of 1 mg/L was proposed (92).

Procedure

Dissolve 5 g of ferrous sulfate heptahydrate in 50 mL of water containing 1 drop of 1:1 sulfuric acid and filter. Place this clear solution in a porcelain dish and dip, one at a time, sheets of Whatman No. 50 filter paper in it, leaving each in for 5 min. Carefully remove the paper and hang to dry in the air. When dry, immerse the paper in ethanolic sodium hydroxide solution prepared by diluting 10 mL of saturated aqueous sodium hydroxide solution to 100 mL with 95% ethanol. When the paper is thoroughly wet, remove it and again allow to air dry. This paper retains its sensitivity for several weeks if stored in a dark, cool, dry place. It should be light green or light tan.

Cut a piece of this prepared paper to fit the glass flanges of the apparatus shown in Figure 30, dampens the piece over steam, and place between the flanges. Fasten the flanges tightly together with rubber bands. Connect one tube to a source of suction.

Control Test. Prepare a potassium ferrocyanide solution by dissolving 0.1 g of the anhydrous salt in 1 L of water and diluting 10 mL of this solution to 1 L with water. Prepare the solution immediately before use.

Place 10 mL of the potassium ferrocyanide solution in a side-arm test tube. Add 10 mg of cuprous chloride and 1 mL of the 1:1 sulfuric acid solution. Fit the side-arm test tube with a one-hole rubber stopper and fit a glass tube to nearly reach the bottom. Connect the test tube to the other side of the flanges as shown in Figure 30. Place the test tube in a hot water bath (about 90°C) and draw air

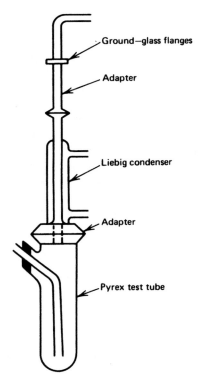

Figure 30. Modified Hubach apparatus for the determination of cyanide and ferrocyanide.

through for 10 min using suction. A rapid stream of bubbles through the liquid is the proper rate of flow. After 10 min remove the paper and dip it in 1:3 hydrochloric acid until it turns white; then dip it in water and air-dry. A blue stain should appear in the center of the paper. Failure to find a blue spot is the result of one of the following causes: the test paper has not been properly prepared; there has been an air leak around the flanges; the steps in the procedure have not been followed; or the chemicals have deteriorated. Repeat until a positive test is obtained.

Clean the apparatus thoroughly to remove all traces of ferrocyanide. To ensure that no trace of ferrocyanide remains in the apparatus, carry out a blank determination using water and following the steps above.

Testing for Cyanide. Place a new prepared paper between the flanges, add 15 mL of the suspected wine sample in the side-arm test tube, assemble the apparatus as above, and carry out the determination as given. If a dark or blue spot develops on the paper, the presence of hydrocyanic acid in the wine has been proved.

Tests for Ferrocyanide. Neutralize 20 mL of the wine sample with 6 N sodium hydroxide solution, then add 1 drop in excess and evaporate on a steam bath to 5 mL. Wash the residue into the side-arm test tube with 10 mL of water, acidify with several drops of 1:1 sulfuric acid, add 10 mg of cuprous chloride, and carry out the determination as given in the control test. A blue stain indicates the presence of ferrocyanide in the wine sample. If the test is faint, repeat it.

A simple, sensitive, and rapid gas chromatographic method is given by Addeo et al. (93). The CN^- is reacted with bromine to form cyanogen bromide, which is then extracted and measured by GLC:

$$CN^- + Br_2 \longrightarrow CNBr + Br^-$$

A concentration of prussic acid as low as 1 μg/L can be detected. Bates and Buick (94) offer a much more complicated method that is less accurate but that can determine as little as 0.2 μg/L. The methods are similar. The method of Addeo et al. (93) is given because of its simplicity and accuracy.

Procedure

Place 10 mL of wine into a 30-mL glass-stoppered centrifuge tube, then add 5 mL of 20% w/w H_3PO_4 and 2 mL of water saturated with bromine. After 5 min add 4 mL of 5% w/w phenol in water and shake. Add 1 mL of chloroform and mix thoroughly for 1 min, then centrifuge for 5 min at 1500 rpm. Inject 5 μL of the chloroform layer into the GLC apparatus. If more than 50 μg/L of cyanide exists in the wine, the extraction need not be made.

Set up the chromatography unit with an electron capture detector. Pack a glass column (1 m × 3-mm i.d.) with Porapak Q, 80/100 mesh. Nitrogen carrier gas flow is 50 mL/min. The injector, oven, and detector temperatures are 120°C,

112°C, and 150°C, respectively. Measure peak heights and compare to a standard curve.

There is also a qualitative screening method for cyanide (95).

A quick photometric method (96) for distilled spirits, sensitive to 0.15 mg/L of HCN, is also an effective method for screening these products.

BETAINE

A compound present in sugar beets and in sugar made from the beet has been suggested as a marker to detect chaptalization of juices by addition of beet sugar to the must (97). Work (98, 99) indicates that the compound

$$CH_3-N\begin{smallmatrix}CH_3\\|\\-O\\CH_2\\|\\CH_3\end{smallmatrix}C=O$$

is present naturally in grapes and hence is not an acceptable choice for detection and chaptalization. The amounts found in 1975 and 1976 French vintages ranged from 0 to 5 mg/L. Determination of betaine by HPLC (100) and GC by derivatization (101) verified that betaine does exist naturally in grapes and varies in amount in both grapes and sucrose from season to season. It is therefore not a satisfactory component to determine adulteration of wines.

β-ASARONE AND OTHER FLAVORS

β-Asarone is a carcinogen and therefore is not an allowable additive for wines; its structural formula is as follows:

$$CH_3O-\underset{CH_3O}{\underset{|}{\bigcirc}}-\underset{OCH_3}{\overset{|}{\bigcirc}}-CH=CH\ CH_3$$

This flavoring agent was commonly added to vermouths as oil of calamus. It was used (and still is in some European countries) as a minor flavorant at or below 5 mg/L and up to 30 mg/L in vermouths (102). Safrole, another flavoring agent, has an LD_{50} of 1950 mg/kg body weight for rats (103). Liddle and de Smedt (104) analyzed 12 vermouths and found thujone (0.01–0.3 mg/L) in 11, β-thujone (0.01–0.08 mg/L) in five, safrole (0.02–0.1 mg/L) in seven, and coumarin (0.2–3.5 mg/L) in two.

Wojtowicz (105) developed a spectrofluorometric method. The β-asarone is steam distilled out of the wine, then extracted into hexane and determined fluorometrically.

Procedure

Place 50 mL of vermouth in a 500-mL distilling flask containing 40 mL of water and 10 mL of saturated sodium chloride solution. Steam distill and collect 150 mL. Transfer the cooled distillate to a 250-mL extraction funnel and extract with 50 mL of hexane. In sequence, wash hexane with 20 mL each of 1 N sodium hydroxide, 1 N hydrochloric acid, and water. Place a portion of the hexane in a 1-cm fluorometric cell and determine emission at 350 nm with excitation at 310 nm. Make a standard curve with a β-asarone concentration range of 0–20 ng/mL in hexane. Compare the unknown to a standard curve. The sensitivity of the method is about 0.1–0.2 mg/L of vermouth.

A gas chromatographic method sensitive to 0.01 mg/L (as well as the determination of α- and β-thujone) has been published (106). Dyer et al. (107), (108) give a routine GLC method suggested for AOAC use. An HPLC method (109) agrees well with GC/MS methods and avoids the occasional high results obtained with GC.

COUMARIN

Coumarin

is found in woodruff, an additive sometimes used to flavor "May" wines. It is a natural product of most plants. It is now necessary to determine coumarin levels in certain beverages to be below specified amounts (110). The toxic oral LD_{50} for rats is 680 mg/kg body weight.

The gas chromatographic method of Dyer and Martin (111) is as follows.

Procedure

Add 10 mL of wine to each of two extraction funnels. Add 1.0 mL of coumarin standard solution (100 mg of coumarin dissolved in ethanol in a 200-mL volumetric flask, brought to volume with ethanol) to one flask, then add 1 mL of ethanol to another flask. Mix both thoroughly and extract with 20 mL of $CHCl_3$. Let the layers separate. Centrifuge sample if necessary to break emulsion. Inject 5 μL into GLC apparatus.

Pack a 6-ft × 2-mm-i.d. glass column with 10% SP-1000 (Carbowax 20M-TPA) on Chromosorb W-AW, 100/120 mesh. The carrier gas is helium, at a flow rate of 30 mL/min. Column temperature is 180°C; injector and flame ionization detector temperatures are 200°C.

Average the results of three extractions and injections. The calculation is:

$$\text{Coumarin}(\text{mg}/\text{L}) = \frac{(\text{PH})(0.5)(1000/100)}{(\text{PS} - \text{PH})}$$

where PH = peak height of sample
PS = peak height of spiked sample
0.5 = coumarin added to 100-mL spiked sample (mg)

ARTIFICIAL WINES

Artificial wines are more widely found in Europe than in the United States. Substitution of grape concentrate for apple or pear concentrate has been some concern, especially at U.S. Customs inspections.

Most reports (112–114) are concerned with detection of artificial coloring agents or with red coloring of white wines. These allow separation by thin-layer or column chromatography of the artificial color compounds. One can differentiate between white wine colored with red pigment extract and real wine based on the ratios of sums of phenol fractions and absorption values of the wine at 420 and 520 nm (113).

By use of certain analyses, pear, apple, and grape concentrates can readily be separated (115). Apple juice is very low in arginine and proline (6 mg/L) and relatively low in sorbitol (4.2 mg/L). Pear juice is high in proline (230 mg/L) and sorbitol (12–20 mg/L) but has only traces of arginine. Grape juice is high in arginine (100–1800 mg/L) and proline (300–2000 mg/L), with only traces of sorbitol.

Rapp et al. (116) found that the presence of 2-hydroxyacetylfuran in wine as a positive indication of invert sugar addition. Caramel use can be detected by the presence of 4-methylimidazole (117) if the caramel was made using NH_4OH. The AOAC (24a, p. 226) gives a method for detection.

The detection of flavor components derived from petroleum sources can be achieved by isotopic analysis. An example of isoamyl alcohol isolated from an artificial wine is given elsewhere (118). General isotopic analysis to verify if the ethanol came from petroleum sources are available (119–121). See also p. 3. Betz (122) gives a simple method of lead precipitation and TLC separation to determine acids present.

POLY(VINYL CHLORIDE)

Poly(vinyl chloride) (PVC) was used widely in some countries for wine containers, but this usage has never been permitted in the United States (123). The monomer of the plastic is a carcinogen:

$$CH_2=CHCl$$
Vinyl chloride

Various types of PVC contain more or less of the monomer. This compound can migrate into the food contained in the PVC package.

Williams and Miles (124) proposed a method, using headspace GLC analysis, capable of detection at 5–10 µg/L. For mass spectral conformation, 50 µg/L is required. Williams (125) confirmed the presence of vinyl chloride in alcoholic beverages stored in PVC at 15 µg/L by detecting bromination of the vinyl chloride to form 1-chloro-1,2-dibromoethane. The 1-chloro-1,2-dibromoethane was analyzed by electron capture detector.

STYRENE

Styrene is a plastic material that is used to construct fiberglass wine containers. The monomer has a very distinctive odor and is very soluble in wine. The threshold in wine is about 0.06 mg/L (126). The oral LD_{50} for rats is 4920 mg/kg body weight (103).

Brun et al. (127) developed a simple extraction procedure that can be used to detect styrene in wine at 10 µg/L: Extract the styrene, then add an internal standard and inject on a gas chromatograph. For wine, the extracting solvent is hexane and the internal standard is cyclohexane. The GLC parameters are as follows: a $\frac{1}{8}$-in. × 7.5-m stainless steel column; packing, 10% Carbowax 1540 on Chromosorb W (60/80), activated at 150°C for 6 hr; carrier gas, nitrogen, 60 mL/min; and temperatures: injector, 180°C; column, 130°C; detector, 150°C.

Procedure

Prepare the internal standard by adding 35 mg of cyclohexane to a 100-mL volumetric flask and bring to volume with hexane. A reference solution is made up with 30 mg of styrene monomer and 35 mg of cyclohexane in 5 mL of hexane.

Extract 200 mL of wine with 10 mL of hexane for 3 or 4 min. Add 1 mL of concentrated acetic acid; shake gently and let the two phases separate. Take 2 mL of the hexane extract and add 2 mL of the internal standard, mix, and inject 5 µL into the GLC apparatus. Inject 1 µL of the reference solution for comparison and calculation, as follows;

$$X = \frac{(A)(H_2)(h_1)(q)}{(B)(H_1)(h_2)(F)}$$

where X = styrene in sample (μg)
       A = styrene in reference solution (μg)
       B = cyclohexane in reference solution (μg)
      H_2 = peak height cyclohexane in reference solution (mm)
      H_1 = peak height styrene in reference solution (mm)
      h_1 = peak height styrene in sample (mm)
      h_2 = peak height cyclohexane in sample (mm)
       q = cyclohexane in sample (μg)
       F = ratio of μL of sample injection to μL of reference injection

POLY(VINYLPYRROLIDONE)

Poly(vinylpolypyrrolidone) (PVPP) is used to adsorb phenol-type materials from juice or wine. The compound itself is insoluble in wine. However, during the production some poly(vinylpyrrolidone) (PVP) is formed, and this compound is soluble to some degree. The PVP can be removed from the wine and measured by the method of Wieczorek and Junge (129). The method involves adsorption of the PVP onto silica gel. The PVP is then reacted with a red dye (vital red). The excess dye is washed off the column. The red-dye–PVP adduct is eluted off the column with dimethylformide, and the red color is determined at 530 nm. Comparisons are made to standard PVP solutions. The sensitivity is about 1 mg/L. PVP and PVPP polymers are shown below:

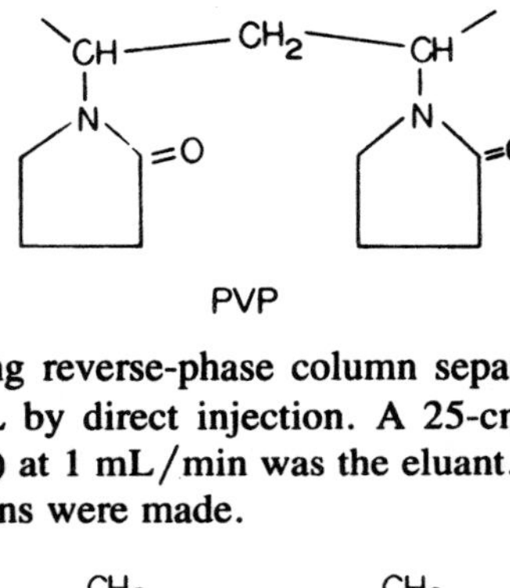

PVP

An HPLC method using reverse-phase column separation (128) could detect styrene at 0.1–2 mg/L by direct injection. A 25-cm column was used, and methanol/H_2O (80:20) at 1 mL/min was the eluant. Detection was by UV at 254 nm; 20-μL injections were made.

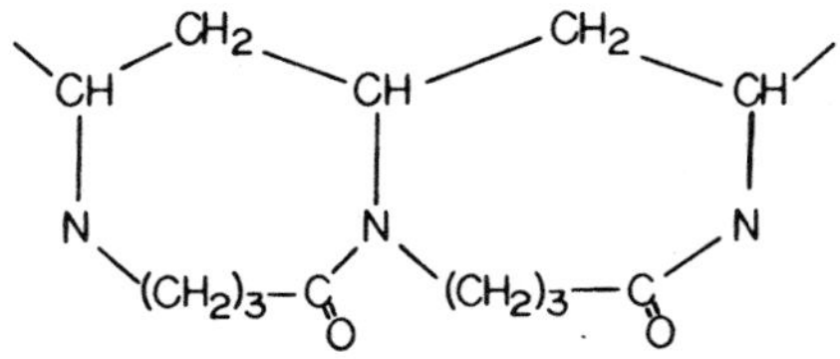

PVPP

AFLATOXINS

Aflatoxins, which are toxic metabolites of fungal origin, are found in many food products. They are invariably of natural origin and occur in foods in their growing state prior to harvest. The formulas of the six common aflatoxins are given below:

[Chemical structures of aflatoxins B_1, B_2, B_{2a}, G_1, G_2, and G_{2a}]

The aflatoxin B_2 has an oral LD_{50} of 85 mg/kg body weight for ducks; for aflatoxin G_2 it is 0.17 mg/kg. For aflatoxin G, the oral LD_{50} for rats is 0.04 mg/kg (103). These compounds are structurally related to other harmful mold products such as patulin.

Lehtonen (130) surveyed 22 European and North African bulk wines for aflatoxin by one-dimensional thin-layer chromatography. None was found in six wines, possible traces were found in 11, and 0.1 µg/L or more was found in five wines. In the wines containing 0.1 µg/L or more aflatoxins, the ranges were as follows: B_1, 0.4–1.0 µg/L; B_2, 0.1–0.2 µg/L; G_1, 0.3–1.0 µg/L; and G_2, 0.1–0.3 µg/L. Imported bottled wines were investigated by one-dimensional thin-layer chromatography and no aflatoxins were found in 11 wines (131). A faint spot suspected of being aflatoxin G_2 was seen in three of the wines. Lemperle et al. (132) were unable to detect aflatoxins in 150 German wines. The sensitivity of their two-dimensional thin-layer chromatography was better than 1 µg of aflatoxin per liter of wine. Takahashi (133) investigated 53 liquid foods and beverages (41 wines) using high-pressure liquid chromatography. The system was sensitive to 0.02 µg/L of the individual aflatoxins. One French sauterne contained 0.3 µg/L of B_{2a}, and one plum wine had 0.05 µg/L of B_{2a}. Otherwise no other aflatoxins were found.

High-pressure liquid chromatography seems to be the best method. For laboratories without HPLC, the thin-layer, two-dimensional method (131) is recommended as an alternative.

ASBESTOS

The use of asbestos as a filtering medium for beverages came to an abrupt halt several years ago when the fibers were found in those beverages that had been filtered through asbestos. Asbestos was an excellent filter medium because of its very high positive charge, due to Mg^{2+} ions associated with the material. Most of the wine-suspended materials have a negative charge. Wale (134) points out that asbestos fibrils can even penetrate a 0.22-µm membrane; thus the only solution was replacement by other material.

The analysis of asbestos requires an electron microscope. Several articles give the technique involved (134–137).

FUNGICIDES AND PESTICIDE RESIDUES

A number of organophosphorus and chlorinated hydrocarbon pesticides have been used on grapes: Dieldrin, Endrin, Parathion, Malathion, Captan, Chlordane, p,p'-DDT, and others. Some of these are now prohibited, or very low and specific tolerances have been established. Enologists, therefore, have the problem of (a) detecting pesticide residues in grapes and in wines and (b) identifying and quantitatively determining any such substances that are present.

The overall problem of pesticide and herbicide residues goes beyond the use on just one food product. Zweig (138) shows the actual and potential problems caused by their uncontrolled use. The World Health Organization (WHO) (139) report lists the acceptable daily intakes for some 34 chemicals. Table 55 gives those that the WHO deems to be acceptable for grapes. Note, however, that the list is subject to change as more toxicological and physiological data are gathered.

A report (140) indicated the relative yeast-growth inhibitive power of diathianon (Delan®), dichlorfluanid (Euparen®), captan (Orthocide 50®), and folpet (Phalton 50®). Each one in this group reacts with thiol groups in the yeast and can delay or prevent yeast growth if present in sufficient amounts. This can cause delayed fermentations. Gnaegi (141) reviewed the problem and suggested

Table 55. WHO Tolerances for Certain Pesticides and Fungicides for Grapes

Compound	Daily maximum acceptable amount (mg/kg of body weight)	Grape residue limit (mg/kg fruit)
Azinphos-methyl®	0.0025	1
Benomyl®	a	10[b]
Bromopropylate®	0.008	5
Captan®	0.01	30 (est)
Carbendozim®	a	10[b]
Demeton-s-methyl sulfone®	0.005	1
Omethoate®	0.005	2
Thiometon®	0.005	0.5
Thiophanate-methyl®	0.08	10
Vamidothion®	a	2[b]

From Reference 139.
[a]More studies needed.
[b]Tentative.

that growers have an obligation to act responsibly in their use of these compounds. By the use GC/EC methods, 6 of 24 grape samples from treated vineyards (with captan and folpet) contained detectable levels of captan and 12 contained detectable levels of folpet. All were below acceptable levels (142). Residues are stable in frozen samples for up to 2 months with less than 10% loss.

Bayleton® has been found (143) to have no inhibitory effects on wine yeasts at 200 mg/L (triadimefon is the active principal). Ronilan® has been reported (144) to remain in the wine for up to 5 months after fermentation but has no ill effects on fermentation or wine quality. A partial list of some fungicides used in the United States on grapes is given is Table 56 (145). In general the LD_{50} for fungicides are quite high and are not a toxic hazard to humans if the regulations are obeyed.

Most of the insecticides are much more toxic. For example, the oral LD_{50} doses of lead arsenate, DDT®, Lindane®, Parathion®, Phosdrin®, and Sevin® (Carbryl®) for rats are 10–50, 113, 88, 13, 3.7–12, and 850 mg/kg, respectively. In general, the herbicides are less toxic. Paraquat® and Piquat® are two of the more toxic, with oral LD_{50} doses of 150 and 231 mg/kg, respectively, for rats. The respective LD_{50} doses of Dalapon®, Duiren®, Linuron®, and Trifluratin® are 9330, 3400, 4000, and 1000 mg/kg of body weight for rats.

Today, analysis for fungicides, herbicides, or pesticides is carried out almost

Table 56. Approved and Experimental Fungicides Used in the United States, Along with Their LD_{50} Values

Fungicide	Use	LD_{50}
Carbamates ($-NH-\underset{\|\|}{\overset{S}{C}}-S-$)		
Maneb®	Foliar protectant	6,750
Ferbam®	Foliar protectant	1,000
Zineb®	Foliar protectant	5,200
Mancozeb®	Foliar protectant	7,500
Other Organics		
Captan® (Cl, N)	Eradicant	9,000
Folpet® (Cl, N)	Protectant	10,000
Dinocap® (NO_2)	Protectant	980
Ronilan® (Vinclozolin®) (Cl, N)	Protectant and contact	10,000
Botran® (Dicloran®) (Cl, N, NO_2)	Protectant and contact	1,500
Roval® (Cl, N)	Protectant and contact	3,500
Rubigan® (Ferarinol®) (Cl, N)	Systemic	2,500
Benomyl (Benolate®) (N)	Systemic	9,590
Bayleton (Triadimefon®) (Cl, N)	Systemic	1,500

From Reference 145.

exclusively by gas chromagraphy. Depending on the type of the chemical, detectors sensitive to halogen (electron capture or microcoulometric detector) or phosphorous (thermionic flame photometric detector) must be used. There has been some attempt to simplify procedures. Thin-layer chromatography has been tested and shown to be suitable for many organochloride, organophosphate, and carbamate compounds when compared to GC methods (146).

The unequivocal identification and quantitative determination of pesticides can be a complicated and lengthy procedure. The procedure given below serves as an example for sample treatment. The gas chromatographic analysis of the concentrate depends on the pesticides present. Methods for determination of pesticides, insecticides, ascaricides, fungicides, herbicides, nematocides, and rodenticides follows similar extraction and cleanup procedures:

Procedure

Evaluate the purity of any solvent used for sample treatment by concentrating 10 mL to 0.1 mL in a Kontes® concentrator tube and injecting a 2-μL aliquot into the gas chromatograph under the conditions used for the gas chromatographic analysis of the sample. The recorder response for the solvent should not be greater than 1 mm for the 2–60-min interval after injection.

In the case of grapes, place 100 g in a Waring blender, add 50 mL of water,

50 mL of ethyl acetate, 10 g of Celite, and blend 2 min at high speed. Filter with suction through sharkskin filter paper in a 12-cm Büchner funnel. In the case of wine or concentrate, proceed directly.

Into a 250-mL separatory funnel, add 100 mL of the grape extract, wine, or wine concentrate, 50 mL of water, and 50 mL of ethyl acetate. Shake for 2 min, venting frequently, then let it set for 10 min. If the layers do not separate, add 25 mL of saturated sodium chloride solution, swirl gently, and allow the layers to separate. Drain off the bottom layer into another 250-mL separatory funnel and repeat the extraction with 50 mL of ethyl acetate. If emulsion forms, break it with saturated sodium chloride solution as above.

Combine the ethyl acetate extracts and pass them through a small filter containing about 30 g of anhydrous sodium sulfate. Rinse with 20 mL of ethyl acetate.

Evaporate the ethyl acetate extracts nearly to dryness, transfer the residue to a graduated conical tube, and continue evaporation just to dryness. Dilute to volume (usually 1 mL) with hexane. The sample is now ready for the gas chromatographic analysis. Additional cleanup of the sample can be accomplished using the Mills procedure of elution through Florisil®.

REFERENCES

1. L. F. Green, *Food Chem.* **1**, 103–124 (1976).
2. B. White and C. S. Ough, *Am. J. Enol. Vitic.* **24**, 148–152 (1973).
3. L. C. Schroeter, *Sulfur Dioxide in Foods, Beverages, and Pharmaceuticals.* Pergamon Press, New York, 1966.
4. M. A. Joslyn and J. B. S. Braverman, *Adv. Food Res.* **5**, 97–160 (1954).
5. E. Peynaud and J. C. Sapis, *Proc. Int. Enol. Symp., 4th, 1975*, pp. 88–93 (1975).
6. F. Medinger, *Weinwirtschaft* **112**, 332–334, 336–338 (1976).
7. H. J. Rehm and H. Wittmann, *Z. Lebensm.-Unters. -Forsch.* **118**, 413–425 (1962).
8. S. M. Hammond and J. G. Carr, *Soc. Appl. Bacteriol. Symp. Ser.* **5**, 89–110 (1976).
9. *Toxicological Evaluation of Some Antimicrobials, Antioxidants, Emulsifiers, Stabilizers, Flour-Treated Agents, Acids and Bases*, FAO Nutr. Meet. Rep. Ser. 40, A, B, and C WHO/Food Additives, p. 67. Food and Agriculture Organization of the United Nations, Rome, 1967.
10. *Toxicological Evaluation of Certain Food Additives with a Review of General Principles and of Specifications*, FAO/WHO 17th Rep. Food and Agriculture Organization of the United Nations, Rome, 1974.
11. *Evaluation of the Health Aspects of Sulfiting Agents as Food Ingredients*, SCOGS-15. U.S. Dep. of Health, Education and Welfare, Washington, DC, 1976.
12. P. Jaulmes, *Bull. O.I.V.* **43**(478), 1320–1333 (1970).
13. R. Shapiro, *Mutat. Res.* **39**, 149–176 (1977).
14. *The Reexamination of the GRAS Status of Sulfiting Agents*, FDA 223-83-2020, Task Order No. 5. Life of Sci. Res. Off., Fed. Am. Soc. Exp. Biol., Washington, DC, 1985.
15. H. Gounelle, C. Boundene, A. Szalsvary, and M. Fauchet, *Bull. Inst. Natl. Sante Rech. Med.* **24**, 1431–1441 (1969).

16. M. A. Amerine, H. W. Berg, R. E. Kunkee, C. S. Ough, V. L. Singleton, and A. D. Webb, *The Technology of Wine Making*, 4th ed. Avi Publishing Co., Westport, CT, 1980.
17. C. S. Ough, in *Antimicrobials in Foods*, A. L. Branen and P. M. Davidson, Eds. Dekker, New York, 1983, pp. 177-203.
18. C. S. Ough, *J. Assoc. Off. Anal. Chem.* **69**, 5-7 (1985).
19. L. F. Burroughs, *Am. J. Enol. Vitic.* **26**, 25-29 (1975).
20. J. Schneyder and G. Vlcek, *Mitt. Hoehere Bundeslehr- Versuchsanst. Wein- Obstbau, Klosterneuburg* **27**, 87-88 (1977).
21. E. Kielhofer and H. Aumann, *Mitt., Rebe Wein, Obstbau Fruechteverwert.* **7A**, 287-296 (1957).
22. F. Paul, *Mitt., Rebe Wein, Obstbau Fruechteverwert.* **8A**, 21-27 (1958).
23. L. F. Burroughs and A. H. Sparks, *Analyst (London)* **89**, 55-60 (1964).
24. J. W. Beuchsenstein and C. S. Ough, *Am. J. Enol. Vitic.* **29**, 161-164 (1978).
24a. *Official Methods of Analysis*, 14th ed. Association of Official Analytical Chemists, Washington, DC, 1984, pp. 391-392.
25. C. R. Warner, D. H. Daniels, F. L. Joe, Jr., and T. Fazio, *J. Assoc. Off. Anal. Chem.* **69**, 3-5 (1986).
26. E. Heike and A. Kreisel, *Z. Anal. Chem.* **285**, 39-42 (1977).
27. J. Ruzicka and E. H. Hansen, *Flow Injection Analysis*, Chem. Anal. Vol. 62. Wiley, New York, 1981.
28. N. Grekas and A. C. Calokerinos, *Analyst (London)* **110**, 335-339 (1985).
29. J. Moller and B. Winter, *Fresenius' Z. Anal. Chem.* **320**, 451-456 (1985).
30. D. List, I. Ruwish, and P. Longhans, *Fluess. Obst* **53**(1), 10-14 (1986).
31. H. O. Beutler and I. Schutte, *Dtsch. Lebensm.-Rundsch.* **79**, 323-330 (1983).
32. H. O. Beutler, *Food Chem.* **15**, 157-164 (1984).
33. G. Schwedt and A. Baurle, *Fresenius' Z. Anal. Chem.* **332**, 350-353 (1985).
34. T. Hamano, Y. Misukashi, Y. Matsuki, M. Ikuzawa, K. Fujita, T. Izumi, T. Adachi, H. Nonogi, T. Fuke, H. Suzuki, M. Toyoda, Y. Ito, and M. Iwaida, *Z. Lebensm.-Unters. -Forsch.* **168**, 195-199 (1979).
35. T. L. C. deSouza, *J. Chromatogr. Sci.* **22**, 470-472 (1984).
36. L. F. Moore, R. P. Bates, and N. R. Marshall, *Am. J. Enol. Vitic.* **38**, 28-34 (1987).
37. D. Barnett and E. G. Davis, *J. Chromatogr. Sci.* **21**, 205-208 (1983).
38. E. G. Davis, D. Barnett, and P. M. Moy, *J. Food Technol.* **18**, 233-240 (1983).
39. L. Reinhard and F. Radler, *Z. Lebensm.-Unters. -Forsch.* **172**, 278-283 (1981).
40. K. Henitze, *Ind. Obst.- Gemueseverwert.* **61**, 555-556 (1976).
41. E. A. Crowell and J. F. Guymon, *Am. J. Enol. Vitic.* **26**, 97-102 (1975).
42. I. F. Gaunt, K. R. Butterworth, J. Hardy, and S. D. Gangolli, *Food Cosmet. Toxicol.* **13**, 31-45 (1975).
43. H. J. Deuel, R. Alfin-Slater, C. S. Weil, and H. F. Smyth, Jr., *Food Res.* **19**, 1-12 (1954).
44. H. Schlotter and A. Stahr, *Weinwirtsch., Tech.* (9), 267-268, 270-271 (1985).
45. H. Schmidt, *Fette, Seifen, Anstrichim.* **61**, 127-133 (1959).
46. L. Eorodog, Z. Jeszenszky, P. Mattyasovszky, and P. Szalka, *Elelmiszervizsqalati Kozl.* **22**, 205-211 (1976).
47. H. W. Van Gend, *Z. Lebensm.-Unters. -Forsch.* **151**, 81-83 (1973).

48. A. Caputi, Jr., M. Ueda, and B. Trombella, *J. Assoc. Off. Anal. Chem.* **57**, 951-953 (1974).
49. A. Caputi, Jr. and K. Slinkard, *J. Assoc. Off. Anal. Chem.* **58**, 133-135 (1975).
50. A. Caputi, Jr. and P. A. Stafford, *J. Assoc. Off. Anal. Chem.* **60**, 1044-1047 (1977).
51. B. Mandrow, E. Roux, and S. Burn, *Ann. Falsif. Expert. Chim.* **68**, 29-48 (1975).
52. J. Monselise, M. Fishman, and R. Hoenig, *Isr. J. Technol.* **10**, 451-452 (1972).
53. F. Eisenbeiss, M. Weber, and S. Ehlerding, *Chromatographia* **10**, 262-264 (1977).
54. H. Woidich and H. Gnauer, *Z. Lebensm.-Unters. -Forsch.* **151**, 109-113 (1973).
55. A. Bertrand and C. Sarre, *Connaiss. Vigne Vin* **9**, 267-272 (1975).
56. Y. Ito, Y. Tonogai, and M. Iwaido, *Eiyo to Shokuryo* **29**, 44-47 (1976).
57. R. G. Coelho and D. L. Nelson, *J. Assoc. Off. Anal. Chem.* **66**, 209-211 (1983).
58. T. Mine and M. Horiuchi, *Shokuhin Eiseigaku Zasshi* **25**, 61-64 (1985).
59. A. Bertrand and C. Sarre, *Connaiss. Vigne Vin* **11**, 345-350 (1977).
60. D. Revuelta, G. Revuelta, and F. Armisen, *An. Quim.* **71**, 179-182 (1975).
61. C. P. Neidig and H. Burre, *Drug Cosmet. Ind.* **54**, 201-207 (1975).
62. L. Chan, R. Weaver, and C. S. Ough, *Am. J. Enol. Vitic.* **26**, 201-207 (1975).
63. *Recueil des Méthodes Internationales d'Analyse des Vins*, 5th ed. Office International de la vigne et du Vin, Paris, 1978.
64. N. Christoph, P. Kreutzer, and K. Hildenbrand, *Weinwirtsch., Tech.* (9), 272, 274-275 (1985).
65. K. Boek, A. Hartmann, and K. Speer, *Dtsch. Lebensm.-Rundsch.* **81**, 275-278 (1985).
66. W. Gilsbach, *Dtsch. Lebensm.-Rundsch.* **82**, 107-111 (1986).
67. F. Bandion and M. Valenta, *Mitt.Hoehere Bundeslehr- Versuchsanst. Wein- Obstbau, Klosterneuburg* **27**, 227-231 (1977).
68. G. G. Valuiko, N. I. Buryan, and L. V. Tyurina, *Vinodel. Vinograd. SSSR* **35**(5), 16-18 (1975).
69. J. Farkas, *Mitt., Rebe Wein, Obstbau Fruechteverwert.* **25**, 279-284 (1975).
70. S. Lafon-Lafourcade, *Connaiss. Vigne Vin* **9**, 177-186 (1975).
71. J. Masquelier, *Ann. Technol. Agric.* **27**, 291 (1978).
72. F. Malik, A. Navara, and E. Minarik, *Mitt. Klosterneuburg* **35**, 45-47 (1985).
73. F. Malik, L. Rudicka, and M. Drdak, *Wein-Wiss.* **38**, 51-57 (1983).
74. C. Junge, *Dtsch. Lebensm.-Rundsch.* **75**, 210-211 (1979).
75. R. Burkhardt, *Dtsch. Lebensm.-Rundsch.* **77**, 97-98 (1981).
76. H. J. Jeuring and A. Brands, *Dtsch. Lebensm.-Rundsch.* **78**, 350-351 (1982).
77. C. S. Ough, in *Antimicrobials in Foods*, A. L. Branen and P. M. Davidson, Eds. Dekker, New York, 1983, pp. 299-325.
78. J. Cuzner, P. D. Bayne, and A. J. Rehberger, *Proc. Am. Soc. Brew. Chem. 1971*, 116-127 (1971).
79. C. S. Ough, L. L. Langbehn, and P. A. Stafford, *Am. J. Enol. Vitic.* **29**, 60-62 (1978).
80. P. Stafford and C. S. Ough, *Am. J. Enol. Vitic.* **27**, 7-11 (1976).
81. P. Cenci and A. Gallerani, *Riv. Vitic. Enol.* **26**, 272-283 (1973).
82. R. Battaglia and J. Mitiska, *Z. Lebensm.-Unters. - Forsch.* **182**, 501-502 (1986).
83. A. Bertrand, *Connaiss. Vigne Vin* **19**, 191-195 (1985).
84. G. Lehmann and J. Ganz, *Z. Lebensm.-Unters. -Forsch.* **181**, 362 (1985).

85. A. Rapp, M. Spraul, and E. Humpfer, *Z. Lebensm.-Unters. -Forsch.* **182**, 419-422 (1986).
86. A. Rapp, L. Engel, and H. Ullemeyer, *Z. Lebensm.-Unters. -Forsch.* **182**, 498-500 (1986).
87. P. Pfeiffer and F. Radler, *Weinwirtsch., Tech.* **(8)**, 234-235 (1985).
88. D. Fuhrling, H. Wollenberg, and S. Littmann, *Dtsch. Lebensm.-Rundsch.* **81**, 325-329 (1985).
89. C. E. Hubach, *Anal. Chem.* **20**, 1115-1116 (1948).
90. B. L. Bates, *J. Assoc. Off. Anal. Chem.* **53**, 775-779 (1970).
91. G. Nidasio, *Boll. Lab. Chim. Prov.* **20**, 1-7 (1969).
92. H. Hoppe and K. Romminger, *Naehrung* **13**, 227-233 (1969).
93. F. Addeo, G. Nota, and L. Chianese, *Am. J. Enol. Vitic.* **29**, 7-10 (1978).
94. B. L. Bates and D. R. Buick, *J. Assoc. Off. Anal. Chem.* **59**, 1390-1395 (1976).
95. B. L. Bates, *J. Assoc. Off. Anal. Chem.* **56**, 840-842 (1973).
96. A. Wurzinger and F. Bandion, *Mitt. Klosterneuburg* **35**, 42-44 (1985).
97. P. Dubois and P. Dupry, *C. R. Acad. Agric. Fr.* **60**, 62-66 (1974).
98. P. Dubois and P. Simand, *Ann. Technol. Agric.* **25**, 337-342 (1976).
99. S. Chaunet and P. Sudraud, *Connaiss. Vigne Vin* **11**, 339-344 (1977).
100. J. Vialle, M. Kolosky, and J. L. Rocca, *J. Chromatogr.* **204**, 429-435 (1981).
101. E. LaNotte, V. A. Liuzzi, and A. M. Leone, *Vignevini* **6**(7-8), 21-24 (1979).
102. J. A. Miller, *Toxicants Occurring Naturally in Foods*, 2nd ed. National Academy of Sciences, Washington, DC, 1973.
103. *The Toxic Substances List*. U.S. Dep. of Health, Education and Welfare, Washington, DC, 1973.
104. P. A. P. Liddle and P. de Smedt, *Ann. Falsif. Expert. Chim.* **69**, 857-864 (1976).
105. E. J. Wojtowicz, *J. Agric. Food Chem.* **24**, 526-528 (1976).
106. E. Merat, G. E. Martin, M. Duret, and J. Vogel, *Mitt. Geb. Lebensmittelungters. Hyg.* **67**, 521-526 (1976).
107. R. H. Dyer, G. E. Martin, and P. C. Burscemi, *J. Assoc. Off. Anal. Chem.* **59**, 675-697 (1976).
108. R. H. Dyer, *J. Assoc. Off. Anal. Chem.* **60**, 1041-1043 (1973).
109. G. Mazza, *Sci. Aliments* **4**, 233-245 (1984).
110. *AFT Ruling 74-10* (May 6, 1974), Bulletin 1974-1975. Bureau of Alcohol, Tobacco and Firearms, Washington, DC, 1974.
111. R. H. Dyer and G. E. Martin, *J. Assoc. Off. Anal. Chem.* **59**, 780-782 (1976).
112. G. E. Martin and D. M. Figert, *J. Assoc. Off. Anal. Chem.* **57**, 217-281 (1974).
113. G. Montedoro and E. Miniati, *Rev. Sci. Tecnol. Aliment. Nutr. Um.* **6**, 177-180 (1976).
114. C. A. Diez de Bethencourt and E. Sanchez Pastor, *Sem. Vitivinic.* **28** (1.394), 1759-1760 (1973).
115. H. Tanner and M. Sandoz, *Schweiz. Z. Obst- Weinbau* **109**, 287-300 (1973).
116. A. Rapp, H. Mandery, and W. Heimann, *Vitis* **22**, 387-394 (1983).
117. P. Mattyasovszky and Z. Jeszenszky, *Borgazdasag* **33**, 105-110 (1985).
118. N. K. McCallum, H. P. Rothbaum, and R. L. Orlet, *Food Technol. Aust.* **38**, 318-320 (1986).
119. D. J. McWeeny and M. L. Bates, *J. Food Technol.* **15**, 407-412 (1980).

120. G. E. Martin, J. E. Noakes, F. C. Alfonso, and D. M. Figert, *J. Assoc. Off. Anal. Chem.* **64**, 1142-1144 (1981).
121. G. J. Martin and M. L. Martin, *J. Chim. Phys.* **80**, 293-297 (1983).
122. R. Betz, *Wein-Wiss.* **36**, 286-292 (1981).
123. *Fed. Regist.* **38**(95), 12931; **38**(134), 18684 (1974).
124. D. T. Williams and W. F. Miles, *J. Assoc. Off. Anal. Chem.* **58**, 272-275 (1975).
125. D. T. Williams, *J. Assoc. Off. Anal. Chem.* **59**, 32-34 (1976).
126. M. Bertuccioli, *Vini Ital.* **18**, 403-406 (1976).
127. S. Brun, M. Giffone, and H. Mattras, *Trav. Soc. Pharm. Montpellier* **37**, 207-216 (1977).
128. W. R. Sponholz and P. Lamberty, *Z. Lebensm.-Unters. -Forsch.* **171**, 451-452 (1980).
129. H. Wieczorek and C. Junge, *Dtsch. Lebensm.-Rundsch.* **68**, 137-139 (1972).
130. M. Lehtonen, *Chem., Mikrobiol., Technol. Lebensm.* **2**, 161-164 (1973).
131. D. M. Takahashi, *J. Assoc. Off. Anal. Chem.* **57**, 875-879 (1974).
132. E. Lemperle, E. Kerner, and R. Heizmann, *Wein-Wiss.* **30**, 82-86 (1975).
133. D. M. Takahashi, *J. Assoc. Off. Anal. Chem.* **60**, 799-804 (1977); see also *J. Chromatogr.* **131**, 147-156 (1977).
134. R. Wale, *Food Technol. Aust.* **34**, 30 (1982).
135. F. R. Albright, D. V. Schumacher, D. S. Sweigart, T. J. Stasny, C. Husack, and K. Boyer, *Food Technol. (Chicago)* **33**(11), 69-70, 72-76 (1979).
136. D. R. Beaman and H. M. Baker, *Anal. Chem.* **52**, 1984-1987 (1980).
137. G. Dufour, P. Sebastien, A. Gaudichet, J. Bignon, and G. Bonnaud, *Ann. Nutr. Aliment.* **32**, 997-1009 (1978).
138. G. Zweig, *Qual. Plant., Plant Foods Hum. Nutr.* **23**, 77-112 (1973).
139. World Health Organization, *W.H.O.* **592**, 1-45 (1976).
140. L. Drobnica, E. Sturdik, E. Minarik, and P. Ragala, *Wein-Wiss.* **35**, 404-413 (1980).
141. F. Gnaegi, *Rev. Fr. Oenol.* **25**(99), 9-13 (1985).
142. D. M. Gilvydis, S. M. Walters, E. S. Spivak, and R.K. Hedblad, *J. Assoc. Off. Anal. Chem.* **69**, 803-806 (1986).
143. A. Tromp and P. G. Marais, *S. Afr. J. Enol. Vitic.* **2**, 25-28 (1981).
144. L. Barbero and P. Gaia, *Vini Ital.* **21**, 95-100 (1979).
145. W. T. Thompson, *Agricultural Chemicals*, Book IV. Thompson Publications, Fresno, CA, 1982.
146. V. Batora, S. L. Vitorovic, H. P. Thier, and M. A. Klisenko, *Pure Appl. Chem.* **56**, 1040-1049 (1981).

Nine
OTHER CONSTITUENTS

The determination of the ash, the alkalinity of the ash, and of cations and anions is of importance to the enologist for a variety of reasons: legal, health, taste, and regional definition.

ASH

The ash is the inorganic matter that remains after evaporation and incineration of a must or wine. Table 57 (1-6) lists data on the ash content reported in various types of wine. In general, the ash amounts to about 10% of the sugar-free extract. Its determination is needed in some enological formulas designed to detect watering, sugar addition, or fortification. Generally, ash content is lower in wines from unripe grapes, from sugared and watered musts, and from the free-run juice; also, white wines tend to have lower ash content than do red wines. The Q-value is the sugar-extract (g/L) − alcohol (%) + ash (g/L). It decreases as the must is watered. Since it varies between varieties, regions, and seasons, it must be determined for each set of conditions. Ashing is usually carried out at 500–550°C. The temperature of ashing should not exceed 550°C lest chlorides be lost. During ashing, the cations are converted to carbonates or other anhydrous mineral salts. Ammonium ion is lost.

An empirical equation relating ash content to conductivity was developed (1). The average difference between the two procedures was ±0.2 g/L. This procedure has been employed (5–8) with multiple regression analysis. Two equations were proposed:

$$\text{Ash (g/L)} = 1.346 \times 10^{-3} L_{20} + 1.684 \times 10^{-2} A + 9.706 \times 10^{-3} E - 2.282 \quad (1)$$

$$\text{Ash (g/L)} = 1.346 \times 10^{-3} L_{20} - 28.549 d_{20/20} + 5.396 \times 10^{-2} R_D^{20} + 25.719 \quad (2)$$

Table 57. Ash Content of Various Types of Wines

Region	Type	Number of samples	Ash (g/L) Min.	Ash (g/L) Max.	Ash (g/L) Average
Argentina	White table	139	1.28	3.08	2.63
Argentina	Red table	208	1.90	5.07	3.47
Austria	Table	151	1.35	4.74	2.31
California	Sherry	4	2.36	3.74	3.02
Czechoslovakia	Table	708	1.60	2.30	2.00
France	Table	64	1.20	3.90	1.84
Germany	Table	434	1.31	3.97	2.05
Germany	Table	32	1.28	4.50	2.20
Hungary	Table	10	1.92	4.46	2.97
Hungary	Table	23	2.33	7.50	3.15
Italy	Table	1168	1.10	4.80	2.06
Italy (Tuscany)	Table	176	1.70	3.16	2.30
Portugal	Table	606	1.00	4.00	2.61
Romania	Table	33	1.33	4.74	2.24
Spain	Sherry, etc.	81	2.06	7.08	4.29
Tunisia	Table	25	1.90	3.40	2.65
Turkey	Table	105	1.02	4.60	2.16
Yugoslavia	Table	170	1.13	3.84	1.76

From References 1-6.

where L_{20} = conductivity (μS) at 20°C
A = ethanol content (g/L)
E = total extract (g/L)
R_D^{20} = refractive index at 20°C measured at the wavelength of sodium doublet.

The absolute deviation (g/L) was ±0.11, and the relative error was ±5.21%.

Rather than determining the ash content from conductivity or directly, it may be calculated from the potassium, calcium, and sodium contents (calculated as carbonates), the magnesium content (as the oxide), and the phosphorous content (as phosphate), then summing. If sulfate or iron pickup occurs, they must be subtracted (9–11). The correlation between the direct ash determination and that calculated was ±0.993, with a reproducibility of ±1.02%. The results were automatically obtained by a computer program from previously stored data.

The ashing procedure given below is essentially that of the OIV (12).

Procedure

Heat a 60-mL broad-bottom platinum dish to 600°C for 10 min, cool in a desiccator, and weigh to 0.1 mg. Pipet 25 mL of the wine sample into the platinum dish and evaporate to dryness, first on a water bath and then in an oven at 100°C. If the extract has been determined by direct weighing, it may be used for ashing. Drying under an infrared light is also satisfactory. Place the dish with the residue in a muffle furnace at 525°C ± 25° for 5 min. Remove from the furnace, cool, add dropwise 5 mL of water, dry in the oven, carefully avoiding splattering, and reheat in the muffle furnace for 15 min. If the residue is still black, cool, add again 5 mL of water, redry, and reash.

Musts and sweet wines are difficult to ash. Adding a drop of olive oil after drying will help prevent undue swelling of the residue in the muffle furnace.

When the ash is gray or white, place the dish in a desiccator to cool and weigh rapidly; the ash is hydroscopic. Reheat to 525°C for 15 min, again cool in the desiccator, and reweigh. The change in weight should be less than 0.3 mg. Express the ash in units of grams per liter of wine.

ALKALINITY OF THE ASH

During the ashing of a wine the free organic acids are destroyed, but their acid salts and the salts of the organic acids are converted into carbonates. The alkalinity of the ash is thus a measure of the quantity of organic acid salts present in the original must or wine. Typical results on the alkalinity of ash of various wines are summarized in Table 58.

The alkalinity of the ash has some theoretical interest in making a balance of the total organic cations present in the wine against the total organic anions. Theoretically the milliequivalents of titratable acid plus the alkalinity of the ash should equal the sum of the milliequivalents of volatile acidity, tartrate, malate, citrate, lactate, succinate, and acid esters. However, in practice, a number of corrections are needed when setting up this balance.

In the determination of the alkalinity of the ash, certain errors occur that must be taken into account. The ammonia that was originally present as ammonium ion with some organic anions is volatilized during ashing. One must therefore determine the ammonia content separately and add the milliequivalents found.

The titratable acidity as originally determined will include one-third the phosphate present in titrating from a pH of about 3.5–8.3. The milliequivalents of phosphate present divided by 3 should be subtracted from the titratable acidity. If the alkalinity of the ash is determined to a methyl orange end point, pH about 4.0, phosphate will not interfere because the phosphate in the ash will be nearly in the same condition as in the wine at the methyl orange end point.

Approximately half of the free sulfurous acid present is neutralized at the normal pH of table wines. During evaporation, the sulfur dioxide is volatilized and an equivalent amount of base is liberated. The percent of free sulfurous

Table 58. Alkalinity of the Ash of Various Types of Wine

Region	Type	Number of samples	0.1 N acid for 100 mL of wine (mL)		
			Min.	Max.	Average
Czechoslovakia	White table	572	10.9	25.9	18.4
France	Table	97	12.4	35.6	21.7
Germany	Table	77	12.3	39.0	20.1
Italy	Table	975	5.0	30.2	12.9
Portugal	Table	885	4.8	60.0	25.5
Romania	Table	33	6.9	30.5	12.8
Spain	Montilla	51	9.1	64.7	21.0
Tunisia	Table	25	11.9	29.4	21.0
Yugoslavia	Table	164	8.8	40.0	19.2

From Reference 3.

acid oxidized to sulfuric acid during evaporation depends on the oxygen content of the wine.

Of the combined sulfur dioxide, about 75% is lost during evaporations. The remainder is oxidized to sulfate (which is bivalent) and thus is equivalent to half the combined monovalent bound sulfur dioxide. At the pH of wine, none of the bound sulfur dioxide is titrated. The sulfate produced thus reduces the alkalinity of the ash. The exact correction that should be made is not important in the case of red wines but is important in certain sweet table wines. As an approximation, one-half of the milliequivalents of total sulfur dioxide present is subtracted from the titratable acidity.

As a conclusion, the balance is to be calculated as follows; all values are to be expressed as milliequivalents:

$$\text{Titratable acidity} + \text{Alkalinity of ash} + \text{Ammonia} - \frac{\text{Phosphate}}{3} - \frac{\text{Sulfur dioxide}}{2} = \text{Acetate} + \text{Tartrate} + \text{Succinate} + \text{Malate} + \text{Acid esters} + \text{Lactate} + \text{Citrate}$$

The alkalinity of the ash is determined by the addition of an excess of acid and back-titration with a base.

Procedure

Place the platinum dish containing the ash in a 250-mL beaker. Add a little water and 25 mL of 0.05 N sulfuric acid. Cover with a watch glass, then heat on a water bath for 15 min. Use a stirring rod to get all the ash in contact with the acid. Rinse off the watch glass and the sides of the beaker with boiled carbon dioxide-free water. Add 5 drops of 0.1% aqueous methyl orange indicator solution and titrate while stirring with 0.05 N sodium hydroxide solution.

$$\text{Alkalinity (meq)} = \frac{(V_1 N_1 - V_2 N_2)(10)(100)}{v}$$

where V_1 = volume of sulfuric acid added (mL)
N_1 = normality of the sulfuric acid
V_2 = volume of sodium hydroxide solution used for titration (mL)
N_2 = normality of the sodium hydroxide solution
v = volume of the original wine sample (mL)

CATIONS

Among the cations present in wine, the following ones are of interest: potassium, sodium, calcium, magnesium, iron, copper, arsenic, lead, cadmium, zinc, and others. Table 59 (13-21) lists the concentrations of six major cations in wines. During alcoholic fermentation, losses in metals occur (22): iron, 25-80%; copper, 75-95%; zinc, 7-66%; manganese, 15-48%; and cadmium, 60-75%. Use of calcium and/or sodium bentonite increases the sodium, calcium, magnesium, aluminum, and iron contents of the wine (23). With calcium bentonite there is a negligible pickup of sodium. Bentonite fining does not significantly change the cadmium, copper, tin, or zinc content, but there is an increase in chromium. Although lead is slightly higher after bentonite fining, even with high levels of bentonite (400 g/mL) the 0.3 mg/L limit is not exceeded. Potassium is only slightly decreased with bentonite fining.

Among the new procedures there is inductively coupled argon plasma–atomic emission spectroscopy (ICAP-AES). It can be programmed for the quantitative determination of as many as 25 metallic ions. The procedure is sensitive, accurate, and rapid, but the special instrument and equipment are expensive (24). It has been used for wines (18). Extraction photometric methods can be employed for copper, iron, nickel, manganese, and zinc (25), but wet ashing is required. Proton-induced X-ray fluorescence (PIXF) has been recommended (26). A comparison of ion chromatography (IC) with post-chromatography derivatization and flame atomic absorption spectroscopy (AAS) gave comparable results (27). IC was preferred because it had no matrix effects. For calcium, iron, magnesium, sodium, and potassium, AAS is currently favored (28).

Table 59. Cation Content of Various Wines

Region	Potassium (g/L)		Sodium (g/L)		Calcium (g/L)		Magnesium (g/L)		Iron (g/L)		Copper (g/L)	
	Range	Average	Range	Average	Range	Average	Range	Average	Range	Average	Range	Average
California[a] (3)	0.28–1.580	0.940	0.010–0.172	0.055	0.006–0.117	0.052	—	—	0.0–35.0	4.89	0.04–0.43	0.11
California[b] (13)	0.732–1.672	1.131	0.009–0.309	0.101	0.031–0.078	0.053	0.069–0.138	0.103	2.3–12.4	8.4	0.07–0.59	0.18
California[c] (14)	0.125–2.040	0.937	0.003–0.279	0.054	0.025–0.310	0.084	0.032–0.245	0.114	0.3–16.1	3.3	0.02–2.40	0.16
France[d] (3)	0.094–1.760	0.654	0.030–0.125	0.062	0.036–0.112	0.091	0.074–0.165	0.123	3.5–26.0	8.81	0.54–1.78	1.28
Germany[a] (3)	0.627–1.293	0.903	0.005–0.043	0.015	0.054–0.115	0.092	0.073–0.091	0.084	2.24–9.89	5.82	0.00–3.68	1.24
Germany[c] (15)	0.090–0.816	0.313	0.004–0.045	0.020	—	—	—	—	0.6–11.4	3.9	—	—
Italy[a] (3)	0.43–1.33	0.84	0.019–0.085	0.026	0.30–0.165	0.103	0.060–0.173	0.109	1.5–90.0	16.00	0.12–1.12	0.36
Italy[d] (16)	1.310–1.838	1.44	0.016–0.047	0.026	0.104–0.152	0.129	0.068–0.107	0.085	7.80–21.20	13.82	Trace–0.80	0.27
Italy[e] (4)	0.43–1.33	0.84	0.019–0.05	0.045	0.030–0.165	0.104	0.060–0.173	0.109	—	—	—	—
Italy (17)	0.340–1.500	0.868	0.006–0.320	0.354	0.047–0.198	0.102	0.052–0.192	0.101	0.26–55.0	10.8	0.5–2.62	0.35
Italy (18)	0.575–1.165	0.888	0.010–0.057	0.026	0.060–0.139	0.094	0.062–0.141	0.094	0.86–10.15	5.9	0.06–12.80	0.18[f]
New Zealand[g] (19)	0.420–1.350	0.882	0.014–0.145	0.033	0.048–0.208	0.085	0.021–0.104	0.065	1.4–22.0	6.3	0.04–1.20	0.22
Romania (20)	0.54–0.80	0.69	0.013–0.030	0.017	0.060–0.134	0.091	—	—	1.15–3.96	2.77	—	—
Spain[h] (21)	0.656–1.60	1.069	0.006–0.023	0.012	—	—	—	—	—	—	—	—

[a] The number of samples used varied widely; thus the results cannot be compared directly without reference to the original reports (3). The numbers in parentheses are reference numbers.
[b] 14 California wines.
[c] 75 German wines.
[d] 12 Trentino wines.
[e] 176 Tuscany table wines.
[f] Omitted 12.80 mg/L sample.
[g] 48 New Zealand wines.
[h] 130 La Mancha wines.

The amounts of potassium, sodium, magnesium, and calcium, as well as potassium/sodium and magnesium/calcium ratios, have been suggested as measures of the authenticity of regional Italian wines (29). Considerable data on varieties, regions, seasons, time of harvest, and methods of production would be needed to unequivocally establish such ratios.

Potassium

Accurate information on the potassium content of wines is desired by modern wine makers to control processing. Prevention of potassium acid tartrate precipitation in bottled wines is considered essential. With a knowledge of the potassium, tartrate, and ethanol contents, along with a knowledge of the pH, the wine maker can predict the potential stability or instability of potassium acid tartrate in the wine. This is particularly important for wineries that use cation exchange to reduce the potassium content of their wines.

The potassium content of the must is a function of variety of grapes, soil and climatic conditions where grown, time of harvest, and other variables. The temperature of fermentation and storage, the time and kind of storage, the pH, the percentage of alcohol, and the use of ion-exchange resions, fining agents, or filter aids, as well as other factors, all influence the potassium content of the finished wine. There is considerable difference in potassium content between varieties, regions, and, to a lesser extent, years. The potassium content decreases during fermentation and aging, primarily because of precipitation of potassium acid tartrate. White table wines usually have lower concentrations of potassium than do red ones. Nutritionists are interested in the high potassium/sodium ratio of wines as a part of hypertension diets.

The OIV (12) recommends precipitation with sodium tetraphenylborate as the standard procedure. Since the wines must be ashed, the procedure is somewhat laborious. Other classic procedures have been precipitation as potassium perchlorate or chloroplatinate. The chemical procedure of precipitation as the acid tartrate has been recommended.

Today, flame emission photometry (the AOAC official method) or AAS at 382.3 nm is used (30). At high potassium levels, a 1:100 to 1:250 dilution of the wine sample is recommended. Olmedo et al. (21) recommended adding 10 mL of concentrated hydrochloric acid to 5 mL of wine and diluting with water to 100 mL. This was further diluted 20- to 50-fold for AAS (31) because of the nonlinear standard curves with the less dilute samples.

Procedure

Dilute 2 mL of wine to 100 mL in a volumetric flask. Then use in the atomic absorption spectrophotometer according to the manufacturer's directions. In the Hill–Caputi (30) procedure a Perkin-Elmer® model 305 equipped with a DCR-1 digital concentration readout device was used. A Boling three-slot burner head turned 90° out of phase was used to reduce air turbulence within the flame. The air-acetylene flame was adjusted in accordance with the manufacturer's directions. A 303-6052 hollow cathode tube was used. The operating conditions were as follows: wavelength, 382.3 nm; lamp current, 30 mA; slit, 4; filter in; noise supplement, 3; average mode, 4x; gas flow rate, acetylane at 11.5 mL/min. For flame photometry a standard curve should be prepared. The potassium content of the wine is determined from the curve. The determination may also be made by the method of addition (see p. 338).

An ion-selective electrode (ISE) has been used for grape juice (32). In one study (33) the sodium, calcium, magnesium, and iron ions did not interfere. The grape juice was diluted 1:20 with a buffer solution (pH 9.2, 0.05 M borax/dm^3). The results were 3.4-13.3% lower than by AAS. For wines the ISE gave low readings at high potassium levels ($>10^{-2}$ M). At high concentrations of other ions, inconsistent readings were obtained, because the ionic strength of the standards and the sample are so different. A response time of 1 min was claimed, but several minutes (up to 10) may actually be needed. Besides the special electrode, a reference electrode and a high-impedance voltmeter ($>10^{12}$ Ω) are needed. A pH-mV meter will be suitable. See also Reference 34.

Sodium

The sodium content was seldom determined on wines until the introduction of ion-exchange resins in wine processing and the recognition that knowledge of sodium intake was of importance to people on low-sodium diets. Cation exchangers are allowed in Italy and the United States but not in France or Germany. Addition of sodium chloride (up to 0.1 g/L) is allowed in Italy and France but not in Germany.

In France there is a legal sodium limit of 160 mg/L. In Spain, sodium chloride may be added to wines to improve clarification, provided the total sodium chloride content does not exceed 1 g/L (21). This is equivalent to 393 mg/L of sodium. Other sources of high sodium are sodium sulfite, sodium metabisulfite, sodium sorbate, sodium bentonite, and (in some countries) sodium sulfide. This is based on the assumption that the natural sodium concentration is 35 mg/L and that as much as 125 mg/L may be picked up: 45 mg/L from sodium hydrogen sulfite (not allowed in France but used in Germany), 40 mg/L from sodium bentonite, and 40 mg/L from sodium sorbate (allowed in France but not in Germany). In Germany the limit is the natural limit, presumably 50 mg/L. Patschky and Schöne (35) indicate a maximum of 30 mg/L, and they believe that the limit should not be greater than 150 mg/L. In 290 ordinary Austrian table wines, only one exceeded 20 mg/L (36). In 152 Austrian wines made from riper grapes ($>19°$Klosterneuburg), only two wines exceeded 30 mg/L. The OIV (12) sets a limit of 0.060 g/L of sodium in excess of the chloride present. This restriction is intended to prevent excessive use of sodium cation exchange, but it may be exceeded (37) if no chloride is added. However, some data (3) cast doubt on this. Obviously better maximum sodium limits are necessary. The potassium/sodium ratio is generally above 10 in wines that have not been treated by ion exchange (38). In treated wines it was less than 10. The 10:1 ratio is generally true with California wines (39).

The usual method of the OIV (12) is flame photometry. They recommended using an artificial wine base for preparing a standard curve. This artificial wine is made by dissolving 481.3 mg of potassium acid tartrate in 500 mL of very hot water, adding 10 mL of ethanol, 700 mg of citric acid, 300 mg of sucrose, 1000 mg of glycerol, 10 mg of anhydrous calcium chloride, 10 mg of anhydrous sodium chloride, and diluting 1 L; this represents a wine sample diluted 1:10. In the United States, atomic absorption spectrophotometry is preferred (36). The sodium ion-selective electrode has been successfully used with dietetic foods and should be equally accurate with wines (40). The electrode should be prepared, calibrated, and cleaned according to the manufacturer's instructions. A standard addition procedure is used (p. 338).

Calcium

The accurate determination of calcium is important because of the problem of precipitation of calcium salts (tartrate, mucate, gluconate, etc.). This precipitation is particularly insidious because it occurs very slowly, usually after the wine is bottled. In a few cases, calcium mucate precipitation has occurred. The calcium content will be influenced by soil conditions, by treatment of the musts with calcium sulfate or calcium carbonate, by use of filter aids and pads, by storage in concrete tanks, by use of fining agents such as calcium bentonite, by ion-exchange treatment, by the concentration of ethanol and other constituents (sulfates, tartrates, etc.) of the wine, by the pH, and by the time and temperature of storage. Table 59 indicates calcium contents of 0.006–0.310 g/L in wines from the United States and Europe. Using AAS for 27 Spanish wines the calcium varied from 28.6 to 79.4 mg/L [average 42.9 mg/L (41)]. No difference between diluted (1:4) and ashed samples was noted.

For the determination of calcium in the ash of wine, the classic procedures such as precipitation as the oxalate and titration with permanganate may be used (42) but is slow, particularly in sweet wines. Other procedures (colorimetric, complexometric, etc.) have also been described (43–46). Ion-selective electrodes for calcium are also now available (47), and they yield satisfactory results. The official procedure of the OIV (12) is based on complexometric determination with EDTA. The interfering compounds can be eliminated by passing the wine through a strongly basic anion-exchange column; interference of aluminum, iron, and manganese is prevented by the addition of triethanolamine (44, 48, 49). The method is slow.

Procedure

Prepare a strongly basic anion-exchange column in the acetate form as follows: Place 100 g of resin such as Merck III in 200 mL of 30% acetic acid for at least 1 day. Fill a 10–11-mm × 30-cm ion-exchange column with the prepared resin, taking care that no air bubbles are trapped in the column. Wash the column thoroughly with 0.5% acetic acid; washing four times with 10–12 mL is suggested.

Add 10 mL of wine at a rate of 1–1.5 drops/sec and collect the eluate in a 100-mL volumetric flask. Wash seven times with 10 mL of water and dilute the content of the volumetric flask to volume with water. Place a 50-mL aliquot of this solution in a platinum dish, evaporate, dry at 100°C, and ash at 525°C.

Dissolve the ash in 5 mL of 2 N hydrochloric acid and transfer quantitatively to a 200-mL Erlenmeyer flask. Add 2 mL of 20% sodium hydroxide solution, 1 mL of triethanolamine solution, and a pinch of Calcon indicator (a mixture of 1 g of sodium-2-hydroxy-1-(2'-hydroxynaphthyl-1'-azo)-naphthalenesulfonic acid-4 and 100 g of sodium chloride), then stir with a magnetic stirrer. Titrate with 0.005 M EDTA solution until the red color changes to blue.

$$\text{Calcium (mg/L)} = \frac{(V)(M)(400.8)(1000)}{v}$$

where V = volume of EDTA solution used for titration (mL)
M = molarity of EDTA solution
v = volume of wine sample (mL)

A satisfactory complexometric method (50) is given as follows. It can be used with wine directly.

Procedure

Prepare the buffer solution, pH 11, by dissolving 3.6 g of orthoboric acid in 900 mL of water and bring to 1 L with ethanolamine. Store in plastic in the refrigerator. To 1 g of 8-hydroxyquinoline add 5 mL of concentrated hydrochloric acid and 40 mg of *o*-cresolphthalein (Complexon®), then bring to 100 mL with water. Keep this color reagent in plastic; do *not* store in glass. To prepare the calcium standard dissolve 2.500 g of calcium carbonate in a minimum amount of hydrochloric acid and bring to 1 L with water (1 mL = 1 mg of Ca). Take 5 mL of this solution and dilute to 1 L with water (1 mg = 5 µg of Ca). Take 50 g of EDTA (disodium salt) and dissolve in 1 L of water. Into a series of tubes add 0, 0.2, 0.3, 0.4, 0.5, and 0.6 mL of the calcium standard solution and dilute each tube to 1 mL with distilled water. These solutions have respective calcium concentrations of 0.0, 1.0, 1.5, 2.0, 2.5, and 3.0 mg/L. Then to the tubes add 1 drop of water, 5 mL of the pH buffer, and 0.5 mL of the color reagent. Mix and let stand for 10 min in a 25°C water bath. Read absorbance against tube concentration.

For white wine analysis dilute a sample 1:20 with water and repeat the procedure above using 1 mL of diluted wine in place of the standard. Prepare a blank using 1 mL of distilled water. Read values from the standard curve and multiply by the dilution factor to get calcium content in milligrams per liter.

For red wine set up two tubes, each containing 1 mL of diluted red wine. Also run a water blank. To one of the tubes with wine add 1 drop of EDTA (and no drop of water). Then proceed as above. Subtract the absorbance value for the tube with EDTA from the tube without. This compensates for the color interference from the red wine pigments. Read values from the standard curve and multiply by the dilution factor.

The purpose of the addition of the 8-hydroxyquinoline is to bind any magnesium that may be present. Magnesium reacts with the color reagent at this pH also. The buffer is to screen out or prevent other metal ions from chelating with the color reagent.

Today, AAS is widely used for the determination of calcium in musts and wines (51, 52). Phosphate interference can be eliminated by the addition of 1% lanthanum chloride (45). The wine is analyzed in a 1:20 dilution. Sodium, phosphates, sulfates, magnesium, glucose, and ethanol do not interfere (53, 54). A dilution of 1:10 was preferred (55) when the calcium content was low (<20 mg/L).

Procedure

Pipet 1 mL of wine and 2 mL of 10% lanthanum (26.738 g of $LaCl_3 \cdot 7H_2O/100$ mL) into a 20-mL volumetric flask. Bring to volume with distilled water. Then use the atomic absorption spectrophotometer according to the manufacturer's directions. Instruments with digital concentration readout devices are preferred. The parameters are as follows: wavelength, 422.7 nm; lamp current, 20 mA; slit 4 (for Perkin-Elmer); filter, average mode; gas flow pressures, acetylene, 4 psi; air, 30 psi.

Put some 1% lanthanum chloride in the flame and adjust to zero. Determine the absorbance of the wine samples. Standard samples of calcium (0, 2, 5, 8, 10 mg/L) in 1% lanthanum are used to prepare a standard curve.

When calcium and magnesium need to be determined at the same time, complexometric titration with ethyleneglycol-bis(2-amino-ethyl ester)N,N'-tetraacetic acid (EGTA) and ethylenediaminetetraacetic acid (EDTA) is recommended (56). A comparison of several methods for determining calcium resulted in recommending AAS as the reference method (57). Direct colorimetry using methyl blue thymol or o-cresolphthalein was also satisfactory. Fluorimetry gave good results with white wines, but red wines have to be decolorized. AAS results were 2–6 mg/L higher compared to the other procedures. From this study it appears that a small part of the calcium appears to be present combined with high-molecular-weight compounds.

Magnesium

Thus far enologists have not been concerned with normal variations in the magnesium content of wines, although it is usually the third most important cation after potassium and calcium. However, there are some indications that it may be of importance in tartrate stability and to the acid taste. Magnesium content is influenced by use of filter aids, by storage in concrete containers, by treatment with fining agents, by use of ion-exchange resins, by the concentration of alcohol and other constituents such as tartrates and sulfates, by the pH, and by the time and temperature of storage. During fermentation the calcium content decreases and the magnesium/calcium ratio increases severalfold (58). Table 59 indicates magnesium contents of 0.021–0.245 g/L. Using AAS in 27 Spanish (Galicia) wines, 26.3–49.4 mg/L (average 35 mg/L) of magnesium were reported (41). The low values compared to those in Table 59 are attributed to the low magnesium content of the soil in this region.

Magnesium can be determined by EDTA titration, after cleanup procedures similar to those given above for calcium. However, the method most frequently used today is atomic absorption spectrophotometry. For a simple colorimetric procedure on diluted wine using xylidyl blue I, see Bonnemaire et al. (59). On 13 musts and wines the average atomic absorption, complexometric, and colorimetric results were 97.0, 96.2, and 97.1 mg/L, respectively. Atomic absorption spectrophotometry at 285.2 nm has been used (60). For a general discussion of methods for determining magnesium in biological materials, see Martinek (61).

Iron

The natural iron content of musts varies widely depending on the iron content of the soil, amount of iron-containing dust on the fruit, and contamination during harvesting, transportation, and crushing. Of 8.9 mg/L in Chilean must, only 1.0 mg/L was naturally present in the berry (62). Much of the must iron is lost during fermentation by absorption and adsorption by yeasts. The loss varies from 25 to 80% depending on the aeration, the presence or absence of polyphenolic compounds, and the relative amounts of yeast present. New wines crushed in stainless steel crushers and not exposed to iron usually contain only 1–2 mg/L. In areas with primitive equipment the iron content may range from 10 to 30 mg/L (or higher) because of iron pickup. In 75 ordinary Spanish table wines the iron content ranged from 6.5 to 25.0 mg/L (average 12.1 mg/L). It is noted that commerical Spanish wines contain only 2–5 mg/L. Commercial

Rioja wines (63) contained 2.4–19 mg/L (average 7.3 mg/L). However, 175 German wines ranged from 0.43 to 10.3 mg/L (average 4.0 mg/L) (64). In five Italian wines, 0.86–10.15 mg/L (average 5.87 mg/L) were found (18). In 82 Baden wines, iron contents of 0.087–7.60 mg/L (average 3.40 mg/L) were reported (65).

Iron is of importance to the wine maker because, when present in excess of about 7–10 mg/L, it may cause cloudiness or color change. Oxidation of wines is also accelerated by the presence of iron. The state of oxidation of the iron in a wine depends on the oxidation condition of the wine. Wines kept out of contact with the air for some time (e.g., bottled wines) have 80–95% of their iron in the ferrous state. Ferric iron appears when wines are aerated. It is the precipitation of colloidal ferric phosphate that is responsible for a milky cloudiness in white wines, known as white casse. Ferric polyphenol compounds precipitate as a blue-black film in red wines. This is known as blue casse. The ability of citric acid to protect against ferric phosphate casse is due to the formation of a ferric citrate complex.

Iron is commonly determined colorimetrically, directly in white wines or by extraction of the color with ethyl acetate, isoamyl alcohol, or amyl alcohol-methanol (2:1). The thiocyanate procedure is widely used (66) and is outlined below. Other colorimetric procedures involve 1-o- or 1,10-phenanthroline, 2,4,6-tripyridyl-s-triazine (TPTZ), bathophenanthroline (4,7-diphenyl-1,10-phenanthroline) (67), 3-(4-phenyl-2-pyridyl)-5,6-diphenyl-1,2,4-triazine (PPDT), 3-(2-pyridyl)-5,6-diphenyl-1,2,4-triazine (PDT) (68, 69), or R-nitroso (70). Better results are obtained if the must or wine is ashed, but this considerably lengthens the time required, especially for musts or sweet wines. However, ashing seems to be essential with dark red wines (71). At low iron content the results can be poor compared to those from atomic absorption spectrophotometry.

Procedure

Prepare a standard iron solution by dissolving exactly 1.000 g of clean iron wire in a few milliliters of hot concentrated hydrochloric acid and diluting to 1 L with water. Dilute 1-, 2-, 5-, and 10-mL samples of this standard to a liter; the iron concentration of the diluted standards are 1, 2, 5, and 10 mg/L, respectively. Into a large test tube add 10 mL of water and to four other tubes add 10 mL of the diluted iron standards. To each add 2 mL of 3 N hydrochloric acid, 1 mL of 20% potassium thiocyanate solution, and 5 drops of 30% hydrogen peroxide. If the blank tube has an appreciable pink color, the water or the reagents contain iron, or the test tube and the pipets were iron-contaminated. Extract the colored Fe(CNS)$_6^{3-}$ complex with 10 mL of a 2:1 amyl acetate-methanol mixture. Measure the absorbance in a photoelectric colorimeter with a blue-green filter or in a spectrophotometer at 490 nm and prepare a calibration curve.

Pipet 10 mL of wine into each of two test tubes marked A and B. To each add

2 mL of 3 N hydrochloric acid and 1 mL of 20% potassium thiocyanate solution. To tube A add 5 drops of 30% hydrogen peroxide and to B add 5 drops of water. Mix and extract with 10 mL of the organic solvent mixture. Separate, dry the extract with anhydrous sodium sulfate, and filter. Measure the absorbance. Use the reading for A to express total inorganic iron and the reading for B to express iron present in the wine in ferric state.

In a review (72) of several colorimetric procedures for iron the ferrocyanide procedure was preferred, primarily because it gave more reproducible results.

Procedure

Prepare 30% nitric acid (solution 1), 30% hydrogen peroxide (solution 2), and 10% potassium ferrocyanide (solution 3). Prepare four test tubes, A, B, C, and D, as follows:

Contents	Test Tube			
	A	B	C	D
Wine or iron standard	2 mL	0 mL	2 mL	0 mL
Water	0 mL	2 mL	1 mL	3 mL
Solution 1	5 mL	30 mL	3 mL	3 mL
Solution 2	2 drops	2 drops	2 drops	2 drops
Solution 3	1 mL	1 mL	0 mL	0 mL

Mix and wait 8 min. Measure color at 700 nm. The color is stable for about 10 min.

Prepare a standard curve with solutions of 0–30 mg/L of iron, plotting absorption of $(A - B) - (C - D)$. Use 2 mL of wine in the same way. Calculate $(A - B) - (C - D)$ and read concentration from the standard curve. Tubes B and D need to be prepared only once in a series, since their values can be used both for the standard curve and for the wine samples.

Differential pulse polarography may be used after ashing (71). Also X-ray fluorescence is available (64). AAS (73, 74) is now used for routine determination of iron in wines. Since the concentration of ethanol and glucose in wine has some influence of the results, correction factors are needed for accurate determination, or the must or wine must be ashed (52). ICAP–AES is available for the determination but is expensive.

Copper

Musts and new wines normally contain very small amounts of copper, about 0.1–0.3 mg/L. However, copper sprays employed in the vineyard may introduce larger quantities into musts. Contact of musts or wines with copper-containing alloys may also result in copper pickup. Much of the must copper is fixed by yeasts during fermentation or precipitated as copper sulfide. As Table 59 indicates, the copper content of a variety of wines ranges from traces to 12.80 mg/L. In 41 Brazilian table wines the copper content ranged from traces to 3.20 mg/L (average 0.56 mg/L) (75). In 175 German wines the range was 0.0–9.5 mg/L (average 0.6 mg/L) (64). In 92 Italian wines for export the range was 0.10–0.86 mg/L (average 0.25 mg/L) (76). In a later study (18) in five Italian wines the average copper content was 0.18 mg/L. In 82 Baden wines the range was 0.005–2.400 mg/L (average 0.310 mg/L) (65). The legal limit of various countries is 1–5 mg/L (77).

Copper cloudiness occurs in wines containing more than about 0.2–0.4 mg/L of copper. This cloudiness was believed to occur in the absence of air; protein-copper compounds are responsible (78–81).

Copper (Cu^+) forms a red color with 2,2′-diquinoline (in pentanol or hexanol). Hydroxylamine in sodium acetate is used to reduce the copper. The alcohol layer contains the color, which is proportional to the copper content of the wine. The color may be compared to standards or to a standard curve (up to 2 mg/L) (82).

Copper can be determined in the ash solution by the color formed with pyrrolidine–dithiocarbamate or diethyl dithiocarbamate through extraction with organic solvents (82, 83). However, at low copper concentration the results can be less satisfactory than those obtained with atomic absorption spectrophotometry.

Procedure

Standard Copper Solution. Dissolve 0.50000 g of reagent quality copper metal in dilute nitric acid and make to a liter with water. Pipet 1, 2, 3, etc., mL of this stock solution into 1000-mL volumetric flasks, then add to each flask 150 mL of 95% ethanol and dilute to volume with water. The copper concentration of these diluted solutions is 0.5, 1.0, 1.5, etc., mg/L.

Hydrochloric–Citric Acid Reagent. Dissolve 75 g of citric acid in 300 mL of water in a 500-mL volumetric flask. Add 50 mL of concentrated hydrochloric acid and dilute to volume with water.

Sodium Diethyl Dithiocarbamate Solution. Dissolve 1 g of sodium diethyl dithiocarbamate in a few milliliters of ethanol, wash into a 100-mL volumetric flask with water, and dilute to volume.

Pipet 10 mL of each of the standard copper solutions into clean test tubes. Add 1 mL of the hydrochloric acid–citric acid reagent to each of the tubes, mix, add 2 mL of 5N ammonium hydroxide and remix, add 1 mL of the sodium diethyl dithiocarbamate solution to each tube, and shake. After a minute add 10 mL of amyl acetate and 5 mL of absolute methanol to each tube. Shake for 30 sec, then let stand until the phases separate. Draw off the lower aqueous phase completely and pour the colored organic layer from the tube into a small beaker. Add anhydrous sodium sulfate to the liquid until any cloudiness disappears. Stir and filter the dry organic phase through a hard filter paper into a clean dry test tube.

Prepare a standard curve by plotting copper concentration versus percentage of transmission in a photoelectric colorimeter with a blue 435-nm filter, or determine the optical density at 435 nm in a spectrophotometer. For the analysis of red or white wine, follow the procedure given above, substituting 10 mL of the wine sample for the standard solution.

Today, atomic absorption spectrophotometry is the preferred method (52, 70, 75, 84–86) if the equipment is available. It is reported (85) that the greatest sensitivity is achieved when the copper is chelated and extracted into ketone. Caputi and Ueda (73) used wine directly in a three-slot burner head but advised use of calibration curves because of the interference of glucose and ethanol.

AAS using ethanol 5% diethylenetriaminepentaacetic acid (ADTP) as the spectrophotometric buffer showed the copper contents of 14 Galician (Spanish) table wines (87) to range between 0.34 and 2.21 mg/L (average 0.88 mg/L). X-ray fluorescence (64) or neutron activation (88) may also be used for determining copper. Activated carbon (free of metals) added to red wines (1–3 g) did not interfere with the absorbance. Ashing avoids interferences but is, of course, slow (especially with musts and dessert wines). ICAP-AES (18) is also available. For the determination of copper, cadmium and lead potentiometric stripping analysis (PSA) has been used (89). The wines were used directly, and the results compared favorably with AAS.

TRACE ELEMENTS

Many old and new techniques have been used in detecting and measuring elements of less important or amounts. Among these are: spectrophotometric, X-ray fluoresence spectrography; flameless AAS (with corrections for ethanol and sugars and pretreatment); neutron activation analysis; ion-sensitive electrodes; flame atomic emission spectrometry; fluorometry; inverse voltammetry; and differential pulse polarography. ICAP-AES was used for antimony, arsenic, barium, cobalt, silver, and tin (18). In the same study AAS was preferred for lithium, potassium, rubidium, and sodium. Sugar and ethanol can cause major errors (90). For a review of trace elements see Reference (91): antimony, beryllium, cadmium, cesium, gold, hafnium, mercury, selenium, tantalum, thallium, and tungsten, as well as the ultrarare europium, lanthanum, samarium, and scandium. The reason for the determinations, except curiosity, is often not clear. The chance of contamination and public health dangers, as well as a small or large relation to wine quality, are possibilities. Recently (28), flameless AAS was recommended for aluminum, chromium, copper, nickel, and manganese,

AES was recommended for lithium and rubidium, and AAS was recommended for zinc.

We briefly describe the current analysis for these materials.

Aluminum

Wines should not be fermented or stored in aluminum vessels because of the danger of aluminum pickup.

Flame emission (13) analysis of 14 California table wines gave a range of 0.67–2.80 mg/L (average 1.48 mg/L) for aluminum. In an unknown number of Bulgarian wines the range was 0.31–8.70 mg/L (92). In 12 Romanian wines the range was 1.86–4.65 mg/L (average 3.15 mg/L) (11). In 51 Italian wines the range was 0.81–5.40 mg/L (average 1.40 mg/L). ICAP-AES was used (18). Techniques recommended are flame emission at 396.15 nm or atomic absorption at 309.3 nm (93).

Antimony and Arsenic

In 75 German wines the antimony content ranged from 1.2 to 9.0 µg/L (average 3.7 µg/l) (15) on ashed samples. Neutron activation was used. In 51 Italian wines the antimony did not exceed 0.06 mg/L (18).

Arsenic is normally present only in trace amounts in musts and wines. However, when arsenic-containing materials were used in wine making, higher amounts were present. The OIV (12) suggested that the maximum limit for wine is 0.1–0.2 mg/L. The U.S. limit of As_2O_3 in grapes is 3.5 mg/kg (94). In one recent study (95), 0.020 (or less) to 0.265 mg/L (average 0.078 mg/L) of total arsenic was reported in 23 wines (origin not given). The authors state that the amounts of arsenic present are so small that chronic consumption of large quantities of contaminated wine would be required before a substantial risk of toxic effects occurred. Campbell and co-workers (96, 97) reported 0.02–0.10 mg/L in European wines and 0.02–0.11 mg/L (average 0.06 mg/L) in nine California wines. They suggest "that no problematical levels (of toxicity) of arsenic exist in wines investigated." Crecelius (98) reported 0.001–0.530 mg/L (average 0.153 mg/L) in 19 California wines. In 18 varieties of Bulgarian wine (numbers of each not stated), traces to 0.20 mg/L were reported (92). In eight Italian wines, only 0.10–0.16 mg/L (average 0.12 mg/L) was found (58). In 51 Italian wines the arsenic content varied from 0.03 to 0.17 mg/L (average 0.08 mg/L) (18).

Numerous procedures for the quantitative determination of arsenic have been developed. The Gutzeit method is the most sensitive chemical method, but obtaining reproducible results is difficult (99). For submicro amounts, spectrophotometric analysis (99), gas chromatographic procedures (100), flameless AAS (95, 101), and proton-induced X-ray fluorescence (PIXF) analysis have been used.

Barium

In 14 California wines, barium concentrations ranging from 0.02 to 0.22 mg/L (average 0.15 mg/L) were reported (13). In 51 Italian wines, they varied from 0.05 to 0.20 mg/L (average 0.11 mg/L) (18).

Beryllium

The element has been determined in wines by a variety of procedures (91). The amounts found in seven Romanian wines varied from 0.2 to 0.5 μg/L (102).

Boron

Boron is seldom determined in musts or wines, although there are maximum limits. In general, 2–112 mg/L concentrations of H_3BO_3 have been reported; but when vineyards are treated with borax, higher amounts (up tp 200 mg/L) are found (103). In 309 musts and wines of various countries, H_3BO_3 concentrations of 5.3–50.5 mg/L were reported (average 17.9 mg/L) (101, 103). In 51 Italian wines, boron concentrations of 3.89–10.10 mg/L (average 5.19 mg/L) were found. ICAP–AES was used for the determination (18). In a variety of 41 European wines the boron content (as H_3BO_3 acid) ranged from 13.7 to 80.3 mg/L (average 22.2 mg/L) (104). These workers used wet ashing with sulfuric acid and hydrogen peroxide, as well as colorimetric determination at 620–630 nm with 1,1'-dianthrimide (105).

An ion-specific electrode has also been satisfactorily used (106), although considerable time (16 hr) is required. In 29 European and African wines the boron content (as H_3BO_3) range was 13.5–44.5 mg/L (average 32.4 mg/L).

Cadmium

Very little cadmium is found in musts (<5 μg/L) and about two-thirds of this is lost in fermentation (22). However, in the older literature, values of up to 4000 μg/L can be found. In the recent reports, 67 German musts and wines ranged from <1 to 49 μg/L (average 6 μg/L) (107). The highest value was believed to be a result of tank contamination. Flameless AAS (228.8 nm) was used for the determination in these studies. In 31 wines, Mack (108) reported traces to 10 μg/L (average 5 μg/L). However, from vines grown near a lead and cadmium pigment factory, higher values (up to 40 μg/L) were reported.

The cadmium appeared to be *in* the fruit not on the surface, as with lead. In 28 apertifs (vermouth, etc.) the range was 0.88–5.56 μg/L (average 1.42 μg/L). The lower cadmium content of recent European wines is attributed to better winery operations leading to less contamination (109). In 82 Baden wines, concentrations of 0.1–7.0 μg/L (average 1.0 μg/L) were reported (65).

AAS (often flameless), by the method of additions, has been used with lanthanum to reduce interference. Extraction of cadmium from ashed solutions of musts or wines by pyrrolidine dithiocarbamate or methyl pentyl ketone has also been recommended (52). Ashing seems to be necessary (110, 111). Details of the wet ashing required are available.

The official AOAC method involves wet digestion, dithizone extraction, and AAS. An alternative procedure using wet digestion, extraction with a chelating ion-exchange resin, eluting with sulfuric acid, and AAS gives comparable results somewhat faster (112). Potentiometric stripping analysis (PSA) was also used, with good concordance with AAS results (89).

Cesium and Chromium

In 74 German wines the range of cesium was 0.2–4.7 μg/L (average 1.7 μg/L). Neutron activation analysis was used (15).

Very few reports on chromium in musts or wines are available. In 75 German wines, chromium ranged from 0.010 to 2.570 mg/L (15). However, only six contained 0.200 mg/L or more. In the 69 remaining wines the average was 0.075 mg/L. Neutron activation analysis was used. Concentrations in four Soviet wines ranged between 0.050 and 0.069 mg/L, with an average of 0.054 mg/L (113).

Use of flameless AAS indicated that the chromium content of wines of 102 European countries varied from 0.010 to 0.152 mg/L (114). It was concluded that the main extraneous source was from asbestos filter pads. Very little was extracted from stainless steel tanks, 0.002 mg/L (115). Possibly the best method is extraction of chromate with methyl isobutyl ketone and flameless AAS using direct injection and a graphite furnace (114, 116). In seven Romanian wines the chromium varied from 0.004 to 0.031 mg/L (average 0.015 mg/L) (102). In 82 Baden wines, concentrations of 0.011–0.252 mg/L (average 0.077 mg/L) were reported. Flameless AAS was employed (65, 102). The chromium content of 51 Italian wines varied from 0.01 to 0.81 mg/L (average 0.06 mg/L). ICAP–AES was used (18).

Cobalt, Europium, and Hafnium

Neutron activation analysis was used to determine cobalt in 75 German wines: The range was 1.2–29.6 mg/L (average 5.8 mg/L) (105). In 51 Italian wines the cobalt content did not exceed 0.01 mg/L (18).

In 73 German wines the reported europium contents were 0.01–0.90 µg/L (average 0.37 µg/L) (15). One anomalous value of 8.2 g/L is omitted. Neutron activation analysis was employed. No reason for this determination is apparent.

In 73 German wines, investigators found hafnium in concentrations of 0.1–5.5 µg/L (average 0.7 µg/L). Neutron activation analysis was the method used (15). The value of the analysis is not known.

Lead

Lead occurs normally in soils and thus is found in plants and animals, particularly in marine life. Because of the public health aspects involved, the amount of lead in wine has been the subject of active research. For example, contamination of grapes with lead from automobile exhaust fumes has been of concern. Wines made from grapes grown near a freeway were higher (about double) than those taken from a greater distance. Much of the lead on grapes is not found in wine (117). Some lead is lost during fermentation (40–70% in small-scale fermentors) (118). Much is lost by blue-fining treatment with ferrocyanide. Formerly, lead arsenate was used as vineyard spray, but this practice has almost (119) disappeared. Use of a piece of lead to stop leaks in barrels is also now rare. Small amounts of lead are picked up from alloys (brass, bronze, etc.), rubber plastics, lacquers, paints, and so on (120), as well as from the lead capsule used to protect corks (121). Wines stored without lead foils apparently contained less than half as much lead. The cork itself contained high concentrations of lead, and this apparently passed between the cork and the glass to reach the wine. This hypothesis agrees with Bonastre (122) but disagrees with another report (123), which found little relationship between time under a lead foil and lead content in the wine. They make the point that even if old wines stored with lead capsules are consumed, their high price tends to restrict repeated excessive consumption.

The amounts found in commercial wines are generally low and are below the suggested legal limits. The data for lead, listed in the accompanying tabulation, are representative of recent results (123–130).

In 100 Italian wines (artisan and from the Naples region), lead contents of 0.07–1.23 mg/L were found (131). Of 560 Hungarian wines, 527 contained less than 0.4 mg/L of lead, 15 contained between from 0.4 and 0.6 mg/L, and 18 contained more than 0.6 mg/L (132).

The preferred method for lead determination is flameless AAS using wet or dry ashing. If dry ashing is employed, the ashing temperature should be below about 480°C—some say below 450°C—to prevent loss of lead (108, 124, 125, 127, 130, 133–135). Although AAS is the usual method, there are at least nine variations in amounts and procedures (136). In a comparative study with nine

Region of origin	Type	Reference	Number of samples	Lead concentration (mg/L)		
				Min.	Max.	Average
California	Wine	123	99	0.002	0.305	0.13
	Wine	14	300	0.001	0.62	0.04
Germany	Must, wine	107	67	0.01	0.12	0.038
	Grape juice	124	9	0.096	0.224	0.15
	Wine	69	82	0.020	0.193	0.045
Germany, etc.	Wine	108, 125	68	0.000	0.38	0.113
Italy	Wine	16	12	0.02	0.24	0.12
	Wine	76	92	0.08	0.25	0.18
	Wine	126	189	<0.1	0.53[a]	0.16
	Wine	127	120	0.03	0.68	0.24
	Wine	128	106	0.000	0.95[b]	0.16
	Wine	129	357	0.04	0.82	0.15
	Wine	19	132	0.1	1.26	0.23
	Wine	18	51	<0.01	0.37	0.06
Romania	Wine	102	11	0.001	0.06	0.02
Various	Aperitifs	130	18	0.04	0.261	0.127

[a]Omits outliers of 1.03, 1.45, and 2.43 mg/L.
[b]Omits one outlier of 3.30 mg/L.

laboratories, some laboratories obtained outlier results because of faulty technique. Without these a reproducibility of 0.083 mg/L and a repeatability of 0.169 mg/L was obtained, and with attention to proper procedures no difficulty with finding wines with over 0.3 mg/L should occur. In wet ashing, a quartz Kjeldahl flask is recommended, with heating not to exceed 320°C (137, 138).

Addition of ammonium pyrrolidine dithiocarbamate (APDC) to form a copper complex, coupled with extraction of the complex with methylisobutylacetone, has been employed to avoid the necessity for ashing (121, 134). The method of measured additions with flameless atomic absorption was used. The official AOAC method, which consists of wet ashing and photometry on carbon tetrachloride-extracted red dithiazone complex, has been widely and satisfactorily used (139, 140). It is faster than AAS. Proton-induced X-ray fluorescence has been employed (97). Precautions concerning the method are noted, and it is generally agreed that all methods for determining lead in biological materials require great analytical care because of the small amounts normally present. ICAP–AES has also been recommended (18). Potentrometric stripping analysis (PSA) was used directly in wines, and the results compared favorably with AAS (89). Polarography with differential inverse anode pulse voltammetry has been employed (137).

Lithium

This element is seldom determined in wines. Values of 0.13–0.26 mg/L were found in 25 wines of various origins (141), but in 27 Spanish wines analysts reported 0.4–2.70 mg/L (average 1.29 mg/L) (142). In 20 young French wines the lithium content varied from 0.002 to 0.013 mg/L. In bottled wines 10–50 years of age of lithium content varied from 0.009 to 0.033 mg/L. The higher lithium was attributed to extraction from the glass. Values of over 0.050' mg/L were considered aberrant. In 51 Italian wines the lithium content varied from 0.005 to 0.030 mg/L (average 0.010 mg/L) (18). In 14 California table wines the range was 0.010–0.110 mg/L (average 0.049 mg/L) (13). The reasons for the wide differences in the averages are not apparent. AAS directly on the wine is favored (142). Flame emission spectroscopy can also be used (13) and was preferred (143) because of its greater sensitivity (5×10^{-4} mg/L) as compared to AAS (1×10^{-3} mg/L).

Manganese

It is known that small amounts of manganese are normally found in plant materials. It is puzzling that the determination of this element should be of much interest to enologists. It may be that a few rare cases of contamination or the possibility of finding some as-yet undisclosed relation of the metal to wine quality have sparked this interest. The old idea that interspecific hybrids (direct producers or French hybrids) were high in manganese appears not to be correct. More manganese is present in wines in warm years in Germany (64). Blue fining (ferrocyanide) markedly reduces the manganese content. The amounts (mg/L) normally present in wines are small: in 25 Spanish (Galician) wines, 0.68–17.40 (average 5.5) (144); in eight Italian table wines, 0.58–0.84 (average 0.68) (84); in 80 German wines, average 1.00 (22); in 175 German wines, 0.24–2.25 (average 1.04) (64); in 12 Romanian red wines, 1.20–5.50 (average 2.39) (20); in 11 Romanian table wines, 0.09–0.33 (average 0.20) (102); in 82 Baden wines, 0.340–2.600 (average 1.270) (65); in 15 Bulgarian wines, 0.14–1.55 (92); in six Yugoslavian wines, 0.37–1.83 (average 0.88) (93); in 14 California wines, 0.65–2.03 (average 1.19) (13, 97); in 51 Italian wines, 0.37–2.44 (average 1.40) (18). In short, the manganese content of wines is about 1 mg/L. content of wines is about 1 mg/L.

AAS following wet or dry ashing with or without addition of ethanol or aminopolycarboxylate compounds gives similar results (52, 84, 110, 144). Colorimetric, polarographic X-ray fluorescent, neutron activation, ICAP–AES (18), and other analytical procedures have also been used (13, 97).

Mercury

Few analyses of wines for mercury are available. Extreme care is needed for analysis for mercury in biological materials because of the possibility of laboratory contamination. Quartz equipment and high-quality chemicals (Merck® suprapur), as well as other precautions, are required. Cold vapor AAS has been employed. In 63 wines (French, German, and Spanish) the mercury content varied from ≤0.02 to 0.65 µg/L (average 0.06 mg/L). These values are lower than those found in the earlier literature (145). There is more mercury in musts than in wines.

Molybdenum

In 51 Italian wines, molybdenum concentrations of 0.01–0.33 mg/L (average 0.05 mg/L) were found. ICAP-AES was used (18).

Nickel

The amounts of nickel found in wines are small, and extraneous sources are rare. The use of nickel alloys in stainless steel is a potential problem but is probably a small one. Use of blue fining (ferrocyanide) reduces the nickel content. In nine California wines the range was 0.00–0.14 mg/L (97). In eight Italian wines, concentrations of 0.05–0.40 mg/L (average 0.13 mg/L) were reported (110). Recently in 51 Italian wines, concentration of 0.01–0.09 mg/L (average 0.03 mg/L) were found using ICAP-AES (18). Proton-induced X-ray fluorescence and AAS with a graphite furnace (114) have been employed. AAS with differential pulse polarography has also been tried (114). Higher nickel values were traced to contact with asbestos, nickel-containing enamels, contaminated lime used for reducing acidity, and strongly reducing wines (high SO_2 and/or ascorbic acid) in tanks with improperly welded joints. Direct-reading emission spectroscopy has been used (13).

Rubidium, Scandium, Selenium, and Silicon

There are few reports of the rubidium content of wines. In nine California wines the range was 0.58–2.81 mg/L (average 1.21 mg/L) (97). In 71 German wines the range was 0.18–2.29 mg/L (average 0.61 mg/L) (15). In 51 Italian wines, concentrations of 1.17–4.99 mg/L (average 3.30 mg/L) were reported. ICAP-AES was used (18). Proton-induced X-ray fluorescence and neutron activation analysis and ICAP-AES have been employed. A possible value to human nutrition has been noted (20).

Scandium is mentioned in one report (15): In 75 German wines it ranged from 0.03 to 1.58 µg/L (average 0.39 mg/L). Neutron activation was used. The significance of the determination is not known.

Only 0.1–1% of the selenium level of the must is found in wines (110). Most is lost in the pomace or in the sediment. In eight Italian wines, the amounts ranged from 2 to 7 µg/L (average <3 µg/L). Up to 33 µg/L is found in the seeds, skins, and stems. In 28 German wines, selenium was not detected at the measurement sensitivity of the methods of 0.001 mg/L (146). A fluorometric micromethod based on the AOAC procedure was used. AAS has also been employed as has direct-reading emission spectroscopy (13). Another report (147) found undetectable (<0.4 µg/L) to 1.0 µg/L of selenium in 28 table wines.

Silicon was determined in 51 Italian wines by means of ICAP–AES. The range was 12.58–28.87 mg/L (average 19.60 mg/L) (18).

Silver, Strontium, Tantalum, and Tin

Very variable amounts (15) of silver were found in 74 German wines: 0.7–270 µg/L (average 29 µg/L). Only five samples had 100 µg/L or more. Neutron activation was used for the analysis. No obvious reason for this determination is apparent. In 51 Italian wines the silver did exceed 0.01 mg/L (18).

Fourteen California wines contained 0.36–1.72 mg/L (13) (average 1.51 mg/L) of strontium (7). In 12 Romanian wines a range of 1.24–1.95 mg/L (average of 1.75 mg/L) was reported (20). In another report (102), a range of only 0.14–0.78 mg/L (average of 0.44 mg/L) was found for 11 Romanian wines.

The strontium content of 51 Italian wines varied from 0.29 to 3.13 mg/L (average 0.82 mg/L) (18). Direct-reading emission spectroscopy and spectral analysis following semimicro separation were used.

Tantalum was determined on 62 German wines (15): Only 0.01–3.8 µg/L (average 0.43 µg/L) was found. Six wines had more than 1 µg/L. Neutron activation was used.

In Italian wines the tin content averaged 0.12 mg/L (18): Seven Romanian wines averaged only 0.009 mg/L (102).

Titanium and Vanadium

In seven Romanian wines, titanium concentrations of 0.25–0.6 µg/L were reported (102). In 51 Italian wines, a range of 0.01–0.22 µg/L (average 0.07 µg/L) was reported by means of ICAP–AES (18).

In seven Romanian wines a range of 0.2–0.7 µg/L of vanadium was found (102). By means of ICAP–AES on 51 Italian wines, a range of 0.01–0.20 mg/L (average 0.05 mg/L) was found (18).

Zinc

There are many recent reports of zinc determination in wines. Use of zinc-containing insecticides and fungicides may be one reason, but one study (148) showed that zinc-containing insecticides did not increase the zinc content of the wine. The possible importance of this element in yeast and in human nutrition should be noted. Zinc-deficient soils are known, and cases of accidental contamination have occurred; there are other reasons for interest in zinc determination. The accompanying tabulation (149-151) gives amounts of zinc found in wines of various regions.

Region of origin	Type	Reference	Number of samples	Zinc Concentration (mg/L)		
				Min.	Max.	Average
Brazil	Table	75	40	Trace	8.80	1.25
Bulgaria	Table	92	18[a]	0.1	2.5	—
California	Table	13	14	0.13	1.13	0.71
	Table, dessert	96	9	0.26	1.67	0.89
	Table	14	300	0.07	5.60	0.54
Germany	Table	15	75	0.21	3.20	0.99
	Table	22	175	0.00	7.80	1.63
	Table	149	24	0.04	1.31	0.38
	Table	65	82	0.015	2.700	1.060
Italy	Table	18	5	—	—	0.81
	Table	19	132	0.35	4.45	1.40
	Table	16	12	1.85	4.40	2.85
	Table	150	67	0.12	6.05	1.21
	Table	76	97	0.26	5.00	1.92
Portugal	Port	151	10	1.0	1.9	1.4
Romania	Table	102	11	0.13	0.92	0.36
	Red Table	20	12	2.50	11.7	5.34
Spain	Red Table	63	25	0.26	1.53	0.67
Yugoslavia	Table	88	6	0.18	2.08	1.32

[a]Minimum and maximum for 18 varieties.

ANIONS

The major anions that may be present in wines, such as acetates, lactates, malates, tartrates, and nitrates, have already been discussed. Besides these, others are of interest: bromide, chloride, phosphate, sulfate, fluoride, and iodide.

In beer, ion chromatography (IC) has been used to determine chloride, nitrate, phosphate, and sulfate (152). The system used combined ion-exchange separation, eluent suppression, and conductometric detection. The IC components included an eluent reservoir, and eluent pump, a sample introduction loop, a guard column, an anion-exchange separator column, a fiber-suppressor column, and a conductivity detector (Dionex® Corporation, Sunnyvale, CA). This avoids wet ashing. Application to wine seems feasible.

Bromide

Very little bromide is found in grapes and wines, usually less than 1 mg/L (146). The soil has a predominant effect. Soils high in chloride are often also high in bromide. Bromine-containing fungicides and insecticides have been used in the vineyard, however. Addition of organic bromine-containing antiseptics, such as monobromacetic acid and its esters, is illegal in the United States. A high bromine content is, therefore, evidence of sophistication. The suggested international limit is 1 mg/L. In nine California wines, bromine concentrations ranged between 0.14 and 0.54 mg/L and averaged 0.34 mg/L (13). In 73 German wines a range of 0.001–0.162 mg/L and an average of 0.052 mg/L were reported (15). Less than 0.15 mg/L was found in 280 German table wines (153), and 0.2 mg/L was suggested as the limit. In another investigation of 175 German wines, 31 had traces and 66 had up to 0.09 mg/L, 63 had 0.1–0.2 mg/L, 13 had 0.21–0.30 mg/L, and two had 0.40–0.42 mg/L (22). Higher amounts were reported (154) in non-German wines, and a 0.5 mg/L limit was recommended. Another report (155) found 0.6–1.8 mg/L in northern European and 0.8–5.55 mg/L is southern European wines. The 5.5 mg/L value strongly suggests fraudulent use of an organic bromine-containing antiseptic agent (156). In 1974, 12 of 32 Spanish wines contained fraudulently added organic bromine (157). This is surely rare today.

Procedures used include proton-induced X-ray fluorescence spectrometry (97), neutron activation analysis (15), and ion-selective electrodes (155).

The method for the determination of total bromine in wines, as suggested by the OIV (12), is to ash the wine under strongly alkaline conditions, add Chloramine-T at pH 4.6 to release free bromine, and then add phenosulfonphthalein. The formed tetrabromphenosulfonphthalein is determined colorimetrically. The procedure given is basically that of Jaulmes et al. (158) with a number of modifications.

Procedure

Prepare a standard bromide solution by dissolving 1.489 g of potassium bromide in 1 L of water and diluting a 25-mL aliquot to 500 mL. The bromide concentration of the diluted solution is 50 mg/L. By further dilution prepare solutions with bromide concentrations between 0.05 and 1.0 mg/L and analyze 100-mL aliquots according to the procedure given below. Prepare a calibration curve by plotting absorbance against concentration.

Pipet 100 mL of wine sample into a platinum dish. Add 7.5 mL of 2 N potassium hydroxide solution; dry first on a water bath, then dry in a 100°C oven; finally, ash at 525°C. Cool, add a few drops of water, redry, and reash at a temperature not exceeding 550°C. Dissolve the residue in 50 mL of hot water and 7 mL of 2 N sulfuric acid, then filter into an evaporating dish. Evaporate to 6–8 mL and transfer to a 20-mL graduated cylinder containing 0.5 mL of 2 N sulfuric acid and 1 drop of phenol red indicator solution (prepared by dissolving 0.15 g of phenol red in 15 mL of 1 N sodium hydroxide solution, then diluting to 1 L with water). Use water to rinse the dish, up to 19.5 mL. Mix well. Add about 5 drops of 2 N sulfuric acid until the color turns yellow; then dilute to 20.0 mL with water.

To a 5-mL aliquot of this solution, add 2 mL of a pH 4.7 buffer containing 30 mL of acetic acid and 68 g of sodium acetate trihydrate in a 1-L aqueous solution and 0.2 mL of the phenol red indicator solution. Mix, add 0.2 mL of a 0.35% aqueous Chloramine-T solution, and let react for exactly 1 min. Add 0.1 mL of a 10% sodium thiosulfate pentahydrate solution, mix, and measure the absorbance at 580 nm.

Organic Bromine. Extract the 100-mL wine sample in a liquid–liquid extractor at pH 1 with 100 mL of diethyl ether. Dry the extract and ash as above, and carry out the determination as above.

Chloride

Increasing interest in the sodium and chloride content of wines has occurred because of the use of anion exchange for tartrate stabilization, the growing of grapes near the ocean, the possibility of the use of chloride-containing additives, and legal restrictions on the chloride content.

There are no specific limits on chloride content in the United States. In European countries the suggested limit of sodium chloride is 500 mg/L, which corresponds to a chloride concentration of 304 mg/L. A limit as low as 152 mg/L has been suggested (35), but wines made from coastal-grown grapes or with grape concentrate exceeded this limit. There are a number of studies of the chloride contents of wines available. These are summarized, together with data on two other anions, in Table 60. The data show that normal table wines contain 30–150 mg/L and that wines made from grapes near the ocean may contain up to 300 mg/L.

The determination of the chloride content of wines is generally straightforward. Coloring material that may interfere with the end point must be removed; this can be done either by separating the chloride as barium chloride or by passing the wine through a strong anion-exchange column. The following procedure is essentially that of the OIV (12).

Table 60. Anion Content of Various Wines[a]

Region	Chloride[b] (g/L)		Phosphate[c] (g/L)		Sulfate[d] (g/L)	
	Range	Average	Range	Average	Range	Average
Algeria	0.042–0.108	0.083	—	—	0.45–0.70	0.58
Austria	0.010–0.210	0.082	0.150–0.600	0.292	0.18–1.53	0.71
California	—	—	—	—	0.07–3.03	1.10
Czechoslovakia	—	—	0.110–0.420	0.260	0.33–0.44	0.39
France	0.003–0.212	0.057	0.039–0.600	0.262	0.31–2.31	0.93
Germany	0.011–0.138	0.043	0.026–0.686	0.276	0.16–0.52	0.33
Hungary	—	—	0.052–0.129	0.084	—	—
Israel	0.046–0.140	0.082	—	—	0.90–1.72	1.47
Italy	0.024–0.170	0.070	0.070–0.637	0.236	0.12–2.92	0.75
Italy[e]	0.028–0.064	0.042	0.220–0.343	—	—	—
	—	—	—	0.407	0.42–0.62	0.55
Portugal	0.017–0.370	0.071	0.080–0.900	0.360	0.11–2.34	0.38
Romania	0.009–0.065	0.021	0.010–0.628	0.303	0.07–1.48	0.65
Spain	0.106–0.596	0.226	0.073–0.527	0.216	1.05–4.39	2.03
Switzerland	0.005–0.031	0.017	0.276–0.468	0.383	—	—
Tunisia	—	—	0.122–0.364	0.284	—	—
Yugoslavia	—	—	0.130–0.820	0.420	—	—

[a] The number of samples and wine types used vary; thus the results may not be compared directly without reference to the original reports (1, 3, 13, 35).
[b] Expressed as Cl.
[c] Expressed as PO_4.
[d] Expressed as K_2SO_4.
[e] 12 Trentino wines.

Procedure

Prepare a strong anion-exchange resin such as Dowex 1 × 2, 50/100 mesh, by treating it with 1 N sodium hydroxide solution, with 1 N nitric acid, then repeating the treatment. Keep under 1 N nitric acid. Place about 15 mL of the treated resin in a 25-mL column and wash with 50 mL of water. Pass 50 mL of wine through at a rate of about 1.5 mL/min, then rinse the column with 50 mL of water. Elute the chlorides from the column by passing 50 mL of 1 N nitric acid through it at the same speed. Collect the eluate in a 250-mL round-bottom flask. The eluate will not be colorless; to remove the color add several milliliters of 1:5 nitric acid and 3–5 drops of saturated potassium permanganate solution, stir, and let stand until the violet color disappears. If it does not completely disappear add a few drops of dilute hydrogen peroxide.

Now add 10 mL of a 15% solution of ferric ammonium sulfate dodecahydrate or 10 mL of 10% ferric nitrate solution, 20 mL of ethyl ether, and 10 mL of 0.1 N silver nitrate solution. Titrate the excess silver nitrate with 0.1 N potassium thiocyanate solution to a brick-red color that persists 5 min.

$$\text{Sodium chloride (g/L)} = \frac{(V_1 N_1 - V_2 N_2)(F)}{v}$$

where V_1 = volume of silver nitrate solution added (mL)
N_1 = normality of silver nitrate solution
V_2 = volume of potassium thiocyanate solution used for titration (mL)
N_2 = normality of potassium thiocyanate solution
F = molecular weight of sodium chloride (=58.45)
v = volume of wine sample (mL)

If the concentration is to be calculated in terms of chloride, use $F = 35.45$.

This procedure may be modified as follows (159):

Procedure

To 50 mL of wine in a 250-mL Erlenmeyer flask add 20 mL of nitric acid (1 + 5) and 10 mL of 0.100 N silver nitrate. Mix and let stand 5 min in the dark. Filter through a Büchner (or Gooch) funnel with a nitric acid-washed asbestos pad. Rinse the Erlenmeyer flask with several small amounts of 2% nitric acid. Bring to volume in a 200-mL volumetric flask. Measure the residual silver by atomic absorption spectrophotometry. This avoids the end point problems of the original method.

A conductometric procedure for beer should be satisfactory for wine (160). With the introduction of the chloride ion-selective electrode it has been used with wines (161, 162). The wine is blended 1:1 with a total-ionic-strength-adjusted buffer (85 g $NaNO_3$ is dissolved in deionized water, and 57 mL of acetic acid and 30 g of NaOH are added; when cool make to a 1 L). Since ethanol reduces the electrode potential, a correction must be applied for ethanol. The method of additions is used, and the results are plotted on a graph. The results agreed well with methods based on titration with silver nitrate or on potentiometric response.

Fluoride

A fluoride-specific electrode is used for the determination of this element (106, 163, 164). Aluminum may interfere. The error is about 10%. The suggested limit of the OIV is 5 mg/L; however, on January 14, 1977, German law reduced the limit in that country to 0.5 mg/L (128). A literature survey (165) showed that only 16 of 351 European grape juices and wines had more than 0.5 mg/L of fluorine and only one exceeded 0.95 mg/L. In 362 German grape juices and wines, at least six exceeded 0.5 mg/L. In 46 non-German grape juices and wines, four had more than 0.5 mg/L. The fluoride content of wines is usually less than 0.5 mg/L. However, wines made from grapes grown near a ceramic factory (166) were higher. (40% had 2–5 mg/L.) Also, liquid sugar may be made with fluoridated water (19). Possibly some fluoride comes from containers or from illegal preservatives. Using the ion-specific electrode (167) in 132 Italian wines the fluoride content was 0.04–1.33 mg/L, but only 6% of the samples exceeded 0.5. This is the present German federal limit.

Iodide

The normal iodide content of wines is small (168): 0.1–0.2 mg/L. In 130 Italian wines it varied from <0.1 to 0.400 mg/L. Wines made from vineyards near the sea were higher in iodides. The ion-selective electrode was used with good reproducibility and repeatability. With an iodide-specific electrode, the iodine content ranged from 0.01 to 0.87 mg/L with 17.4% of the samples above 0.5 mg/L (167).

Phosphate

The phosphate content of wines is normally derived entirely from that present in the musts. Occasionally phosphate is added to musts that show a sluggish fermentation or which have "stuck." The phosphate is present in both the inorganic and organic forms. The latter is of considerable significance in certain phases of alcoholic fermentation. The phosphate content is also significant in the formation of ferric phosphate casse.

The belief that high-phosphate wines are high in quality does not appear to be substantiated by the data now available. Certain wines do appear to be much higher in phosphates than others, largely because of variations in soil conditions. The phosphate content normally varies from about 50 to 900 mg/L (as PO_4), with white wines containing somewhat lower amounts as compared to red wines. Table 60 summarizes data for a number of wines of different origin.

The recommended reference procedure of the OIV (12) is to precipitate the phosphate with ammonium phosphomolybdate and then titrate the precipitate. Better recovery was obtained (168) by ashing at 600–650°C, removing the silica, and precipitating as quinolein phosphomolybdate. Ion-selective electrodes have not been as successful for phosphate as for other anions and cations. See, however, p. 353.

The procedure given below consists of ashing, removing the silica by dehydration in hydrochloric acid, and developing a blue color using a molybdate-aminonaphtholsulfonic acid solution. This gives the total phosphate concentration.

Procedure

Prepare an aminonaphtholsulfonic acid solution in the following way. Into a 1-L beaker add 120 g of sodium bisulfite or 110 g of sodium metabisulfite, 24 g of sodium sulfite, and 2.5 g of purified 1-amino-2-naphthol-4-sulfonic acid. Gradually add water to form a paste. Stir to break up lumps and add up to 800 mL of water. Transfer to a 1-L volumetric flask and dilute to volume. Filter and store in a brown bottle.

Prepare an ammonium molybdate solution by dissolving 25 g of ammonium molybdate in 200 mL of water, adding 500 mL of 10 N sulfuric acid, and diluting to 1 L with water.

Prepare standard phosphate solutions by dissolving 25 g of potassium dihydrogen phosphate, dried for 2 hr at 105°C, in 1 L of water. The phosphate concentration of this solution is 0.306 mg/L expressed as PO_4, or 0.1 mg/L expressed as phosphorus.

Pipet a 10-mL aliquot of the wine or must sample into a platinum or porcelain crucible, evaporate to dryness on a steam bath, place in a 100°C oven for 1 hr, and ash at not over 550°C in a muffle furnace. Heat only as long as necessary to obtain a white ash. Leave the lid off the crucible and open the muffle furnace door slightly to facilitate oxidation. If necessary, cool, add 2-3 drops of water, dry, and reash.

Now add 30 mL of 1:5 hydrochloric acid to the ash and carefully evaporate to dryness on a hot plate, being careful to prevent spattering. Continue heating for 10-12 min after the hydrochloric acid is entirely evaporated. Cool and repeat the treatment. Usually two treatments are sufficient to render the silicic acid insoluble. Finally, add 30 mL more of 1:5 hydrochloric acid and bring to a boil. Filter through hard filter paper directly into a 100-mL volumetric flask. Wash the filter with hot water. Cool and dilute the solution in the flask to volume with water.

Pipet an aliquot from the solution prepared in the preceeding paragraph, containing 0.3-1 mg of phosphorus, into another 100-mL volumetric flask. Usually 25 or 50 mL is sufficient. Dilute to about 75 mL with water, then add 10 mL of the ammonium molybdate solution and 5 mL of the aminonaphtholsulfonic acid solution, dilute to volume with water, and shake. A blue color, proportional to the amount of phosphate present, will develop. Since the color will continue to darken for a few minutes, color comparisons should be made after a specified time, usually 30 min.

Parallel to the determination, pipet 1, 2, and 3 mL of the standard phosphate solution into 100-mL volumetric flasks, add 5 mL of 1:5 hydrochloric acid, about 70 mL of water, 10 mL of the ammonium molybdate solution, and 5 mL of the aminonaphtholsulfonic acid solution; finally dilute to volume with water.

After the specified time determine the absorbance of the standards and unknown at 600 nm, plot a calibration curve, and establish the concentration in the unknown from the curve.

A modification of this procedure adds an excess of ammonium molybdate and determines the excess molybdenum by AAS after the ammonium phosphomolybdate precipitate is separated (169). Results to 2% accuracy with the proper dilution were obtained. A molybdenum blue colorimetric procedure on grape juice or wine gave comparable results (correlation coefficient 0.9837) to a volumetric procedure using ashed samples (170).

Sulfate

There are sulfates in normal musts, but the amount is small and varies from vineyard to vineyard, depending on soil conditions. These sulfates are partially carried over into the wines. Some is utilized by yeasts. Some is converted to sulfide or sulfur dioxide during fermentation. Yeasts differ in their ability to form sulfite or sulfur dioxide (usually reciprocally). Obviously, low-sulfide-producing strains are preferred. Additional sulfates in wines may arise from the oxidation of sulfur dioxide or from the addition of calcium sulfate in the treatment known as *plastering* (*platrage* in France, *enyesado* in Spain). This treatment is legal in the sherry district of Spain and in the production of flor sherry in the United States. The result of plastering is the formation of tartaric acid, which reduces the pH of the must. This has a salutary influence in keeping the fermentation free of bacterial activity. The overall reaction is

$$CaSO_4 + 2KH(C_4H_4O_6) \longrightarrow Ca(C_4H_4O_6) + K_2SO_4 + H_2(C_4H_4O_6)$$

Potassium sulfate has a slightly bitter taste that is undesirable in white table wines but that may have some value in sherries.

The legal limit of the sulfate content is 2 g/L in the United States and 3 g/L in certain other countries, calculated as potassium sulfate. Unplastered California white wines have a sulfate concentration of 0.5–1.0 g/L, expressed as potassium sulfate, whereas unplastered red wines give a concentration of 0.5–2.0 g/L, but an occasional sample exceeds the 2 g/L limit. This indicates that the continuous use of high concentrations of sulfur dioxide and the storage of such wines in warm cellars leads to an accumulation of sulfate that approaches the legal limits. For data on various wines, see Table 60.

Cases of sophistication of low-acid wines using sulfuric acid are known. The sulfate and pH determinations are obviously helpful in such cases. Sulfur dioxide and sulfide determination have been given previously. Where many sulfate determinations are regularly needed, an automated turbidimetric procedure is possible (171, 172). Another rapid procedure precipitates the sulfate with excess barium chloride and, after filtration, determines the excess barium by AAS. The coefficient of variation was ±4.5% (137). With barium sulfate precipitation there is always the question as to how stoichiometric the reaction is and whether there are occluded ions. The procedure given below is a modification of the classic procedure (173, 174). Good results with this procedure (175) have been obtained in comparison with OIV (121) reference procedure.

Procedure

Prepare 50 mL of desulfited wine by adding 1 mL of concentrated hydrochloric acid to the wine sample, placing it on a steam bath, and bubbling carbon dioxide through while reducing the volume to about 25 mL. Add dropwise 5 mL of a 10%

barium chloride solution in 5:95 hydrochloric acid and continue heating for 5 min. Cover with a watch glass and allow to set at least for 6 hr in a warm place. Test for the completeness of the reaction by adding a drop of the barium chloride solution.

During this period stir up some asbestos fibers in water and pour into a Gooch crucible under suction until a fine pad is developed. Place the crucible in a muffle furnace and heat at about 650°C for 20 min. Cool in a desiccator and weigh.

Now place the crucible into holder, place on a suction flask, apply suction, and moisten with water. Decant the contents of the beaker into the crucible and finally transfer the precipitate to the crucible with hot water, using a rubber policeman. Continue washing until the filtrate is free of chloride. Be sure the filtrate is perfectly clear; if not, recombine it with the unfiltered solution, reheat, and allow to set an additional period of time before refiltering.

Dry the precipitate in the crucible at 100°C and heat in a muffle furnace at about 650°C for 30 min. Cool in a desiccator, weigh, and reheat until constant weight.

$$\text{Sulfate, as } K_2SO_4(g/L) = \frac{(W)(0.747)(1000)}{v}$$

where W = weight of barium sulfate precipitate (g)
v = volume of wine sample (mL)

REFERENCES

1. S. Ferenczi and I. Tuzson, *Borgazdasag* **17**, 87–89 (1969).
2. E. O. Flores, *Univ. Nac. Cuyo, Fac. Cienc. Agrar. Rev.* **16**, 131–136 (1970).
3. M. A. Amerine, *Adv. Food Res.* **8**, 113–224 (1958).
4. M. P. Sabatelli, *Atti Accad. Ital. Vite Vino, Siena* **23**, 103–121 (1971).
5. F. Bandion and M. Valenta, *Mitt. Klosterneuburg* **35**, 96–107 (1985).
6. T. Müller and G. Würdig, *Weinwirtsch., Tech.* **121**, 356–360 (1985).
7. M. Procopio and L. Laporta, *Vini Ital.* **6**, 197–200 (1964).
8. A. S. C. Garcia and M. I. S. Garcia, *Cien. Tec. Vitiv.* **2**, 15–27 (1983).
9. J. Barna and F. Grill, *Mitt. Klosterneuburg* **30**, 247–249 (1980).
10. I. Szabo and K. Kerényu, *Szolesz. Boraszat* **7**, 16–19 (1985).
11. J. Barna and F. Grill, *Mitt. Klosterneuburg* **32**, 122–123 (1982).
12. *Receuil des Méthodes Internationales d'Analyse des Vins.* Office International de Vigne et du Vin, Paris, 1978.
13. R. J. Cox, R. R. Eitenmiller, and J. J. Powers, *J. Food Sci.* **42**, 849–850 (1977).
14. C. S. Ough, E. A. Crowell, and J. Benz, *J. Food Sci.* **47**, 825–828 (1982).
15. H. Siegmund and K. Bächmann, *Z. Lebensm.-Unters. -Forsch.* **164**(1–7), 298–303 (1977–1978).
16. G. Marghesi, D. Tonon, and C. Mattarei, *Vini Ital.* **19**, 301–308 (1977).

17. B. Cerutti, S. Mannino, and A. Vecchio, *Riv. Vitic. Enol.* **34**, 145-156 (1981).
18. F. S. Interesse, F. Lamparelli, and V. Aloggiò, *Z. Lebensm.-Unters. -Forsch.* **178**, 272-278 (1984); see also *ibid.* **181**, 470-475 (1985).
19. J. M. Robertson, B. L. Kirk, and A. C. Crum, *Rep. N.Z. Dep. Sci. Ind. Res., Chem. Div.* **CD-2247**, 1-43 (1976).
20. D. Davidescu, M. Jacob, and N. Anghelide, *Lucr. Stiint. Inst. Agron. "Nieolae Balescu," Bucuresti* **12**, 95-104 (1969).
21. R. G. Olmedo, P. Puertas, T. A. Masoud, and M. C. Diez, *An. Bromatol.* **29**, 281-304 (1977).
22. K. G. Bergner and B. Lang, *Wein-Wiss.* **26**, 186-193 (1971).
23. W. Postel, B. Meier, and R. Markert, *Mitt. Klosterneuburg.* **36**, 20-27 (1986).
24. G. Charalambous and K. J. Bruckner, *Tech. Q. Master Brew. Assoc. Am.* **14**, 197-208 (1977).
25. M. Karvánek and G. Janiĉek, *Sb. Vys. Sk. Chem.-Technol. Praze [Oddil]* **E35**, 75-79 (1971).
26. J. A. Zee, I. M. Szöghy, R. E. Simard, and M. Desmarais, *Am. J. Enol. Vitic.* **32**, 93-99 (1981), see also *ibid.* **34**, 152-156 (1983).
27. D. Yan, E. Stumpp, and G. Schwedt. *Z. Anal. Chem.* **322**, 474-479 (1985).
28. E. Martin, V. Iadaresta, J. C. Giacometti, and J. Vogel, *Mitt. Geb. Lebensmittelunters. Hyg.* **77**, 528-534 (1986).
29. G. Pallotti, B. Bencivenga, and C. Botré, *Rass. Chim.* **27**, 293-305 (1975).
30. G. L. Hill and A. Caputi, Jr., *Am. J. Enol. Vitic.* **20**, 227-236 (1969).
31. M. Lepage, R. Brisson, J. Daniel, S. W. Frey, E. Heger, R. L. McAdam, M. Moll, D. Thompson, D. Whitney, and S. T. Likens, *J. Am. Soc. Brew. Chem.* **35**, 133-136 (1977).
32. H. Greve and J. Rehbein, *Fluess. Obst* **46**, 48-50, 55 (1979).
33. N. Kh. Grinberg and E. I. Kesel'brener, *Konservn. Ovoshchesush. Prom-st.* (3), 37-38 (1984).
34. G. Scollary and C. Skepper, *Aust. Grapegrower Winemaker* **244**, 74, 76, 78-79 (1984).
35. A. Patschky and H.-J. Schöne, *Mitt., Rebe Wein, Obstbau Fruechteverwert.* **20**, 432-435 (1970).
36. F. Bandion and M. Valenta, *Mitt. Klosterneuburg* **33**, 73-74 (1983).
37. M. Astier-Dumas and H. Gounelle de Pontanel, *Ann. Hyg. Lang. Fr. Med. Nutr.* **8**, 95-102 (1972).
38. A. Tamborini and A. Magro, *Riv. Vitic. Enol.* **23**, 89-94 (1970).
39. H. W. Berg, M. Akiyoshi, and M. A. Amerine, *Am. J. Enol. Vitic.* **30**, 55-57 (1979).
40. F. G. McNerney, *J. Assoc. Off. Anal. Chem.* **59**, 1131-1134 (1976).
41. A. González, F. Bermejo, and C. Baluja, *Rev. Agroquim. Tecnol. Aliment.* **24**, 233-238 (1984).
42. J. Ribéreau-Gayon, E. Peynaud, P. Sudraud, and P. Ribéreau-Gayon, *Traité d'Oenologie*, Vol. **1**, Dunod, Paris, 1976.
43. J. Blouin, L. Llorca, and P. Leon, *Connaiss. Vigne Vin* **5**, 99-107 (1971).
44. H. Bieber and K. Wagner, *Mitt., Rebe Wein, Obstbau Fruechteverwert.* **16**, 104-106 (1966).
45. J. P. Bonnemaire, S. Brun, J. C. Cabanis, and H. Maltras, *Trav. Soc. Pharm. Montpellier* **31**, 245-252 (1971).

46. T. I. Semenova, R. I. Sukhomlin, and V. F. Yas'ko, *Vinodel. Vinograd. SSSR* **37**(6), 30-31 (1977).
47. A. Casanova and L. Mora, *Ann. Technol. Agric.* **23**, 403-410 (1974).
48. B. Weger, *Riv. Vitic. Enol.* **21**, 441-443 (1968).
49. R. Franck and C. Junge, *Weinanlytik, Untersuchungen von Wein und ähnlichen alcoholischen Erzeugnissen sowie von Fruchtsäften.* Carl Heymanns Verlag, Cologne, 1970-1983.
50. M. T. Hagnet and J. C. Cabanis, *Trav. Soc. Pharm. Montepellier* **35**, 49-54 (1975).
51. M. C. Polo, M. D. Garrido, C. Llaguno, and J. Garrido, *Rev. Agroquim. Tecnol. Aliment.* **9**, 600-605 (1969).
52. K. G. Bergner and B. Lang, *Dtsch. Lebensm.-Rundsch.* **67**, 121-124 (1971).
53. M. R. de L. A. Peres, *An. Inst. Vinho Porto* **25**, 157-167 (1974).
54. F. Griselli, O. Colagrande, and S. Bonsi, *Sci. Tecnol. Aliment.* **4**, 167-169 (1974).
55. L. Hart, G. Mylonas, and G. Scollary, *Aust. Grapegrower Winemaker* **244**, 30-32 (1984).
56. A. J. Courtoiser, F. Forestier, and A. Rey, *Ann. Technol., Agric.* **26**, 1-23 (1977).
57. J. Blouin and P. Bonnin, *Rev. Fr. Oenol.* **23**, 3-15 (1983).
58. A. Corrao, A. M. Gattuso, and G. Fazis, *Agric. Intal. (Pisa)* **69**, 369-381 (1969).
59. J. P. Bonnemaire, S. Brun, J. C. Cabanis, and H. Mattras, *Trav. Soc. Pharm. Montpellier* **31**, 253-260 (1971).
60. M. R. de L. A. Peres, *An. Inst. Vinho Porto* **25**, 143-155 (1974).
61. R. G. Martinek, *J. Am. Med. Technol.* **36**, 241-253 (1974).
62. A. Lavin, J. P. Sotomayor, and F. Martin, *Agric. Tec. (Santiago)* **36**, 86-88 (1976).
63. G. Chinchetru, *Sem. Vitivinic.* **30**, 2765, 2767, 2769 (1975).
64. K. G. Bergner and B. Lang, *Mitt., Rebe Wein, Obstbau Fruechteverwert.* **20**, 281-295 (1970).
65. H. D. Mohr, *Wein-Wiss.* **37**, 275-284 (1982).
66. F. Bermejo, C. Baluja, and J. A. Ravina, *Analusis* **3**, 157-163 (1975).
67. A. O. Calagrande and A. del Re, *Ind. Agrar.* **7**, 206-209 (1969).
68. A. A. Schilt and P. J. Taylor, *Anal. Chem.* **42**, 220-224 (1970).
69. B. Mandrou, H. Tabre, and H. Mattras, *Ann. Falsif. Expert. Chem.* **69**, 219-228 (1976).
70. O. Mitoseru, *Ind. Aliment. (Bucharest)* **24**, 143-145 (1973).
71. G. Bohm, G. Sontag, and G. Kainz, *Mikrochim. Acta* **1**, 311-318 (1977).
72. M. Vidal and J. Blouin, *Rev. Fr. Oenol.* **70**, 49-55 (1978).
73. A. Caputi, Jr. and M. Ueda, *Am. J. Enol. Vitic.* **18**, 66-70 (1967).
74. M. K. Meredith, S. Baldwin, and A. A. Andreasen, *J. Assoc. Off. Anal. Chem.* **53**, 12-16 (1970).
75. G. R. Biancalana, L. D. Moretto, and R. Tokoro, *Arq. Inst. Biol., Sao Paulo* **39**, 311-315 (1972).
76. S. Martina, R. Caravell, and R. M. Barbagallo, *Riv. Vitic. Enol.* **26**, 509-512 (1973).
77. J. Schneyder, *Mitt., Rebe Wein, Obstbau Fruechteverwert.* **24**, 129-134 (1974).
78. C. E. Kean and G. L. Marsh, *Food Technol.* **10**, 335-359 (1956).
79. M. A. Joslyn and A. Lutkon, *Food Res.* **21**, 384-396 (1956).
80. A. Lutkon and M. A. Joslyn, *Food Res.* **21**, 456-476 (1956).
81. R. G. Peterson, M. A. Joslyn, and P. W. Durbin, *Food Res.* **23**, 518-524 (1958).
82. M. Boese, *Beckman Rep.* **2**, 13-15 (1973).

83. M. A. Amerine and M. A. Joslyn, *Table Wines, The Technology of Their Production*, 2nd ed. Univ. of California Press, Berkeley and Los Angeles, 1970.
84. A. Amati, R. Rastelli, and A. Minguzzi, *Ind. Agrar.* **6**, 630–634 (1968).
85. A. G. Cameron and D. R. Hackett, *J. Sci. Food Agric.* **21**, 535–536 (1970).
86. D. H. Struck and A. A. Andreasen, *J. Assoc. Off. Anal. Chem.* **50**, 334–337 (1967); see also pp. 338–339.
87. F. Bermejo, C. Baluja, and P. Sanchez, *Papers Int. Congr. At. Absorp. At. Fluorescence Spectrom. 3rd, 1971* Vol. 2, pp. 469–486 (1973).
88. J. Stuper, J. Korosin, Z. Kerin, and D. Kerin, *Hrana Ishrana* **18**, 322–332 (1977).
89. S. Mannino, *Riv. Vitic. Enol.* **35**, 297–304 (1982).
90. J. C. Méranger and E. Somers, *J. Assoc. Off. Anal. Chem.* **51**, 922–925 (1968).
91. H. Eschnauer, *Z. Lebensm.-Unters. -Forsch.* **134**, 13–17 (1967).
92. V. Lechev and Ts. Ganeva, *Lozar. Vinar.* **24**,(8), 33–41 (1975).
93. M. Ihnat, *Anal. Biochem.* **73**, 120–133 (1976).
94. C. F. Jelinek and P. E. Corneliussen, *Environ. Health Perspect.* **19**, 83–87 (1977).
95. D. D. Siemer, R. K. Vitek, P. Koteel, and V. C. Houser, *Anal. Lett.* **10**, 357–369 (1977).
96. J. L. Campbell, B. H. Orr, and A. C. Noble, *Nucl. Instrum. Methods* **142**, 289–291 (1977).
97. A. C. Noble, B. H. Orr, and J. L. Campbell, *J. Agric. Food Chem.* **24**, 532–536 (1976).
98. E. A. Crecelius, *Bull. Environ. Contam. Toxicol.* **18**, 227–230 (1977).
99. J. Kellen and B. Jaselskis, *Anal. Chem.* **48**, 1538–1540 (1976).
100. G. Schwedt and H. A. Russel, *Chromatographia* **4**, 242–245 (1972).
101. K. C. Tam, *Environ. Sci. Technol.* **8**, 734–736 (1974).
102. D. Davidescu, *Ann. Technol. Agric.* **27**, 53–54 (1978).
103. R. Wohler and B. Holbach, *Mitteilungsbl. GDCh-Fachgruppe Lebensmittelchem. Gerichtl. Chem.* **30**, 89–91 (1976).
104. H. E. Haller and C. Junge, *Aktuel. Ernaehrungsmed. Klin. Prax.* **2**, 51–55 (1977).
105. C. Junge and H. E. Haller, *Weinwirstchaft* **113**, 140–143 (1977).
106. A. R. Deschreider and R. Meaux, *Rev. Ferment. Ind. Aliment.* **29**, 75–80 (1974).
107. S. Wallrauch, *Fluess. Obst* **41**, 134–135 (1974).
108. D. Mack, *Dtsch. Lebensm.-Rundsch.* **72**, 431–432 (1975).
109. H. Eschnauer, *Bull O.I.V.* **55**, 592–597 (1982).
110. M. Giaccio, *Boll. Lab. Chim. Prov.* **26**, 260–268 (1975).
111. H. S. Billig, H. Dreger, and H. Trepton, *Fluess. Obst* **42**, 369–375 (1975).
112. R. A. Baetz and C. T. Kenner, *J. Assoc. Off. Anal. Chem.* **57**, 14–17 (1974).
113. I. F. Gribovskara, A. V. Kariakin, and G. Sh.-V. Damadamaev, *Vinodel. Vinograd. SSSR* **31**(7), 24–27 (1971).
114. K.-G. Bergner and G. Braun, *Mitt. Klosterneuburg* **34**, 64–67, 73–80 (1984).
115. G. Kehry and K.-G. Bergner, *Mitt. Klosterneuburg* **35**, 7–15 (1985).
116. B. Medina and P. Sudraud, *Connaiss. Vigne Vin* **14**, 79–96 (1980).
117. G. Basile and V. Tarallo, *Boll. Lab. Chim. Prov.* **25**, 185–188 (1974).
118. M. Cotrau, M. Proca, G. Danela, G. Iftoda, A. Cotrau, and V. Cotea, *Rev. Med.-Chir.* **81**, 69–73 (1977).
119. G. Hamelle, *Ann. Falsif. Expert. Chim.* **69**, 101–105 (1976).

120. G. Botta, *Ind. Bevande* **5**, 109-112 (1976).
121. B. Medina, G. Giumberteau, and P. Sudraud, *Connaiss. Vigne Vin* **11**, 183-193 (1977).
122. J. Bonastre, *Ann. Technol. Agric.* **8**, 388-437 (1959).
123. M. A. Edwards and M. A. Amerine, *Am. J. Enol. Vitic.* **28**, 239-240 (1977).
124. B. Boppel, *Z. Lebensm.-Unters. -Forsch.* **153**, 345-347 (1973).
125. D. Mack, *Dtsch. Lebensm.-Rundsch.* **71**, 71-72 (1975).
126. F. C. Ramusino, A. Stacchini, R. Giordano, P. Ravagnan, P. Tandoi, and P. D. Femmine, *Riv. Soc. Ital. Sci. Aliment.* **5**, 307-313 (1976).
127. A. Amati, A. Minguzzi, and R. Rastelli, *Riv. Sci. Tecnol. Alimenti Nutr. Um.* **6**, 39-41 (1976).
128. C. Stella and M. P. Sabatelli Gellini, *Atti Accad. Ital. Vite Vino, Siena* **27**, 43-57 (1975).
129. R. Stefani, *Vini Ital.* **20**, 343-345 (1978).
130. W. Ooghe and H. Kastelyn, *Ann. Falsif. Expert. Chim.* **69**, 351-367 (1976).
131. B. Biscardi, R. de Fusco, and A. Parella, *Nuovi Ann. Ig. Microbiol.* **25**, 381-389 (1974).
132. L. Nagy, J. Posta, and L. Papp, *Z. Lebensm.-Unters. -Forsch.* **160**, 141-142 (1976).
133. H. E. Haller, *Dtsch. Lebensm.-Rundsch.*, **71**, 430-431 (1975).
134. P. Petrino, M. Cas, and J. Estienne, *Ann. Falsif, Expert. Chim.* **69**, 87-99 (1976).
135. M. Edwards, S. Sakellariadis, and R. G. Burau, *Conf. Adv. Innov. Soil Plant Water Anal. Berkeley, 1973* pp. 75-78 (1974).
136. R. Ristow and M. Bernau, *Dtsch. Lebensm.-Rundsch.* **78**, 125-130 (1982).
137. H. R. Brunner, *Schweiz. Z. Obst.-Weinbau* **122**, 164-168 (1986).
138. M. R. de L. A. Peres and J. Pereira, *An. Inst. Vinho Porto* **25**, 181-190 (1974).
139. P. Jaulmes, G. Hammelle, and J. Roques, *Ann. Technol. Agric.* **9**, 189-245 (1960).
140. Y. Tep, A. Tep, and S. Brun, *Trav. Soc. Pharm. Montpellier* **33**, 65-72 (1973).
141. M. Ney, *Ann. Falsif. Expert. Chim.* **58**, 263-266 (1965).
142. F. Bermejo and C. Baluja, *Acta Cient. Compostelana* **9**, 123-138 (1972); **11**, 3-12 (1974).
143. B. Medina and P. Sudraud, *Ann. Falsif. Expert. Chim. Toxicol.* **72**, 65-71 (1979).
144. C. Baluja, M. R. de A. Peres, and A. Gonzalez, *Quim. Anal.* **30**, 295-305 (1976).
145. M. Stoeppler, K. May, R. Enkelmann, and H. Eschnauer, *Heavy Met. Environ. Int. Conf. 4th, 1983* vol. 1, pp. 245-248 (1983).
146. J. Schneyder and W. Kain, *Mitt., Rebe Wein, Obstbau Fruechteverwert.* **22**, 106-108 (1972).
147. K. G. Bergner and H. Ackermann, *Mitt., Rebe Wein, Obstbau Fruechteverwert.* **24**, 135-148 (1974).
148. J. Stupar, *Nova Proizvod.* **26**, 189-192 (1975).
149. R. H. Dyer and A. F. Ansher, *J. Food Sci.* **42**, 534-536 (1977).
150. A. Castelli, A. Cavallaro, G. Cerutti, and M. Fittipaldi, *Riv. Vitic. Enol.* **27**, 247-257 (1974).
151. H. de Almeida and M. R. de A. Peres, *An. Inst. Vinho Porto* **22**, 27-40 (1971).
152. J. C. Jancar, M. D. Constant, and W. C. Herwig, *J. Am. Soc. Brew. Chem.* **42**, 90-93 (1984).
153. R. Woller and B. Holbach, *Dtsch. Lebensm.-Rundsch.* **68**, 1-5 (1972).
154. B. Holbach and R. Woller, *Wein-Wiss.* **28**, 210-222 (1973).
155. J. E. Graf, T. E. Vaughn, and W. Kipp, *J. Assoc. Off. Anal. Chem.* **59**, 53-55 (1976).
156. F. Moreno Martin, M. Xirau, and W. H. Perxas, *Circ. Farm.*, **242**, 101-116 (1974).

157. F. Moreno Martin, M. Xirau, and M. A. Perxas, *Medicamenta, Ed. Farm.* **43**, 103-108 (1974).
158. P. Jaulmes, S. Brun, and J. C. Cabanis, *Chim. Anal. (Paris)* **44**, 327-330 (1962).
159. M. D. Garrido, C. Llaguno, and J. Garrido, *Am. J. Enol. Vitic.* **22**, 44-46 (1971).
160. G A. Howard and P. Gertsen, *Brauwissenschaft* **30**, 19-21 (1977).
161. J. L. Bernal, Ma. J. del Nozal, A. J. Aller, and L. Deban, *Rev. Agroquim. Tecnol. Aliment.* **23**, 137-142 (1983).
162. Ma. J. del Nozal, J. L. Bernal, and Ma. M. Pedrero, *An. Bromatol.* **35**, 51-65 (1983).
163. W. Dostel, A. Gorg, and U. Guvenc, *Brauwissenschaft* **31**, 71-76 (1978).
164. J. L. Hidalgo, M. Pastor, and J. A. Pérez-Bustamente, *An. Bromatol.* **35**, 67-77 (1983).
165. R. Woller and B. Holbach, *Wein-Wiss.* **33**, 71-76 (1978).
166. F. Mammi, M. Manfredini, and E. Iogono, *Boll. Chim. Unione Ital. Lab. Prov.* **2**, 190-204 (1976).
167. A. Vecchio, C. Finoli, S. Germani, and G. Cerutti, *Technol. Aliment.* **4**, 10-14 (1981).
168. L. Cava and A. Cavallaro, *Boll. Chim. Unione Ital. Lab. Prov.* **5**, 140-147 (1979).
169. F. Hernandez, D. Garrido, and M. D. Cabezudo, *An. Quim.* **77**, 370-374 (1981).
170. J. Barna and F. Grill, *Mitt. Klosterneuburg* **30**, 117-119 (1980).
171. A. C. Hautman and C. S. Du Plessis, *Agrochemophysica* **7**, 21-25 (1975).
172. C. S. Du Plessis and A. C. Hautman, *Agrochemophysica* **7**, 47 (1975).
173. H. Rebelein, *Mitteilungsbl. CDCh-Fachgruppe Lebensmittelchem. Gerichtl. Chem.* **23**, 107-109 (1969).
174. J. Schneyder, *Mitt., Rebe Wein, Obstbau Fruechteverwert.* **20**, 278-279 (1970).
175. C. Junge and D. Olschimke, *Dtsch. Lebensm.-Rundsch.* **67**, 152-153 (1971).

Ten

GASES

Changes in the oxidation–reduction potential of a wine or grape juice depend mainly on the presence or absence of dissolved oxygen. The gases involved in either removing or protecting the wine or juice from contact with oxygen are nitrogen and carbon dioxide. The presence of budding yeast cells will rapidly remove oxygen. Changes that occur, such as the formation of hydrogen sulfide, can also be an indication of the oxidizing or reducing condition of the wine. The changes in the measurable oxidation–reduction potential can be relatively slow but do respond to the conditions listed above as well as to the contents of sulfur dioxide, phenol, ascorbic acid, iron, copper, and other substances. The oxidation–reduction potential is seldom measured because of the more meaningful analyses available for various factors that cause such changes. Table 61 (1) summarizes the oxidation–reduction potential values of a number of half-reactions that are of interest in wine.

OXYGEN

A positive as well as a negative value for wine results from exposure to oxygen. Young wines in a highly reducing state may require small amounts of oxidation, especially if hydrogen sulfide has formed. Also, for aging of red wine a certain amount of oxygen must penetrate the wooden barrels and as much as 1000 mg/L of oxygen or more is absorbed (2) for the slow oxidation and polymerization of the phenolic compounds in order for the wine to age properly. At a certain point in the life of a wine it becomes advantageous to limit or prevent contact with oxygen. With the white wines it is generally conceded that the oxygen content should be kept to a minimum during aging. The solubilities of oxygen at saturation for various conditions are given in Table 62 (3, 4). The need to monitor oxygen pickup in wine to determine the source of oxygen has been well documented (5, 6). Once the problem points in the winery operation have been identified and the proper steps have been taken to reduce oxygen pickup to a minimum, measurements can be made less frequently. Oxygen ab-

Table 61. Standard Reduction Potentials of Some Half-Reactions found in Wine[a]

Half-reaction	E_0 at pH 7.0 V
$\frac{1}{2} O_2 + 2 H^+ + 2 e^- \rightarrow H_2O$	0.816
$Fe^{3+} + 1 e^- \rightarrow Fe^{2+}$	0.771
$\frac{1}{2} O_2 + H_2O + 2 e^- \rightarrow H_2O_2$	0.30
Dehydroascorbate + 2 H^+ + 2 $e^- \rightarrow$ ascorbate	0.06
Fumarate + 2 H^+ + + 2 $e^- \rightarrow$ succinate	0.03
Acetaldehyde + 2 H^+ + 2 $e^- \rightarrow$ ethanol	−0.163
Acetyl-CoA + 2 H^+ + 2 $e^- \rightarrow$ acetaldehyde + CoA	−0.41
Acetate + 2 H^+ + 2 $e^- \rightarrow$ acetaldehyde	−0.60

From Reference 1.

[a]The standard conditions are unit activity for all components with the exception of H^+, which is 10^{-7} M, and the gases are at 1 atm pressure.

Table 62. Saturation Concentration of Oxygen Under Various Conditions in Alcohol-Water Solutions and in Wine at 20°C and 760 Torr Pressure (3, 4)

Ethanol (vol %)	Oxygen (mg/L)
0	9.2
5	8.5
10	8.3
11.2[a]	8.25
15	8.2
20	8.1
20.5[a]	7.8

[a]In wine.

sorption can occur at leaks in seals of centrifugal pumps, as well as during cold stabilization, racking, fining, mixing, and bottling. Wines undergoing cold stability will be saturated to 12 mg/L of oxygen (7).

The absolute necessity to prevent oxidation of grape juices has not been fully established. The affinity for oxygen by polyphenol oxidase in grape juice has been noted (8) as well as the impracticability of sparging grape juice with nitrogen to remove oxygen. Only 50 mg/L of sulfur dioxide was sufficient to inhibit grape polyphenol oxidase in grape juice up to 6-7 hr (9). The rapid uptake of oxygen in juice is a function of variety (10). Also it was found that oxidized juices, when fermented, gave wines with color about equal to the same juice fermented without prior oxidation.

The classic chemical method for determining oxygen in wines (11) is based on the measurement of the oxidative capacity of a wine sample after removal of oxygen by nitrogen stripping, then that titer is compared to the amount of sodium sulfite required to reduce the nonstripped wine. The red wine color interferes with the oxidation–reduction indicator used (indigo carmine); thus only white wine can be analyzed in this way. The method is time-consuming and erratic.

The Clarke membrane oxygen electrode was first applied to the measurement of dissolved oxygen in wine by Ough and Amerine (12). Further tests (13, 14) have verified the utility of the oxygen electrode for wine oxygen measurements. The electrode consists of a platinum or gold cathode covered with a plastic membrane that is permeable to oxygen. A silver anode is enclosed within the electrode. The two poles are connected by a saturated potassium chloride bridge. A voltage of about 0.6 V is applied to the system. Oxygen passes through the membrane and is reduced. Current flows between the two poles and is measured by a readout device. Most electrodes require the liquid being measured to be circulating over the membrane, otherwise the oxygen in the immediate area of the surface of the membrane will be depleted, giving low results. High levels of carbon dioxide or hydrogen sulfide can cause erroneous reading or poisoning of the electrode.

The partial pressure of the oxygen in the solution is determined. At any given temperature, saturation values depend on the molecular makeup of the liquid. See Table 63 for values.

Figure 31 shows the system in which the membrane-type electrode is used. Today, commercial electrodes with self-contained stirrers for direct insertion into wine bottles are also available. The membrane electrode is calibrated against air. The instrument should be temperature compensated, or steps should be taken to maintain a constant temperature.

Procedure

Obtain a juice or wine sample with the absolute minimum of aeration. (One acceptable method is to attach a hose to the tank and with a slow, gentle liquid

Table 63. Concentration (mg/L) of Oxygen Under Atmospheric Pressure in Alcohol–Water Solutions at Several Alcohol Concentrations and Temperatures

Alcohol (% v/v)	Temperature (°C)						
	0	5	10	15	20	25	30
0	14.6	12.8	11.3	10.15	9.2	8.4	7.7
10	12.6	11.3	10.1	9.2	8.3	7.7	7.2
20	11.7	10.6	9.6	8.9	8.1	7.5	7.1

From Reference 3.

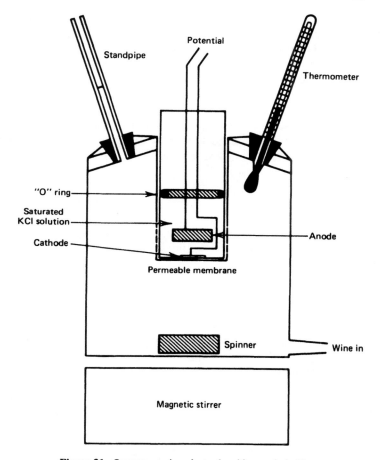

Figure 31. Oxygen-sensing electrode with sample holder.

flow, purge the line of air. Then bottom fill a sample bottle previously filled with nitrogen.) No significant delay is allowable between sampling and start of analysis. Carefully tranfer sample to measuring receptacles, which can be similar to the receptacle shown in Figure 31. Before doing this, however, calibrate the electrode with air for 100% reading at the temperature of the wine or juice. Maintain the wine or juice at the sampling temperature of the tank. After the wine has been transferred to the measuring receptacle, all undissolved gas expelled, and the system closed except for the standpipe vent, start the magnetic stirrer. Allow the system to come to equilibrium, and take a reading of the percentage of saturation. Multiply this value by the saturation value of wine (at 20°C, 8.2 mg/L). See Tables 62 and 63 for values at different temperatures and alcohol values.

In the beer industry the dissolved oxygen is of great concern, and various methods for determining it are available. Some applications appear possible in the wine industry. For dissolved oxygen an in-line procedure is available (15). It consists of a sensor (Orbisphere model 2110 contained in a model 2950 flow chamber) and a model 2606 display sensor showing the dissolved oxygen content at various temperatures. The sensor is of the polarographic type, that is, it consists of an oxygen-detecting cathode of gold, a silver anode or reference electrode, and an alkaline KCl electrode. The cell is enclosed and separated from the flowing beer by a fluorinated plastic membrane. Various membranes are available for different purposes. The response is not influenced by carbon dioxide. The membrane is cleaned daily and did not deteriorate with a year's use.

Concentrations of oxygen and nitrogen were measured by GC (16) in the headspace and, by use of Henry's and the ideal gas law, in the beer. Again one can envisage that this may be of interest in the wine industry, especially in sparkling wine. The equipment is not simple.

CARBON DIOXIDE

Fermentation produces carbon dioxide in large volumes, but most of it is lost during the yeast fermentation. The quantity remaining depends primarily on the alcohol concentration of the finished wine and the temperature of the storage. A significant amount of carbon dioxide is evolved during the malolactic fermentation. The residual amount depends mainly on the same factors. The biochemical source of carbon dioxide is from decarboxylation of organic acids by decarboxylase enzymes. Table 64 gives the calculated saturation values for carbon dioxide in table wines (17).

Sometimes mixtures of nitrogen and carbon dioxide are used to protect wine or are used at bottling to ensure some level of residual carbon dioxide in the

Table 64. Saturation Concentrations of Carbon Dioxide in Wines at Two Ethanol Concentrations and Four Temperatures

Temperature (°C)	Carbon dioxide (g/100 mL)	
	Ethanol (10% v/v)	Ethanol (12% v/v)
0	0.312	0.307
10	0.218	0.214
20	0.150	0.147
30	0.108	0.106

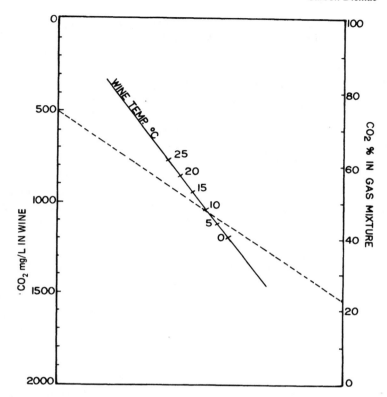

Figure 32. Nomograph relating the percentage of carbon dioxide in gas mixtures of carbon dioxide and nitrogen applied to wine at 1 atm pressure (over the temperature range 0–25°C) to the concentration of carbon dioxide in the wine at the temperature of application. Reproduced with permission of Lonvaud-Funnel and Ribéreau-Gayon (18).

wine. An approximation of the carbon dioxide remaining can be estimated from a nomograph (see Figure 32) (18). In the case of mixtures of gases, the amount of carbon dioxide at saturation depends primarily on the partial pressure of the carbon dioxide in the gas mixture and the temperature of the wine. The taste threshold for carbon dioxide in white and red wines for the "spritz" or gas detection level was found to be 0.05–0.06 g/100 mL (17). Below this level the effect of carbon dioxide in aroma and flavor enhancement is negligible. For white wines a level above 0.06 g/100 mL increased the "freshness" of the aroma. Levels above 0.10–0.12 g/100 mL were sometimes harmful to the quality of the dry wines. However, in Bordeaux wines some were below the desired levels of carbon dioxide.

During the production of sparkling wines carbon dioxide is one of the main products. The solubility of carbon dioxide in the wine is again a function of the temperature and of the partial pressure of the carbon dioxide in the headspace of the bottle or tank. Tables are available (19) for solubility and pressure values under various conditions.

Carbon dioxide is a very toxic gas if the concentrations reach certain levels. At 3–5% carbon dioxide, breathing becomes extremely rapid, and dizziness and ringing in the ears occur because of hyperventilation. For an extended time, 0.5% is the maximum allowable. More than 10% carbon dioxide in the air is deadly, especially over extended periods. At 18% carbon dioxide concentration an immediate cessation of breathing occurs and death results (4). Extreme care should be taken in or about fermenting wine tanks. One cause of winery deaths is the entrance of workers into improperly ventilated areas that contain fermentation gases, mainly carbon dioxide. The use of a candle or a kerosene lantern is not sufficient warning, since combustion can continue at carbon dioxide levels that cause unconsciousness. Modern detection systems should be used in dangerous areas.

A number of systems are in use for analysis of carbon dioxide in wine. A vacuum manometric system measures the volume of carbon dioxide released from an acidified sample with vigorous agitation by measuring the amount of mercury displaced (20, 21). This method is one of the official AOAC methods (22), 11.062; the equipment is expensive and subject to breakdown that necessitates tedious and time-consuming troubleshooting. A source of error was found with apple wine, and a modification has been offered (23).

Another AOAC procedure, 11.067, approved for use with wine is the volumetric method (24), involving the acidification of wine acid, usually phosphoric acid, and the distillation of the freed carbon dioxide into three traps, the last containing barium ion. The barium carbonate formed is titrated with a standard acid. Various improvements of this method have been described (25–28).

The enzymatic method (29, 30) did not prove workable in practice (31). However, the use of the carbonic anhydrase was combined successfully with a direct volumetric titration technique (32, 33). The principle of this reaction is the titration of bicarbonate ion with standard acid between pH 8.6 and 4.0. In this range the bicarbonate ion is completely coverted to carbon dioxide gas (equation 1):

$$HCO_3^- + H^+ \longrightarrow H_2O + CO_2 \qquad (1)$$

Essentially all the carbon dioxide is in the bicarbonate form at the initial pH. The use of the carbonic anhydrase ensures that the solubility of carbon dioxide in alkaline solutions is rapid and complete. When the standard acid is added the tendency to form carbon dioxide is small but significant (equation 2), whereas the reverse equilibrium (equation 3) is slow without the carbonic anhydrase.

The second equilibrium reaction (equation 4) at the alkaline pH is rapid and complete.

$$CO_3^{2-} + 2H^+ \longrightarrow H_2CO_3 \longrightarrow CO_2 + H_2O \qquad (2)$$

$$CO_2 + H_2O \longrightarrow H_2CO_3 \qquad (3)$$

$$H_2CO_3 + OH^- \longrightarrow HCO_3^- + H_2O \qquad (4)$$

Procedure

Cool sample to be analyzed to ~5°C, open, and add equivalent of 5 mL of NaOH reagent (50% w/w) per 375 mL of sample. Immediately recap, mix, pipet 10 mL into 40 mL of H_2O, and add three drops of carbonic anhydrase solution [0.1 mg/mL of H_2O (when refrigerated solution is stable for 1 week)]. Titrate with 0.04545 N H_2SO_4 to pH 8.6. This normality was selected so that 1 mL is equivalent to 20 mg of CO_2 per 100 mL.

Refill buret and continue titration to pH 4.0. Record the titer between pH 8.6 and 4.0.

Obtain a blank as follows: De-gas a 25-mL duplicate of the sample with agitation for about 1 min under ≥28 in. of vacuum in a 500-mL filter flask with 3 drops carbonic anhydrase solution. Add 0.33 mL of NaOH solution. Pipet 10 mL of this sample into a 150-mL beaker containing 40 mL of H_2O. Stir and titrate as above.

Calculate using the following equation:

mg of CO_2 per 100 mL = (mL of sample − mL of blank) × normality H_2SO_4

× 44.0 × 10 = (mL of sample − mL of blank) × 20

To correct for dilution with NaOH, use the equation

mg of CO_2 per 100 mL = (mL of sample − mL of blank) × 20 × 1.013

A specific ion electrode for determination of carbon dioxide in wine has been described (34). The electrode consists of a Teflon membrane which allows the gas to pass through, depending on the partial pressure of the carbon dioxide in the wine, into a saturated bicarbonate solution. The carbon dioxide causes an equilibrium shift in the bicarbonate system:

$$CO_2 + H_2O \rightleftharpoons H_2CO_3 \longrightarrow HCO_3^- + H^+$$

forcing the equilibrium to the right. A glass electrode is in contact with the bicarbonate solution and reacts to the change in [H^+]. The change is proportional to the changes in the carbon dioxide concentration of the wine. A sensitive potentiometer is required. As with other membrane electrodes, good circulation over the membrane is necessary. The method was tested against the standard and classic methods and was found to be adequate (31). The method is simple and direct, but the electrode is costly. The directions accompanying the electrode should be carefully followed.

The carbon dioxide content of wines has been measured by gas chromatography (35). A procedure has been described (36) for the analysis of carbon dioxide, nitrogen, and oxygen in beer and headspace analysis, with the sample taken directly from the capped bottle. Complete analysis took 30 min.

The use of sensitive radioactive detection devices to measure the amount of ^{14}C present in carbon dioxide has allowed the detection of carbonated wines (37). If the source of carbonation is inorganic, as is the case with artificial carbonation, the ^{14}C percentage is appreciably lower. For further information see Reference 38.

A unique determination using the principle of infrared absorbance of carbon dioxide was developed (39). The method is useful for rapid on-stream measurements. The results are comparable to those from the other methods, and a commercial model is available.

The legal limit for CO_2 in still wines in the United States is 0.392 g/L. Above this the wines are considered sparkling and are taxed at a higher rate. Thus, a rapid and accurate method is needed by control agencies and, of course, by still wine producers who produce wines with a carbon dioxide content that approaches this limit. An automatic procedure (39a) uses a lactic acid reagent (containing surfactants) to liberate CO_2 or carbonates from the wine. The gas is forced into a fixed volume space where a thermal conductivity detector responds proportionally to the volume of gas in a reference cell open to ambient air. A carbon dioxide analyzer (Corning 960®) is used.

Procedure

Chill a bottle of wine to be analyzed in an ice–salt bath before opening. Carefully open the bottle and add 1.0 mL of 50% w/w NaOH for each 100 mL of wine. The pH should be ~11. Close and gently invert the bottle several times. Calibrate the analyzer using 60 mmol of sodium carbonate standard per liter (5.04 g $NaHCO_3$ diluted to 1 L). Inject wine into the carbon dioxide analyzer with a 100-μL pipet. Repeat until satisfactory calibration is obtained. Now invert the bottle of wine a few times and with the micropipet inject 100 μL and record the amount of millimoles per liter of CO_2 from the display reading. Convert to milligrams per liter by multiplying by 4.40. To correct for the dilution of adding the NaOH, multiply by (volume wine and base used)/(volume of wine used). The instruction manual should be followed (Corning Medical, Medfield, MA 02052). Each determination requires less than 1 min. The limit of sensitivity is 400 mg/L, but by proper dilution it can be used for wines with over 400 mg/L.

HYDROGEN SULFIDE

Hydrogen sulfide (H_2S) is never purposely present in wine in detectable amounts. The main source of this gas is the reduction of elemental sulfur used for dusting of the grape vines. Many wine yeasts will rapidly convert the sulfur to sulfide:

$$S + 2H \longrightarrow H_2S$$

This reaction took place in about equal amounts with all 12 wine yeast strains studied by Eschenbruch et al. (40).

The second source of hydrogen sulfide is from the yeast reduction of SO_3^{2-} and SO_4^{2-}. Yeasts can be divided into low and high SO_3^{2-} producers. High SO_3^{2-} producers usually give low S^{2-} (40). Addition of L-cysteine with or without pantothenoic acid causes increased S^{2-} only in some strains (low sulfite producers) (41). It is conceivable that grapes low in methionine and fermented with a pantothenate-deficient yeast would be more likely to form hydrogen sulfide via the SO_4^{2-} route. The abbreviated pathway appears in Figure 33. Sulfate is the usual starting source, but other sources exist: SO_3^{2-} and chemical vineyard sprays containing elemental sulfur (42). Benomyl has been shown (43) to produce only limited amounts of hydrogen sulfide when the residue is fermented in grape juice—less than 0.7 μg/L from fruit sprayed four times and with a residue of 2.25 mg per killogram of grapes.

Other volatile sulfur compounds are related to the formation of hydrogen sulfide. The formation of mercaptan and disulfide has been reviewed (44).

The threshold level of hydrogen sulfide depends somewhat on the wine type. A generally accepted detectable level is about 1-2 μg/L.

Hydrogen sulfide is a highly poisonous gas. Levels of 600 mg/L air for 30 min can be fatal (45). Levels associated with wine are much less than this, and wines sold commercially should have below sensory threshold levels.

Various chemical methods have been tried for the determination of hydrogen sulfide in beer (46). Some of these procedures proved to be adequate for beer analysis, but they were not successful when applied to wine.

One of the more promising methods is the use of an ion-specific electrode that can selectively measure sulfide ion activity (47). The electrode consists of a solid silver sulfide membrane separating a silver nitrate internal reference solution and a silver wire from the sample solution. The membrane allows only the flow of silver ions. The distribution of the silver ions between the inner and outer solutions develops a voltage that is dependent on the silver ion concentration in the sample solution. In turn the silver ion concentration depends on the sulfide ion solution, on the pH, and on the ionic strength of the solution. By the use of standard curves and knowledge of the solution pH, the concen-

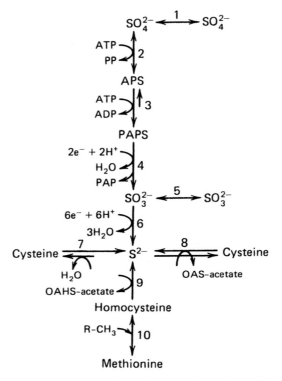

Figure 33. Pathway for the formation of S^{2-} by wine yeast from amino acids, SO_4^{2-}, SO_3^{2-}, and elemental sulfur. 1, Sulfate permease; 2, ATP-sulfurase; 3, APS-kinase; 4, PAPS-reductase; 5, sulfite permease; 6, sulfite reductase; 7, serine sulfhydrase; 8, OAS-sulfhydrase; 9, OAHS-sulfhydrase; 10, methyl transferase. APS, adenosine-5-phosphosulfate; PAPS, 3′-phosphoadenosine-5′-phosphosulfate; OAS, O-acetylserine; OAHS, O-acetylhomoserine.

tration of the hydrogen sulfide can be calculated. Variations in the method have been suggested (47).

Investigation shows a loss of sensitivity of these electrodes with time (48). Films deposited on the surface of the membrane at the membrane/solution interface were metallic silver. This caused Ag_2S and Ag_2O films to form that gave mixed potentials and non-Nerstian responses. On the other hand, it was shown that the only real limitation to sensitivity in micrograms per liter was the solubility of the membrane material and manipulative and interference conditions (49). See Chapter 12 for a general discussion of ion-specific electrodes.

The free sulfide at the pH of wine (3.0–4.0) is essentially all present as hydrogen sulfide, with only traces present as the bisulfide (HS^-) and much less as the sulfide (S^{2-}). Certain other forms may also be present as complexes with proteins or other substances. These forms may be partially or wholly titrated with silver nitrate. The sulfide forms are titrated rapidly and completely. The electrode can be used to determine the end point of the titration. The measurement can be considered an estimate of the amount present and of the potential capacity of the sample to readily form hydrogen sulfide.

NITROGEN

Nitrogen is a nontoxic gas that is harmless unless it is used to exclude oxygen and causes suffocation. It is used to some extent to exclude air in wine making. It also is useful to strip oxygen from the wine. It is more soluble than oxygen and less soluble than carbon dioxide. It normally is not measured in wine because its presence is neither a fault nor an advantage.

PURIFICATION OF GASES

Molecular sieves (50), namely zeolites, are commonly used to clean up gases for analytical purposes. They separate molecules not only according to size but also according to polarity, with the more polar molecules being absorbed more readily. No chemical reactions take place. The sieves can be purged readily by heat or vacuum.

It is important to select the proper size sieve: In cleaning water from ethanol, the pore mesh should be about 3 Å. Molecules up to that diameter will be absorbed. Fine angstrom pore size sieves are used to dry and remove random hydrocarbons from carrier gases for gas chromatographic work. If extreme purity is required, the carrier gas should be passed through a liquid nitrogen trap. These molecular sieves are available in 3–5- and 10-Å pore sizes.

REFERENCES

1. I. H. Segal, *Biochemical Calculations*. Wiley, New York, 1968.
2. J. Ribéreau-Gayon and E. Peynaud, *Traité d'Oenologie*, Vol. 2. Béranger, Paris.
3. J. V. Mejane, M. Debailland, and J. Lecerf, *Ind. Aliment. Agric.* **90**, 719–727 (1973).
4. B. C. Rankine, *Aust. Grapegrower Winemaker* **11**, 18–19 (1974).
5. H. Müller-Spath, *Weinwirtschaft* **113**, 144–146, 148–150, 152–157 (1977).
6. R. R. Cant, *Am. J. Enol. Vitic.* **11**, 164–169 (1960).
7. G. Prass and J. Vingo, *Food Technol. Aust.* **28**, 475–477 (1976).
8. F. Radler and E. Torokfalvy, *Z. Lebensm.-Unters. -Forsch.* **152**, 38–41 (1973).
9. F. Neradt, *Weinberg Keller* **17**, 519–526 (1970).

10. B. B. Wite and C. S. Ough, *Am. J. Enol. Vitic.* **24**, 148-152 (1973).
11. H. W. Berg, *Wines Vines* **32**(5), 23 (1951).
12. C. S. Ough and M. A. Amerine, *Food Res.* **24**, 744-748 (1959).
13. K. Henning and A. Lay, *Weinberg Keller* **10**, 165-169 (1963).
14. B. C. Rankine and K. F. Pocock, *Aust. Wine, Brew. Spirit Rev.* **87**(11), 44-52 (1969).
15. R. Mitchell, J. Hobson, N. Turner, and J. Hale, *J. Am. Soc. Brew. Chem.* **41**, 68-72 (1983).
16. T. J. Wick and K. J. Siebert, *J. Am. Soc. Brew. Chem.* **44**, 72-78 (1986).
17. A. Lonvaud-Funel and P. Ribéreau-Gayon, *Connaiss. Vigne Vin* **10**, 391-407 (1976).
18. A. Lonvaud-Funel and P. Ribéreau-Gayon, *Connaiss. Vigne Vin* **11**, 165-182 (1977).
19. M. A. Amerine, H. W. Berg, R. E. Kunkee, C. S. Ough, V. L. Singleton, and A. D. Webb, *The Technology of Wine Making*, 4th ed. Avi Publishing Co., Westport, CT, 1980.
20. U.S. Internal Revenue Service, *Carbon Dioxide Test Procedures*, Ind. Circ. No. 59-47. Office of the Commissioner of Internal Revenue, Washington, DC, 1959.
21. M. J. Pro, A. Etienne, and F. Feeny, *Vacuum System of the Determination of Carbon Dioxide*, Doc. No. 5157 (5-59). Internal Revenue Service, Washington, DC, 1959.
22. *Official Methods of Analysis*, 14th ed. Association of Official Analytical Chemists, Arlington, VA, 1984, pp. 220-230.
23. A. Caputi, Jr., *J. Assoc. Off. Anal. Chem.* **56**, 843-845 (1973).
24. A. D. Etienne and A. P. Mather, *J. Assoc. Off. Agric. Chem.* **39**, 844-848 (1956).
25. J. E. Beck, *J. Assoc. Off. Agric. Chem.* **43**, 652-654 (1960).
26. K. Hennig and A. Lay, *Weinberg Keller* **9**, 202-205 (1962).
27. E. Capt, J. F. Schopfer, and A. Dufour, *Mitt. Geb. Lebensmittelunters. Hyg.* **60**, 114-120 (1969).
28. W. Postel, *Dtsch. Lebensm.-Rundsch.* **66**, 185-190 (1970).
29. R. L. Morrison, *J. Assoc. Off. Agric. Chem.* **45**, 627-629 (1962).
30. R. L. Morrison, *J. Assoc. Off. Agric. Chem.* **46**, 288-289 (1963).
31. C. Barrère, M. Ducasse, C. Vigne, M. Vidal, and A. Tallec, *Vignes Vins* **273**, 31-38 (1978).
32. A. Caputi, Jr., M. Ueda, P. Walter, and T. Brown, *Am. J. Enol. Vitic.* **21**, 140-144 (1970).
33. A. Caputi, Jr., *J. Assoc. Off. Anal. Chem.* **54**, 782-784 (1971).
34. P. Ribéreau-Gayon and A. Lonvaud-Funel, *C. R. Acad. Agric. Fr.* **262**, 491-497 (1976).
35. H. L. Ashmead, G. E. Martin, and J. A. Schmidt, *J. Assoc. Off. Agric. Chem.* **47**, 730-734 (1964).
36. F. Drawert, W. Postel, and C. Kurer, *Brauwissenschaft* **23**, 217-224, 258-265 (1970).
37. P. Resmini and G. Volonterio, *Sci. Tecnol. Aliment.* **2**, 39-46 (1972).
38. G. E. Martin, H. W. Krueger, and J. M. Burggraff, *J. Assoc. Off. Anal. Chem.* **68**, 440-443 (1985).
39. A. Caputi, Jr. and M. Ueda, *Am. J. Enol. Vitic.* **24**, 116-119 (1973).
39a. J. D. Mitchell and C. R. Benjamin, *J. Assoc. Anal. Chem.* **64**, 547-549 (1981).
40. R. Eschenbruch, P. Bonish, and B. M. Fischer, *Vitis* **17**, 67-74 (1978).
41. H. H. Dittrich and T. Staudenmayer, *Dtsch. Wein-Ztg.* **104**, 707-709 (1968).
42. R. Eschenbruch, *Wynboer* **482**, 22-23 (1971).
43. P. G. Marias and M. M. Kruger, *Argochemophysica* **8**, 61-64 (1976).
44. H. J. Niefind and G. Spath, *Eur. Brew. Conv., Proc. Congr.* **14**, 297-308 (1973).

45. W. B. Deichmann and H. W. Gerade, *Toxicology of Drugs and Chemicals*. Academic Press, New York, 1969.
46. M. W. Brenner, J. L. Owades, and R. Golyzniak, *Proc. Am. Soc. Brew. Chem.* **1953**, 83-89 (1953).
47. *Instruction Manual, Silver Ion Activity Electrode, Sulfide Ion Activity Electrode*, Model 94-16. Orion Research Inc., Cambridge, MA, 1970.
48. J. Gulens and B. Ikeda, *Anal. Chem.* **50**, 782-787 (1978).
49. D. J. Crombie, G. J. Moody, and J. D. R. Thomas, *Anal. Chim. Acta* **80**, 1-8 (1975).
50. D. W. Breck and J. V. Smith, *Sci. Am.* **220**, 85 (1959).

Eleven
WINE COLOR

To the novice, wine is red, pink, or white and is either dark or light. However, to the enologist, color is a much more complex sensation. The most difficult problem with color measurement is the determination of meaningful parameters that are reproducible at a later time. In the early days, "tintometers" and other comparative devices were used. Generally they compared only one attribute: either the intensity of color or the hue. As spectroscopy developed, so did the color evaluation techniques. In 1931, with the culmination of the work and the report of the International Commission on Illumination (CIE—Commission Internationale de l'Eclairage), it became possible to define the color of a liquid such as wine to a more exact degree. A number of abridged methods have also become popular. For details see References 1–3.

Color is defined by a number of terms. These terms can be based on the measure of radiation, on the luminous energy that is gathered by the eye, or on the color sensation that is formed by the mind. The equivalent terminology (4) for the color characteristics or attributes is given in Table 65. Among the three color sensation terms, "hue" is the quality factor defining what the ordinary individual means by color (whether it is blue, red, etc.), "brightness" is the quantitative factor defining the intensity of the color, and "saturation" represents the percentage of hue in a color—whether it is "pale" or "deep" in common terms.

The eye cannot distinguish the separate components of color but sees the blend only. A number of pigments can be present in a wine, but the viewer sees only the one hue and its brightness, as well as the saturation that results. The effect of color on the discrimination of sweetness in table wine (5), the preference for certain red colors in rosé wines (6), and subject variability to color preference (7) have also been tested.

Winkler and Amerine (8) showed the necessity of using tristimulus values to define a wine color clearly. However, in practice the color values of white wines have been restricted to the simple measurement of the absorbance of a sample at around 420 nm. In the general area of 400–440 nm any significant increase in the brown color of white wines is detectable, even on relatively

Table 65. Equivalent Terminology of Three Fields of Science on Color

Physics: radiant energy (characteristic)	Psychophysics: luminous energy (characteristic)	Psychology: color sensation (attribute)
Radiance	Luminance	Brightness
Relative spectral composition	Dominant wavelength	Hue
Radiant purity	Purity	Saturation

insensitive spectrophotometers. Comparisons to standard color solutions made up from stable dyes in the color comparator have been used to estimate the relative brightness of both white and red wines. Cobalt chloride hexahydrate ($CoCl_2 \cdot 6H_2O$, 5–40 g/L) gives an absorption curve suitable for comparison with anthocyanin solutions (9).

For routine control measurement of white wine color, the ordinary colorimeter (such as the Klett-Summerson® or Bausch & Lomb® Spectronic-20) is adequate. The change in absorption at 420 or 430 nm can be followed and closely related with browning changes in the wine. The procedure suggested is used more or less routinely for white wines.

Procedure

Take a 20-mL sample of the white wine and filter through a 0.45-μm membrane filter. Zero the colorimeter with a water blank also filtered through the filter. Read the absorbance. With the Klett-Summerson colorimeter use a No. 42 filter and with the Bausch & Lomb Spectronic-20 set at 420 nm. There is no need for dilution. The ranges used are in the optimum areas of response for the equipment.

For the determination of color values for wine that are objective and that can be used to define the wine color completely, tristimulus measurements are superior. They are the preferred method of the OIV (10). The theory of tristimulus colorimetry is discussed in detail in References 1 and 11. For the application of this technique for wine color measurement, see References 10 and 12–14. Recently (15), for white wines (completely clear), 13 optical density measurements between 400 and 700 nm were used in conjunction with a computer program (BASIC) to calculate the color parameters (brightness, hue, and saturation).

There are a number of reports using the method of Sudraud (16) to approximate color differences in red wine. In his method the absorbance at 420 and 520 nm is determined, hue changes are approximated as the ratio of absorbance at 420 nm to absorbance at 520 nm, and the sum of these two absorbance values is used as a value to correlate with percentage brightness. As shown by Ough et al. (17), the method is adequate for the detection of most color changes in a red wine. However, the human eye can judge differences (18) in a finer detail by using wine glasses with measured quantities of the two wines than by this procedure. The method can still be a useful routine measurement.

Procedure

Filter the red wine through a 0.45-μm membrane filter and determine its pH. Then dilute the wine 1:10 with water whose pH has been adjusted to the same value as that of the wine. This can be done by mixing proportional amounts of weak citric acid and phosphate buffer solutions.

Place the diluted samples in the standard colorimeter tubes and determine the absorbance at 420 and 520 nm. Record the ratio of the absorbances at 420 and 520 nm as the hue and the sum of the absorbances as the brightness. If a spectrophotometer is available that can use 1-mm light path cells, the dilution is unnecessary.

A simpler procedure for blending red wines uses the absorbance value at 520 nm to measure the depth of color (19). The wine is diluted (1:10) to pH 3.30 using a buffer solution (10.21 g of potassium hydrogen phthalate and 6.10 mL of 2 N HCl diluted to 1 L). A 1-cm cuvette was used.

The color of white grape juice and very light-colored white table wine, white special natural wines, and certain fruit wines may need to be determined when activated charcoal is used in their production. Federal regulations set a minimum limit on the amount of color that can remain after treatment. Formerly this was done with Lovibond slides or by comparing the color at 430 nm to a potassium dichromate solution. Little (20) rejected spectrophotometry because of errors introduced by subliminal turbidity and developed (20, 21) an instrument based on transreflectometry. An AOAC method (11.003) using this instrument is now official (22–24). White wines treated with activated charcoal should have a color not more than 95% transmission.

Procedure

A special white wine colorimeter (a double-beam filter photometer utilizing a tungsten incandescent lamp with Corning S-61 high-pass filter, selenium photocells, and 2.5-cm light path test and reference cells) is required (available from Technical Products®, P.O. Box 291, Felton, CA 95018).

Prepare a 0.0002059 M standard potassium dichromate solution by dissolving a 0.0400 g of K_2CrO_4 (primary standard) in 0.05 N potassium hydroxide and dilute to 1 L with 0.05 N potassium hydroxide.

Allow instrument to warm up for 2 hr. Standardize the instrument using the standard chromate solution according to the manufacturer's instructions. Fill the reference and test cells with water and place in the colorimeter. Set the indicator knob to zero, and null the colorimeter by adjusting the zero set knob. Remove the test cell and replace with the zero set cell. Null the meter by adjusting the indicator. The indicator should read about 98.5 on duplicate tests. Repeat the standardization each hour or every 10-15 samples. With the zero set cell in place, set the indicator to the value (about 98.5) determined above. Null the meter with the zero set knob. Replace the zero set cell with the test cell containing the wine sample. Null the meter by adjusting the indicator. Read the percentage transmission on the indicator.

Similar results can be obtained by using a Klett-Summerson colorimeter.

The problem of blending wines to a certain color using tristimulus colorimetry has been studied (25). Blending diagrams were developed. The reciprocal log functions of the measured red, green, and blue transreflectometric readings were plotted against a sliding scale abscissa, from which the composition of blends starting from two different wines could be determined.

Correlation between overall sensory quality and certain color parameters has been shown for Australian red wines (26). For young Beaujolais red wines, significant linear correlation was reported between overall sensory scores and the total pigments, total anthocyanins, and colored anthocyanins. For 1 year, correlations were reported between color density and other factors (27). For information on the influence of tannin and anthocyanins on color see Glories (28).

REFERENCES

1. G. Mackinney and A. Little, *Color of Foods*. Avi Publishing Co., Westport, CT, 1962.
2. G. Mackinney and C. O. Chichester, *Adv. Food Res.* **5**, 301-351 (1954).
3. A. Chapanis, *Am. Sci.* **53**, 327-346 (1965).
4. A. Maery and M. R. Paul, *A Dictionary of Color*, 2nd ed. McGraw-Hill, New York, 1950.
5. R. M. Pangborn, H. W. Berg, and B. Hansen, *Am. J. Pyschol.* **76**, 492-495 (1963).
6. C. S. Ough and M. A. Amerine, *J. Food Sci.* **32**, 796-811 (1968).
7. C. S. Ough and M. A. Amerine, *Percept. Mot. Skills* **30**, 395-398 (1970).
8. A. J. Winkler and M. A. Amerine, *Food Res.* **3**, 429-438 (1938).
9. M. Beaumont, M. Bourzeix, P. du Breil de Pontbriand, and G. Marteau et Salgues, *Ann. Falisf. Expert. Chim.* **69**, 189-195 (1976).
10. *Recueil des Méthodes Internationales d'Analyse des Vins*, 4th ed. Office International de la Vigne et du Vin, Paris, 1978.
11. A. C. Hardy, *Handbook of Colorimetry*. Technology Press, Massachusetts Institute of Technology, Cambridge, MA, 1936.
12. A. G. Little, C. O. Chichester, and G. Mackinney, *Food Technol. (Chicago)* **12**, 403-409 (1958).
13. J. R. Hudson, *Brygmesteren* **28**, 100-102 (1971).
14. M. A. Amerine, C. S. Ough, and C. B. Bailey, *Food Technol. (Chicago)* **13**, 170-175 (1959).
15. A. Piracci, *Riv. Vitic. Enol.* **37**, 139-150 (1984).
16. P. Sudraud, *Ann. Technol. Agric.* **7**, 203-208 (1958).

17. C. S. Ough, H. W. Berg, and C. O. Chichester, *Am. J. Enol. Vitic.* **13,** 170–175 (1962).
18. H. W. Berg, C. S. Ough, and C. O. Chichester, *J. Food Sci.* **29,** 661–667 (1964).
19. J. Schneyder and A. Wurzinger, *Mitt. Klosterneuburg* **30,** 245–246 (1980).
20. A. C. Little, *Am. J. Enol. Vitic.* **22,** 138–143, 144–147 (1971).
21. A. C. Little and J. R. Simms, *Am. J. Enol. Vitic.* **22,** 203–209 (1971).
22. H. L. Wildenradt and A. Caputi, Jr., *J. Assoc. Off. Anal. Chem.* **59,** 777–779 (1976).
23. H. L. Wildenradt and P. A. Stafford, *J. Assoc. Off. Anal. Chem.* **60,** 739–740 (1977).
24. *Official Methods of Analysis*, 14th ed. Association of Official Analytical Chemists. Arlington, VA, 1984, pp. 220–280.
25. A. C. Little and M. W. Liaw, *Am. J. Enol. Vitic.* **25,** 79–83 (1974).
26. T. C. Somers and M. E. Evans, *J. Sci. Food Agric.* **25,** 1369–1379 (1974).
27. M. G. Jackson, C. F. Timberlake, P. Bridle, and L. Vallis, *J. Sci. Food Agric.* **29,** 715–727 (1978).
28. Y. Glories, *Connaiss. Vigne Vin* **18,** 195–217 (1984).

Twelve
GENERAL CHEMICAL AND EQUIPMENT INFORMATION AND THEORY

There are many texts on the practice and theory of chemical and instrumental analysis (1–11). Those cited in this chapter are by no means the only ones available, nor are they the most complete books, but they do cover analyses at a suitable level for practical understanding. This chapter discusses only the material pertinent to wine analysis covered in the previous chapters, including: measuring; extraction; distillation; ion exchange; visible, UV, and fluorescent spectroscopy; flame photometry; atomic adsorption spectroscopy; chromatography; detectors; specific-ion electrodes; complexometric measurements; and enzymatic reactions.

MEASUREMENT EQUIPMENT

The careful use of pipets and volumetric flasks is essential to accurate and reproducible results for volumetric analysis. For all analytical work the measuring glassware (indeed all glassware used) should be clean. This can be accomplished in most cases by a thorough detergent wash and water rinse followed by dichromate–sulfuric acid cleaning solution treatment with a thorough water rinse, followed by an acetone rinse and oven drying. When a glass must be "degreased," alcoholic sodium hydroxide solution may be required, followed by a rinse and then dichromate–sulfuric acid and water washes and acetone rinses. The solutions required are made as follows: For dichromate–sulfuric acid, take 35 mL of saturated $Na_2Cr_2O_7 \cdot 2H_2O$ (made by dissolving 300 g of the $Na_2Cr_2O_7 \cdot 2H_2O$ in 100 mL of water) and add 1 L of concentrated sulfuric acid to the 35 mL of saturated dichromate.

Table 66. Tolerances for Pipets and Volumetric Flasks

Pipet, volumetric			Pipet, Mohr			Flask, volumetric		
Size (mL)	Tolerance (±mL)	(%)	Size (mL)	Tolerance (±mL)	(%)	Size (mL)	Tolerance (±mL)	(%)
1	0.006	0.06	1	0.01	1.0	10	0.008	0.08
5	0.01	0.2	5	0.02	0.4	20	0.015	0.08
10	0.015	0.15	10	0.03	0.3	25	0.015	0.06
20	0.02	0.1	25	0.05	0.2	50	0.030	0.06
25	0.025	0.1	—	—	—	100	0.050	0.05
50	0.035	0.07	—	—	—	200	0.080	0.04
100	0.05	0.05	—	—	—	500	0.140	0.03
—	—	—	—	—	—	1000	0.190	0.02

From Reference 12.

For the alcoholic sodium hydroxide, dissolve 120 g of sodium hydroxide in 120 mL of water and bring to 1-L volume with 95% v/v ethanol. The solution is most effective when hot. Care should be exercised in handling these solutions.

Equipment used to measure solutions must be within certain tolerances. Some expected variations for pipets and volumetric flasks appear in Table 66 (12).

Automated devices should give reproducible and accurate measurements. In certain cases where internal standards are in the sample, absolute accuracy or reproducibility is not as critical. However, if volume is critical these standards should have the same or better tolerances than the glass pipets or flasks.

All weighing devices should be chosen to weigh the material to a desired degree of precision. When a procedure calls for 10 g of a material, do not use an analytical balance because there is no need to determine the weight to four places. Likewise, if the requirement is for accuracy to ±0.5 mg, use the analytical balance. Keep all balances clean and dry. Be sure they are in good working order, zeroed properly, and calibrated at reasonable intervals.

EXTRACTION

If two nonmiscible solvents are placed together, any substance (or any substances) put into either solvent will distribute between both solvents, depending on the degree of solubility in each. This can be represented by the partition coefficient C_d:

$$C_d = \frac{S_1}{S_2}$$

where S_1 is the amount of solute in solvent 1, and S_2 is the amount in solvent 2. The completeness of extraction depends on two factors: the coefficient of partition C_d and the volumes of each solvent. The fraction of compound transferred in one extraction is given by

$$F = \frac{C_d V_2}{V_1 + C_d V_2}$$

where F = fraction transferred
C_d = partition coefficient
V_1 = volume of original solvent
V_2 = volume of added solvent

The use of several extractions rather than one large extraction is more efficient. With the same volume of solvent, more of the compound can be extracted. The principle of partition applies to liquid–liquid chromatography as well as to paper and gas–liquid chromatography.

Differences in the solubility of a given solute in various solvents depend mainly on the polarity of the solvents. Table 67 (13) lists some common solvents and their properties. The dielectric constant is related to the polarity of the solvent.

EVAPORATION AND DISTILLATION

Each substance has a vapor pressure, which can be defined as the pressure exerted outward from the substance (confined) at a given temperature. As the boiling point of a liquid is reached, the vapor becomes equal to the barometric pressure. Normally this is about 760 torr. If the area above the liquid is subjected to a vacuum, the boiling point will be at whatever torr is generated by the system; thus the boiling point will be lowered accordingly (Table 68) (14). For each twofold pressure reduction, the boiling points are lowered by about 15°C. Boiling requires not only that the proper temperature be reached but also that sufficient energy (heat) be applied to furnish the heat of vaporization. Then

Table 67. Some Common Solvents and Some of Their Physical Properties

Solvent	Molecular weight	Density at 25°C	Refractive index	Boiling point (°C)	Dielectric constant
Water	18.02	0.9971	1.3329	100	78.54
Methanol	32.04	0.7866	1.3265	64.7	32.70
Acetonitrile	41.05	0.7766	1.3416	81.6	37.50
Acetaldehyde	44.05	0.7780	1.3311	20.4	21.10
Formamide	45.04	1.1334	1.4475	210.5	109.00
Formic acid	46.03	1.2141	1.3694	100.6	58.50
Ethanol	46.07	0.7850	1.3594	78.3	24.55
Acetone	58.05	0.7900	1.3587	56.3	20.70
n-Methyl formamide	59.07	0.9988	1.4300	182.5	182.40
Acetic acid	60.05	1.0492	1.3719	117.9	6.15
n-Propanol	60.10	0.8038	1.3856	97.2	20.33
Cyclopentane	70.13	0.7454	1.4065	49.3	1.96
Pentane	72.15	0.6262	1.3579	36.1	0.0
Methyl acetate	74.08	0.9342	1.3614	56.3	6.68
n-Butanol	74.12	0.8060	1.3973	117.7	17.51
Ethyl ether	74.12	0.7076	1.3495	34.5	4.34
Carbon disulfide	76.14	1.2700	1.6319	46.2	2.64
Benzene	78.12	0.8737	1.4979	80.1	2.28
Pyridine	79.10	0.9782	1.5075	115.3	12.40
Dichloromethane	84.93	1.3148	1.4211	39.8	8.93
Hexane	86.17	0.6548	1.3723	68.7	1.89
Ethyl acetate	88.11	0.8946	1.3698	77.1	6.02
Toluene	92.14	0.8623	1.4941	110.6	2.38
Chloroform	119.38	1.4799	1.4429	61	4.81
Carbon tetrachloride	153.82	1.5844	1.4574	76.7	2.24

From Reference 13.

amounts required depend on the liquid being used. Care must be taken to be sure all the gas is rapidly removed from the space above the boiling liquid. To ensure this, the piping to the condensers must be of adequate size and the condenser cannot hold up the condensed liquid. If vacuum is being used, an adequate pump capacity must be available also, to maintain the desired vacuum once boiling commences.

Distillation is a common method to purify solvents or to separate miscible mixtures of liquids. Unless the boiling points of the liquids are quite dissimilar (25–50°C difference), more than a simple distillation is required. Yost (15) has described the good and bad points of various laboratory still designs. These are summarized as follows:

Still type	Throughput	Holdup	Separation efficiency	Pressure drop
Classic (1-plate)	High	Low	Low	Low
Bubble cap	High	High	Low (laboratory size)	Low
Packed (glass beads)	Low	High	High	High
Teflon spinning band	High	Low	Very high	Moderate

Table 68. Boiling Points (°C) of Some Liquids at Pressures up to 760 Torr

Substance	Pressure (torr)				
	100	200	400	600	760
Acetone	7.3	21.8	38.7	49.5	56.2
Benzene	21.8	43.0	61.1	73.2	80.1
Carbon tetrachloride	22.3	39.1	57.0	68.9	76.8
Chloroform	9.5	25.2	42.2	53.8	61.1
Ethanol	34.9	48.2	63.1	72.4	78.3
Ethyl ether	−8.0	1.9	17.5	28.4	34.6
Ethylene glycol	138.1	156.1	176.0	188.9	195.9
Methanol	20.9	34.3	49.4	59.0	64.7
Water	51.0	60.7	80.3	93.6	100.0

From Reference 14.

For corrosive materials or those with very close boiling points, the Teflon spinning band still is the one that is preferred. If there is a 15–20°C difference in boiling points of the solvents to be separated, a 10-plate still is adequate.

A glossary of pertinent terms follows:

Azeotrope. A mixture of two or more compounds that boils at one temperature regardless of the individual boiling points. The vapor given off has a fixed composition. Ethanol and water form such an azeotrope—78.2°C is its boiling point. The composition is 95.6% v/v ethanol and 4.4% v/v water.

Distilland. The material to be distilled.

Distillate. The separated material taken off the head of the distillation system.

Flooding. Excess liquid forced up the column because of excessive heat input into the pot.

Fraction. A discrete portion of the distillate.

Holdup. The liquid portion, at any one time, that is in the still columns.

Pressure drop. The difference between the pressure in the pot and that in the head of the still.

Rectification. The interaction of returning condensed vapors with fresh vapor and the enrichment or purification that results.

Reflux. Condensation of vapor and return of this liquid to the pot.

Theoretical plates. Units used to measure separating efficiency. The more plates per column the better the separating efficiency. Height equal to a theoretical plate (HETP) is found by calculating the theoretical plates and dividing by the length of the column.

Throughput. Rate of flow at which efficient distillation can be done.

ION EXCHANGE

The ion-exchange matrix is usually composed of cross-linked styrene polymers. To obtain ion-exchange capabilities, the matrix is chemically altered. Sulfonation results in a strong-cation resin, and chloromethylation and amination result in a strong-anion resin. Figure 34 gives the chemical formula. Other substitutions can be made to alter the resins. Some of the possible variations are as follows:

Basic		Acidic		
Strong	Weak	Strong	Intermediate	Weak
Quaternary ammonia	Primary amine	Sulfonic	Phosphonic	Carboxylic
Sulfonium	Secondary amine	Methylene sulfonic	Phosphonous	Phenolic
Phosphonium	Tertiary amine		Phosphoric	

Choosing the proper resin for a separation requires some knowledge of the capabilities of the various resins. Some desirable information is given in the next paragraph. For more details see Reference 14.

Strong-base anion exchangers in the hydroxide form split neutral salts and remove acids (even very weak acids). The chloride form removes SO_4^{2-}, HCO_3^-, organic acid, and so on. Weak-base anion resins have a chemistry sim-

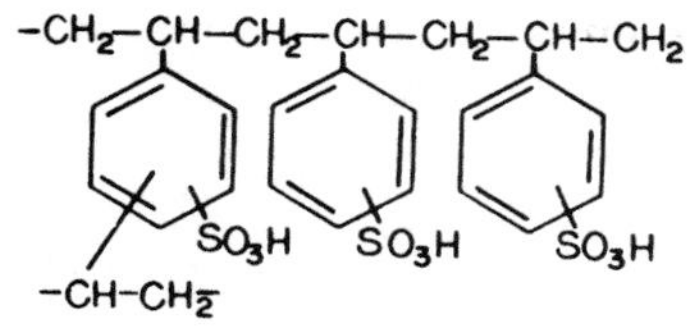

CATION EXCHANGER

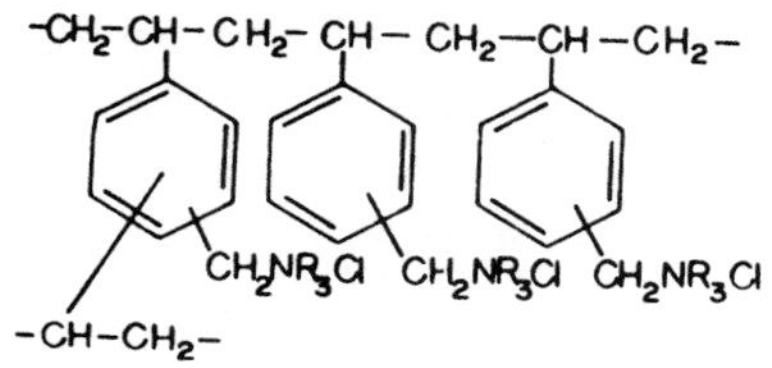

ANION EXCHANGER

Figure 34. Structural formaulas for ion-exchange resins.

ilar to that of ammonia. The free-base form absorbs strong acids. In the acid form they liberate free acids in contact with water. The weak-base resins can be stoichiometrically regenerated with a strong base. Acids can be fractionated depending on the degree of ionization of the acid. Strong-acid cation resins split neutral salts. The chemical activity is similar to that of sulfuric acid. Nitrogen bases, including amines, amino acids, and so on, can be readily absorbed on the hydrogen form. Weak-acid cation exchangers have chemical properties resembling those of acetic acid. They are effective in a pH range of 6–14. Nearly stoichiometric amounts of strong acid are required to convert to the hydrogen form. Intermediate-acid cation resins have a chemistry similar to that of phosphoric acid. They have an efficient pH range of 5–14. Also, they can be regenerated with near stoichiometric amounts of acid.

The application of ion-exchange resins can be divided into a number of categories. A partial list and description of these uses is as follows:

Category	Description
Ion exchange	Replacement of one ion in solution with another
Elimination	Removal of unwanted ions from solutions
Fractionation-chromatography	Ions captured on ion-exchange column are selectively eluted
Neutralization	Addition of acid- or alkali-charged resin to solution

Resins must be pretreated before use. For cation resin fill the column, backwash, and drain. Wash with 2 bed-volumes of 2 N hydrochloric acid, 5 bed-volumes of water, 2 bed-volumes of 1.5 N sodium hydroxide, and 5 bed-volumes of water. (Keep the resin covered with liquid at all times.) Check the effluent on the last wash with phenolphthalein to ensure removal of excess sodium hydroxide. Wash with 2 bed-volumes of ethanol, then wash with 2 bed-volumes of water, which should be neutral. The resin is now in the sodium form. To put into the hydrogen form, repeat the 2 N hydrochloric acid wash, then rinse. For anion resin, follow the same procedure but start with the 1.5 N sodium hydroxide wash; then rinse with 2 N hydrochloric acid. The final wash should be negative to the silver nitrate test (no silver chloride precipitate). The resin ends up in the chloride form. To get in the hydroxide form, repeat first step.

To determine the capacity of the strong-cation resin, first condition the resin and put into the hydrogen form, then treat with excess sodium hydroxide (5 g/100 mL). Wash with water until the last of the effluent is pH neutral. Titrate the effluent with 2.00 N sodium hydroxide. Regenerate the resin, transfer to a graduated cylinder, and determine the resin volume.

Calculate the resin capacity:

$$\frac{\text{meq}}{\text{mL}} = \frac{\text{mL NaOH} \times N \text{ NaOH}}{\text{mL resin in H form}}$$

Anion resin capacity can be done similarly with the resin in the hydroxide form and 1.00 N hydrochloric acid used to titrate:

$$\frac{\text{meq}}{\text{mL}} = \frac{\text{mL HCl} \times N \text{ HCl}}{\text{mL resin in OH form}}$$

For further information see Reference 16.

Some of the standard resins used in the laboratory are given in Table 69, with the equivalent trade names.

SPECTROPHOTOMETRY

Colorimetry is usually considered to refer to analysis that involves the response of the human eye, whereas in spectrophotometry an instrument is used to measure differences in light intensity. Most laboratories now use spectrophotometric measurements.

Light is absorbed as it passes through a liquid. The amount of light absorbed can depend on several factors: wavelength of the light, type and thickness of

Table 69. Some Equivalent Ion-Exchange Resins

Active group	Trade name			
	Dowex	Duolite®	Amberlite	Permutit®
$-SO_3Na$	50	C-20	IR120	Q
$-SO_3Na$	50W	ES-28	IR121	—
$-COOH$	ECR-1	EC-3	IRC84	H-70
$-NH(R)$	3	A-2	IR45	W
$-NR_3^+$	1	A-101D	IRA400	S-1
$-NR_3^+$	2	A-102D	IRA410	S-2

walls of container, material in the container, and thickness of the container. This assumes that light is not scattered by particles in the liquid. The usual transmission loss from reflection from the container walls is about 2% (up to 4%). The internal transmittance of the sample T_i is related to the absorbance A_i by the following formula:

$$A_i = -\log_{10} T_i$$

It is usual to use a solution blank to zero the instrument. This accounts for reflectance and other-than-sample absorbance.

The bandpass of the instrument refers to how narrow and specific a portion of the spectra can be directed to the sample. For normal colorimetric analysis a very narrow bandpass is not always required. If spectral transmission curves are desired, however, a narrow bandpass becomes essential in most instances.

The absorption of a sample dissolved in a solvent is directly proportional to the concentration and to the cell depth. This is described in the Beer–Bouger equation:

$$T_i = 10^{-a_i b c}$$

where T_i = internal transmittance of the sample
  a_i = absorbance index sample (varies with wavelength)
  b = sample path length
  c = concentration of the absorbing component

Figure 35 shows a simple single-beam spectrophotometer. A dual-beam spectrophotometer would have a dual optical system and would split the light between a reference and a test sample.

Some sources of sample error are: hydrogen ion changes, changes in molecular structure, temperature changes, interfering ions, turbidity, fading (caused

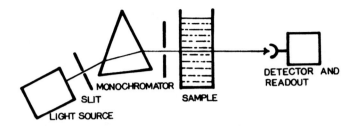

Figure 35. Schematic diagram of a single-beam spectrophotometer.

by irradiation), and fluorescence. Other sources of error are those of the instrument: wavelength setting, stray light, slit width too wide, nonlinear photometric system, wrong transmission area (the optical density of the material should be within range of the instrument), wrong cell window material, and wrong path length. Table 70 (17) lists light sources and their wavelength ranges.

Filters or some other device are required to isolate the proper or desired wavelengths. Simple spectrophotometers use bandpass filters, which are glass plates that incorporate certain inorganic material to get the desired bandpass. These are useful in fluorometry and nephelometry. Cutoff filters, which completely eliminate part of the spectra, are useful and necessary for fluorometry as secondary filters to eliminate the primary spectra.

Interference filters are designed to transmit only a narrow portion of the spectra. Monochromators are either prisms or gratings. Quartz is the most useful material for prisms in the UV region. Gratings are very efficient means of obtaining narrow bandpasses.

Cuvettes are made of various materials. Depending on the desired light path, cells ranging from 1 to many millimeters in light-path distance can be acquired. Table 71 gives types of cell material and the range of wavelength transmission.

Table 70. Light-Source Wavelength Ranges

Light source	Wavelength range (nm)
Tungsten	320–2000
Medium-pressure mercury	254– 578 (254, 313, 366, 405, 436, 546 are the major singlet or doublet lines)
Low-pressure mercury	254 (with phosphor coatings other wavelengths possible)
Deuterium	220– 620
Xenon	220–1100 (used for spectrofluorometry)

From Reference 17.

Table 71. Type of Cuvette Material at Usable Wavelength Range

Cuvette material	Wavelength range (nm)
Mirror glass of Pyrex	400–2500
Optical glass	320–2200
Fused quartz	220–1000
Fused synthetic silica	180– 700

From Reference 17.

Dirty cuvettes can be cleaned gently with lens-cleaning tissue. First they should be rinsed several times with a solvent, then with ethanol. Fused (not cemented) cuvettes can be soaked in sulfochromic acid or a detergent solution. Never use any abrasive material to wipe or otherwise touch the surface of a cuvette.

Calculation of results from standard curves can be achieved by reading directly or by use of a factor. For the latter, multiply the optical density of each measurement by the reciprocal of the slope of the standard curve. Plot optical density on the y axis and wt/vol on the x axis; then the slope $b = (Y_2 - Y_1)/(X_2 - X_1)$ and $1/b = (X_2 - X_1)/(Y_2 - Y_1)$, with the weight/volume units divided by optical density units. Optical density sample reading multiplied by this value gives weight/volume of the test sample. If the standard curve does not go through zero, a plus or minus adjustment must be added. If the x intercept is negative, add the intercept value; if it is positive, subtract it.

TURBIDIMETRY AND NEPHELOMETRY

Figure 36 compares turbidimetry and nephelometry, which have some application in the wine industry. Turbidimetry is the measure of light transmitted after loss by scattering. Nephelometry is the measure of light scattered at 90° angle from the incident beam. If the particle sizes are 0.001–1 μm in diameter, they will cause UV or visible light to scatter. This scatter is termed the *Tyndall effect*.

Turbidity measurements are best when significant amounts of the incident beam is scattered. If only a small amount is scattered, then nephelometry is superior. Most spectrophotometers can be used for turbidity measurements (such as monitoring the growth of yeast); however, a wavelength must be used that is not absorbed by any changing compound in the medium. Many spectrophotometers have nephelometric attachments.

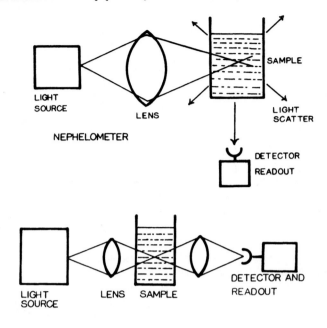

Figure 36. A comparison of nephelometric and turbidimetric optical systems.

FLUOROMETRY

Fluorometry is the measure of emited light from a compound that has been irradiated with a shorter wavelength light. Many compounds, when subjected to high-energy UV light, will emit longer wavelength radiations as the electrons of the molecule return from an excited state to the ground state. Figure 37 depicts a fluorometer setup for the usual analytical purpose. The light source must radiate significant amounts of radiation at the wavelength that will excite the molecule in question. A selection of fluorescent bulbs, mercury vapor lamps, and xenon tubes is available to cover the desired ranges. The primary filter serves to exclude as much of the undesirable emissions as possible. The desired wavelength radiation impinges on the compound and excites certain electrons to a higher π-energy level. The molecule returns to the ground state while releasing a portion of the absorbed energy as a characteristic photon of light (with a wavelength somewhat longer than the exciting wavelength). The emitted photons are given off in all directions. The right-angle emission to the incident beam is usually selected for measurement. A secondary or cutoff filter is placed between the sample holder and the detector to eliminate undesired radiation.

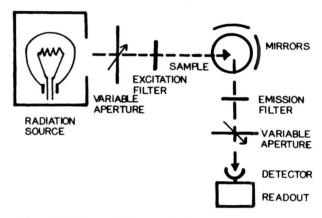

Figure 37. Schematic diagram of a single-beam fluorometer.

Certain other compounds can absorb the emitted secondary radiations. This is called the *quenching effect*.

The fluorescence intensity is usually linear for changes in dilute concentrations of fluorescent materials. Levels as low as 10 µg/L of certain compounds can be detected.

FLAME PHOTOMETRY AND ATOMIC ABSORPTION SPECTROMETRY (AAS)

Two methods of measuring metals in trace amounts in wine, juice, and plant tissue (and soils) have become extremely important: Flame photometry and AAS both depend on the capacity of the outer electrons of elements, particularly the metals, to be raised to higher energy levels by various energy sources and, in the process, to absorb or emit light of a specific wavelength.

The outer-shell electrons have specific absorption wavelengths in the 10^5–10^3-Å region. Figure 38 shows the excitation potential for each of three commonly measured elements.

Some characteristics of energy transitions in atoms are as follows: An atom in the natural ground state will accept energy depending on its specific quantum characteristics; each element has different characteristics; and the energy may take a thermal, physical electromagnetic radiation, or other form. When an electron accepts energy, it is in the excited state. As it returns to the ground state, the same energy absorbed is emitted as a quantum of light of a specific wavelength.

The principle of emission spectroscopy can be used to analytically measure

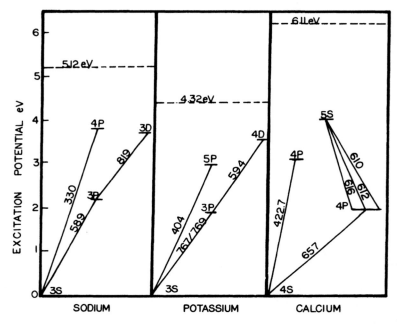

Figure 38. Energy level diagram for sodium, potassium, and calcium: wavelengths are in nanometers; the dashed lines are the individual ionization potentials.

the concentration of a specific element. The intensity of emitted light is proportional to the number of atoms N in the excited state. This, in turn, is proportional to the concentration of ground-state atoms N_0 in the flame (multiplied by a weighting integer) and the temperature of the flame and the ionization potential I of the element:

$$N = N_0 (\text{factor})\, e^{KT/I}$$

where K = the Boltzmann constant
T = absolute temperature (°K)

In AAS (Figure 39), light having a specific wavelength is sent through an area containing atoms, in the ground state, which will absorb that light at the resonant wavelength of the element. The absorption of the light obeys the Beer–Lambert law:

$$\frac{I_0}{I} = e^{kLN_0}$$

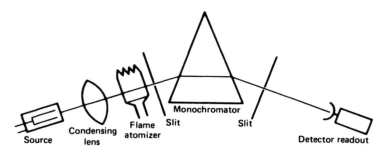

Figure 39. Schematic diagram of an atomic absorption spectrophotometer.

where I_0 = light passing through flame when $N_0 = 0$
N_0 = concentration of the ground state elements in the flame
I = light intensity passing through the flame when $N_0 > 0$
k = constant
L = length of light path through the flame

Putting the elements into the ground state is usually a fairly inefficient process. One of the most common methods is to suction up a liquid sample and introduce it continuously as an aerosol into a flame. The flame energy causes the solution to evaporate and the molecules to dissociate and be put into the ground state. As an example, consider potassium in wine:

$$2K^+ + SO_4^{2-} \xrightarrow[\Delta]{\text{Evaporation}} K_2SO_4 \xrightarrow[\Delta]{\text{Dissociation}} 2K° + SO_4°$$

Once the atom is in the ground state, it can be raised to the excited state in the flame:

$$K° \xrightarrow{\Delta} K^*$$

Then it can degenerate to the ground state again:

$$K^* \longrightarrow K° + h\nu$$

In the case of flame photometric spectroscopy (Figure 40), the amount of light given off is measured. In the case of absorption spectroscopy the amount of abosrbed light is measured.

Flameless atomic absorption differs from the conventional flame unit by using a high-temperature electric oven with quartz windows, to put the atoms in the

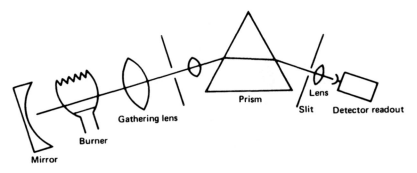

Figure 40. Schematic diagram of a flame photometer.

ground state (Figure 41). In principle there is no difference. In practice a gain in sensitivity is achieved.

To furnish a source of light of a specific wavelength, the hollow cathode tube (Figure 42) was developed. This unit consists of a cathode and an anode. The tube is filled with argon gas and is lined with the metal to be measured, or an alloy of it. A potential is put on the tube, and electrons move and collide with gas molecules, causing them to be ionized. These heavier particles collide with gas molecules, which become ionized. The heavier particles collide with the wall after being accelerated by the voltage difference between anode and cathode. This causes atoms of the metal to be put into motion within the tube. These atoms are then energized by other collisions to the first excitation state, then they degenerate to give off a photon ($h\nu$) of the specific wavelength desired. Again, efficiency is poor. The photons are not directionally oriented, so only those hitting the quartz window at the proper angle are useful. In some cases several metals may be incorporated into a single tube if their emission spectra do not interfere.

Some of the problems, besides general inefficiency, include incomplete dis-

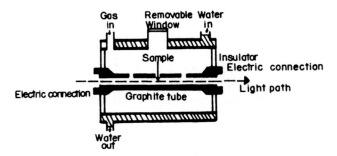

Figure 41. Cross-sectional view of a heated graphite atomizer.

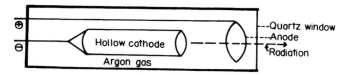

Figure 42. Schematic diagram of a hollow-cathode, specific radiation source.

sociation in the flame and difficulty in dissociating compounds (such as calcium phosphate—this can be eliminated by the use of La^{3+}, which will competitively form phosphate compounds and free the calcium). Another solution is the use of chelating agents to complex with the desired cation to form easily vaporized compounds.

The flame temperature affects the percentage of dissociated atoms that are in the ground state (Table 72).

Some of the fuel gases available, as well as oxidants, are listed in Table 73, with their flame temperatures.

Other interferences can come from other cations, such as potassium interference when measuring sodium. This can be overcome by use of cooler flames.

Table 72. Dissociated Atoms in the Ground State

Element	Resonance, line (nm)	Temperature of flame		
		2000°K	3000°K	4000°K
Cesium	852	99.96	99.28	97.02
Sodium	590	99.9999	99.94	99.56
Calcium	423	99.99999	99.996	99.94
Zinc	214	99.9999999999	99.99999994	99.99995

Table 73. Some Fuel Gases and Flame Temperatures with Air and Oxygen

Fuel	Oxidant	Flame temperature (°C)
Hydrogen	Oxygen	2800
Hydrogen	Air	2100
Acetylene	Oxygen	3000
Acetylene	Air	2200
Propane	Oxygen	2800
Propane	Air	1900
Butane	Oxygen	2900
Butane	Air	1900

Table 74 gives spectral emissions used for some common elements measured with flame photometry. The emission wavelengths employed for absorption spectrophotometry appear in Table 75 (18).

For both these instrumental methods (as well as others) the data can be calculated from standard curves or by the method of addition. The method of addition is most useful when standard curves can vary from linearity, when sample interference is a problem when reagents change continuously, or when only a few samples are to be measured. This technique involves reading of the sample, then adding a known amount of a compound to the sample and reading it a second time.

$$\frac{x}{m} = \frac{x + a}{n}$$

$$nx = mx + ma$$

$$nx - mx = ma$$

$$x(n - m) = ma$$

$$x = \frac{ma}{n - m}$$

where
- x = sample concentration of the first sample (wt/vol)
- a = added amount (wt/vol)
- $x + a$ = total concentration of the second sample (wt/vol)
- m = reading for first sample—optical density
- n = reading for second sample—optical density

Since all the values on the right-hand side are known, x can be calculated. The value of a should be in the same general amount as x. If $a \gg x$ or $a \ll x$, the optical density will be too large to be read accurately or will cause such a small difference in optical density that the accuracy is lost.

Table 74. Flame Spectral Lines of Some Elements Used with Flame Photometry

Element	Wavelength (nm)	Sensitivity (mmol/L)
Calcium	423	1×10^{-5}
Potassium	767	7.7×10^{-6}
Magnesium	285	2.5×10^{-4}
Sodium	589	1.75×10^{-5}

Table 75. Flame Spectral Lines for Some Elements Used with AAS

Element	Wavelength (nm)	Best flame	Analytical sensitivity (μg/mL for 1% abs.)	Detection limit (μg/mL)
Aluminum	309.3 (396.2)[a]	Acetylene–nitrous oxide	1.0	0.2
Arsenic	193.7	Hydrogen-argon	1.5	0.5
Calcium	422.7	Acetylene–nitrous oxide	0.02	0.005
Chromium	357.9	Acetylene-air	0.1	0.01
Copper	324.7	Acetylene-air	0.05	0.005
Iron	248.3	Acetylene-air	0.01	0.005
Potassium	766.5	Acetylene-air	0.05	0.005
Magnesium	285.2	Acetylene-air	0.005	0.001
Manganese	279.5	Acetylene-air	0.05	0.005
Molybdenum	313.3	Acetylene–nitrous oxide	0.5	0.1
Sodium	589.0	Acetylene-air	0.05	0.005
Nickel	232.0 (341.5)[a]	Acetylene-air	0.1	0.01
Lead	283.3 (217.0)[a]	Acetylene-air	0.5	0.05
Strontium	224.6 (286.3)[a]	Hydrogen-air	0.5	0.05
Zinc	213.8	Acetylene-air	0.02	0.002

[a]Values in parentheses are secondary, but useful, absorption wavelengths.

CHROMATOGRAPHY

Chromatography consists of two phases: The first, or moving, phase carries the compounds in question over and around the second, or stationary, phase. The moving phase can be either liquid or gas. If it is liquid, the compounds to be separated must have certain and different solubilities in the liquid. If it is gas, the separation depends primarily on the stationary-phase attractions. The moving and stationary phases must be reasonably insoluble or unreactive to each other. The stationary phase should not form products with compounds to be separated.

The degree of attraction for the compounds to be separated depends on solubility, polarity, hydrophobic attraction, and other forces. These can generally be divided into several main types.

Adsorption is the strong attraction for the compound to the surface of the solid stationary phase. This can be hydrophobic, ionic, or hydrogen binding. The compounds to be separated have different degrees of attraction for the solid stationary phase, and as they are carried along by the liquid or gas moving phase, they spend different amounts of time attached to the stationary phase. They then are eluted by the moving phase at different intervals or appear on a

TLC plate at different R_f values. An example of this type of chromatography is silica gel using nonpolar moving phases or solid phase (Chromosorb 100 series) for gas chromatography.

Common adsorbants listed in decreasing order of adsorption power are: aluminum silicate > charcoal > activated alumina > Florisil > silica gel > calcium oxide > magnesium oxide > talc > starch > powdered sugar.

The distribution of the compounds to be separated between a moving liquid or gas phase and a liquid stationary phase is called *partition*. The separate usually depends on the differences in the solubilities of the compounds in the two phases. Paper chromatography is one type of partition chromatography. The cellulose fibers act as a solid support but attract water molecules, and this thin layer of water is the liquid stationary phase. The degree of separation is achieved because of the solubility differences between this water layer and the solvent. The solvent is a two-compound (nonpolar and polar) system. In some cases silica gel systems become "partition" rather than "adsorption" if the level of polarity of the solvent system is high. Gas–liquid chromatography is a partition system. The gas carries along the volatilized materials to be separated. The amount of time each spends dissolved in the liquid phase determines the retention time or how long it takes to go through the column.

Paper and Thin-Layer Chromatography

The theory of paper or thin-layer chromatography is given in a number of texts (19–21). As just noted, paper chromatography is a form of partition chromatography; special grades of paper are available for this purpose. The usual method of separation is by either descending or ascending solvents. The smaller and more concentrated the spots, the better the separation achieved. Choice of solvents depends on the material being separated. The distance the material being separated moves in comparison to the distance the solvent front moves is called the R_f value (it is never more than unity or less than zero):

$$R_f = \frac{\text{Distance traveled by material}}{\text{Distance traveled by solvent front}}$$

Since many factors affect the R_f of a compound, it is more satisfying to run a standard of the material in close proximity of the test sample.

Several conditions can cause a spot to become diffuse, mainly molecular diffusion and eddy diffusion. Choice of good paper, reasonably fast times, and small original spots minimize this problem. Tailing or streaking can be caused by overloading, too fast a solvent flow, change in solvent composition, or strong absorption in the paper. There are other spot distortion problems as well (19).

Paper chromatography is seldom used for quantitative analysis. Thin-layer chromatography, however, is a most useful tool for both qualitative and quantitative separation and measurement of compounds. An inert absorbant is usually slurried in water and, with or without a binder ($CaSO_4$), is spread onto a glass plate. The absorbant can have a fluorescent material added if desired. (The spot will appear black on a fluorescent background under UV light.) A selection of adsorbants is available (see page 340). Silica gel is the most popular adsorbant. The thickness of the adsorbant is 300–500 μm. The plates are usually heated about 105°C, to activate them (i.e., to remove most of the water). If very nonpolar compounds are being separated and the solvents are specially treated to be free of water, longer heating and higher temperatures (3 hr at 150°C) may be beneficial. The plates must be uniform. A portion along the edge should be stripped off to prevent uneven running of the solvent.

Spotting the samples should be done carefully to minimize spot size. The spots should be dry before the chromatography is run. Choice of solvents depends on the material being separated.

Once the sample has been run, it becomes necessary to dry the plate and identify the compound. As in paper chromatography, the R_f values are only approximately correct. It is better to run a known sample alongside the unknown to determine exact location.

Once the spot has been identified, several options are open for quantifications: (a) The spot can be removed by scraping off the silica gel (or other adsorbant), eluting the material, and making the determination spectrophotometrically or by some other analytical means, or (b) the intensity of the spot can be measured by a photometric scanner. The second method can be used on either visible or fluorescent spots.

Gas Chromatography

Gas chromatography depends on an inert solid material to support the liquid phase. An exception is solid absorption materials such as Porapak that require no inert support. For most quantitative work, packed glass, stainless steel or fused silica capillary columns are used.

Solid Supports

A solid support should be inert, of uniform size, resistant to fracture, and temperature resistant.

Diatomaceous earth approaches these characteristics but is fragile. Heating with sodium carbonate to 1600°C causes the diatomaceous earth to fuse into a white crystaline substance, "critobalite," commonly marketed as Celite® or Chromosorb W by Johns-Mansville®. If the heating is carried out without the carbonate, a pink material results—"firebrick" or Chromosorb P. The latter form has a higher density than Chromosorb W and holds more liquid phase because of smaller pore size Chromosorb G is a high-density, very inert form and requires about $2\frac{1}{2}$ times less liquid phase than Chromosorb P or W. It is 4–6 times less fragile. Maximum loading for Chromosorb G is about 5%, whereas for Chromosorb W or P it is about 12.5% (wt/wt). Other supports made by other companies are similar.

Tailing is broadening of peaks or unintended holdup of compounds to be separated; attractive forces, usually from the solid support, are responsible. A typical structure for a solid support is as follows:

$$\begin{array}{c} | \quad\quad | \\ -Si-O-Si- \\ | \quad\quad | \\ O \quad\quad OH \\ | \\ -Si-OH \\ | \end{array}$$

The OH groups (silanol) have strong proton-donor capabilities, and the Si—O—Si groups (siloxane) have proton-binding tendencies.

To improve nontailing properties of the solid support, the material is acid washed to remove any alkali material. Then a series of compounds called *silanizing reagents* can be used to derivatize the silanol groups. Commonly used compounds are dimethyldichlorosilane and hexamethyldisilazane.

Mesh size is the number of holes per square inch that are used in the screen to separate the material. With a 60/80 mesh solid support, for example, a material may pass through a 60 hole/in.^{2} screen but not through an 80 hole/in.^{2} screen. The standard mesh sizes used for solid support material are 60/80, 80/100, and 100/120.

New glass capillary columns and, more recently, fused silica columns have replaced the older metal or large bore glass columns for many applications. These columns have a much smaller internal diameter, and the walls act as the solid support for the liquid phase. In some instances the liquid phase is chemically bonded to the fused silica. Although the capacity of these columns is much less than that of the larger bore columns, the separation is far superior. With the use of internal standards and methods to separate away large portions of the solvent they can be used effectively for quantification as well as peak identification.

Liquid Phases

Liquid phases that coat the solid support should be liquid at the oven temperatures used in the gas chromatograph; also, they should be capable of separating the material to be tested. Ability to coat the solid support thoroughly, inertness, and a low vapor pressure (so as not to "bleed" off the column into the detector) are other necessary properties.

The choice of liquid substrate for any particular job is very important. Separation of the compounds can be very simple or very difficult, and the liquid phase is by far the most important controlling factor for success or failure. The number of liquid phases is numerous. Lists of those available are easily obtained from commercial suppliers.

Gas Chromatograph

The instrument consists of an injector, oven, column, detectors, and so on.

In the last few years, great strides have been made in the computerization of all instruments. Samples can be loaded onto the autosampler and are automatically injected; the retention window (time) of each compound in question is preset; and the peak areas are electronically measured, compared to internal standard, and the data are stored on disks or printed out for immediate use in the desired format. Figure 43 is schematic drawing of a gas–liquid chromatograph.

The inject is a device that allows a sample to be placed into the apparatus while a certain gas pressure exists in the column. It consists of a tube (either glass or stainless steel) with a septum over one end. The septum is self-sealing and penetrable by a syringe needle. The other end of the tube is connected to the column. The injector area is usually temperature controlled.

The temperature in the oven, an insulated area designed to contain the column, must be very accurately controlled. The oven can be operated at a set temperature (isothermally). In some ovens, however, the temperature can be proportionally increased either linearly or at some predetermined pattern (temperature programming). The oven must be well insulated but must have a low

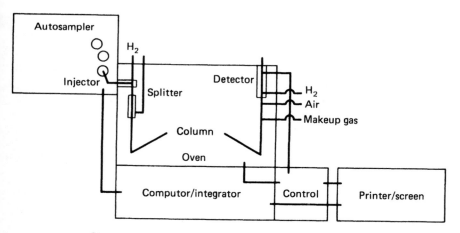

Figure 43. Schematic diagram of a gas–liquid chromatograph.

heat-holding capacity to permit rapid temperature changes. Heating is accomplished by high-volume air circulation around the column with electrical coils as a heat source.

Columns

The outer material of the column can be fused silica, glass, stainless steel, copper, aluminum, or Teflon. Metals tend to cause oxidation of organic materials. The column tube must be very clean before it is coated or filled (22). This can be ensured by passing through some concentrated nitric acid, washing well with water, then giving a final wash with methylene chloride and drying at 110°C with a gentle stream of nitrogen passing through.

To fill the column attach an end, plugged with about $\frac{1}{2}$-in. of glass wool, to a vacuum line. Cover the end with a piece of cheesecloth. Apply a vacuum (aspirator). Attach a small funnel to the other end of the column and slowly add the prepared column-packing material. A mechanical vibrator can be used to assist in the filling. Excess vibration can cause exposure of solid support, resulting in poor separation of compounds. When the column is full, remove the funnel, plug that end with about $\frac{1}{2}$-in. of glass wool, and connect to the injector. Turn on the inert carrier gas, set the oven at the proper curing temperature, and precondition the column—usually overnight. This removes the "bleed" or easily volatile materials before use and saves on detector fouling. Then connect the column to the detector end and it is ready for use.

Techniques for preparing packing material vary. One simple method is to weigh out the solid support first, then the liquid phase. Dissolve the liquid phase in the recommended solvent. Make a slurry of the solid support in excess solvent (the same one). Mix together in a round-bottom flask. Attach to a rotary evaporator and turn on for about 10 min without vacuum. Then turn on vacuum and evaporate off the solvent. Use minimal heat. When the packing is completely dry and free flowing, it can be further dried in an oven until no odor of the solvent remains. It is then ready to load into the column. Special problems arise with certain coatings and substrates. The proper selection of material for a specific job is an extensive area.

The new fused silica capillary columns usually come coated. The flexibility of the material makes them simple to handle and easy to install. Some problems that damage these columns are: (a) overloading with solvent (washes off the coating), (b) overheating (stay within manufacturers' recommended ranges), and (c) failure to have adequate gas flow (columns easily damaged if gas flow is off or too low because of elevated temperature). Because of the low flows involved, dead spaces must be kept to a minimum. Use of makeup gas to the detector is essential. A splitter or direct on-column injection equipment must be used and conditions must be optimized for best results. For further information see specific journal articles or texts on the subject.

Headspace or Gas Extraction Analysis

Headspace analysis is primarily a qualitative tool as it is used today. The main concern is that the techniques used by various workers are not standardized with regard to (a) volume of headspace gas collected, (b) temperatures of the test material during sampling, and (c) the method of obtaining the headspace gas. The technique does allow for great sensitivity and gives an indication, for volatile compounds, of a measure of what level one may be sensing by olfactory techniques. However, the effects of ethanol on the various components change the vapor pressures rather unpredictably and sometimes cause values greatly in excess of predicted amount by normal calculations. Thus another variable results. The many factors that affect the vapor pressure, as well as the complexity of the sampling methodology, lead to large standard deviations in any analytical work. Collection methods vary, ranging from adsorption on active absorbants, such as carbon or Tenax, to more complex in-line collection systems. Generally, when absorbants are used the ethanol absorbed must be removed before the other organic volatiles can be put onto a GC column for separation. This is accomplished by back flushing the trapping column at a slightly elevated temperature. The material trapped on the column must then be flushed off and trapped either in a capillary tube for syringe injection or in a chilled device that can be rapidly heated, and the material must be flushed rapidly, directly onto the GC column. There are numerous systems for collection and handling. One that would seem particularly effective was discussed by Maarse (23). A diagram of the system is shown in Figure 44. Carrier gas is used to purge the sample of volatiles. The ethanol and water are frozen out in the $-15°C$ trap; then the other volatiles are trapped in a fused silica column by the use of a $-120°C$ (by bubbling air through liquid nitrogen) air condenser. When the purging is complete, a switching valve transfers the carrier gas to the column inlet and the $-120°C$ airflow is shut off. The area of the column where the collected volatiles are trapped is heated, by the preheating oven, and the volatiles are flushed on through the capillary column for GC separation. This type of instrument would tend to give better analytical results, because it strips out most of the ethanol and water, but it does not fairly represent what the nose detects.

Detectors

The detector senses compounds that are eluted from the column of a gas–liquid chromatograph. Usually they respond to all organic material. Specific detectors are extremely useful for certain jobs. Four types of detector are listed below.

THERMAL CONDUCTIVITY. The carrier gas and the eluted compounds pass over a "hot wire" (with an electric current passing through it). The presence of the compounds in the carrier gas raises the heat conductivity capacity of the mixture, and more heat is removed from the wire. As the temperature of the wire

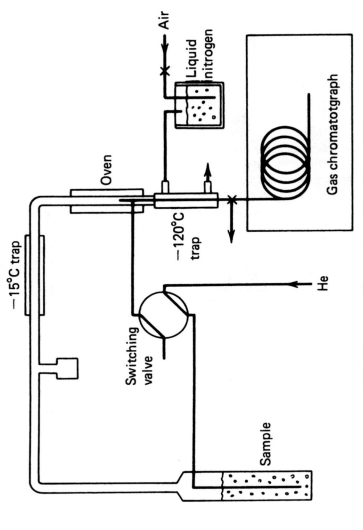

Figure 44. A purge-and-trap system for GC headspace analysis.

is lowered by the removal of this extra heat, the current flow is increased because of decreased electrical resistance:

$$I = \frac{1}{T}$$

The increased current flow is measured electronically, amplified, and recorded.

The method has a distinct advantage in that the sample is not destroyed. However, it is not extremely sensitive, and the presence of water in a sample will cause very severe problems because it is also detected.

FLAME IONIZATION. The most commonly used detector consists of a small jet supplied with hydrogen and air; a small flame burns. In flame ionization detection the effluent passes through the flame, and as it is burned, ionized particles are formed. An electric potential is maintained between the jet and the metal sleeve (collector ring or sleeve). As the ionized particles pass over the collector, they are attracted and neutralize part of the potential. The change in voltage is amplified and recorded. This is a very sensitive method and is not bothered significantly by water. Most combustible compounds that are eluted are measured.

THERMIONIC DETECTORS. The thermionic detectors are very similar to flame ionization detectors, but certain salts can be added to the flame area to catalyze desired reactions. For example, in the nitrogen (thermionic) detector a rubidium bromide crystal is inserted into the flame area, and the following reactions are postulated:

$$RuBr \underset{Heat}{\rightleftharpoons} Ru^+ + Br^-$$

$$2Ru^+ + H_2 \longrightarrow 2Ru^+ + 2H^+$$

$$Ru + RNH_2 \rightleftharpoons Ru^+ + CN^-$$

Under usual flame ionization conditions, nitrogen compounds are burned and ionized only to about 0.001%, but in this case the effective ionization is 1.0%, which is an increase of 10,000-fold. Other carbon compounds are still detected, but the nitrogen compounds are magnified in proportion. One of the problems associated with thermionic detection is extreme sensitivity to flow changes. Certain other elements may also interfere. The equipment is sensitive in the same range as the flame ionization unit.

EMISSION DETECTOR. The emission approach is an extremely sensitive detection method, particularly for sulfur and phosphorus. The effluent is burned in a jet, similar to other jets, and a percentage of the elements ends up in the ground state with the emission of a photon of a specific wavelength. Appropriately designed mechanical and optical systems filter these elements and feed them to a photomultiplier, and measurements are made. Since the wavelength is specific, the recording measures only that compound and is unresponsive to other compounds or to column bleed. The emission wavelengths used for sulfur and for phorphorus are 3939 and 5260 Å, respectively. Sensitivity of the system can be as low as 0.2 ng per injection. At present, its use is limited to measuring sulfur and phosphorus.

COULSON ELECTROLYTIC CONDUCTIVITY. The Coulson® electrolytic conductimeter measures only nitrogen, with some interference from halides. The effluent from the column is mixed with hydrogen gas and is passed over a nickel catalyst in a heated quartz tube (800–900°C). All the nitrogen-containing compounds are reduced to ammonia or to non-nitrogen compounds. Any acidic components are trapped in a neutralizing scrubber (a piece of glass wool soaked in strontium hydroxide). The rest of the gases, including the ammonia, pass out the end and through a Teflon tube. A constant-head system for uniform washing of the gas traps part of the ammonia. This ammonia, dissolved in the water, passes through a glass conductivity cell. In the presence of the ammonium ion the following reaction occurs:

$$NN_3 + H_2O \rightleftharpoons NH_4^+ + OH^-$$

The ions formed allow current to flow. This current is amplified and measured. Sensitivity is about 1 μm per injection. This detector is especially suited for measuring small amounts of nitrogen compounds in biological solutions.

The Hall detector is an improved electrolytic conductivity system that can be adapted to capillary columns.

MASS SPECTROMETRY. Recently the use of single-ion monitoring (SIM) of mass spectrographs and the use of capillary columns of gas chromatographs have proven to be most sensitive and reliable methods of identification and of quantification of trace amounts. Specific ions that are reasonably unique for the compound in question can be measured.

For further theory on gas chromatography, see the textbook by Grob (24). A recent textbook on capillary columns is also recommended (22).

Supercritical Chromatography

An emerging technology is supercritical chromatography. This technique involves the use of a solvent under pressure so that it is neither a gas nor a liquid as the mobile phase. Solvents that are useful include carbon dioxide because of its low critical temperature (31.3°C) and its use in thermally sensitive compounds; others include nitrous oxide and ammonia for separation of polar compounds and n-pentane for high-molecular-weight hydrocarbons. Compatibility with sensitive detectors is also essential.

Three essential units of a system are: (a) a high-pressure pumping system; (b) a high-precision oven equipped with an injector system, capillary column, and microcomputer control; (c) appropriate detector; and (d) good computer-operated control.

The usefulness of this system is that both volatile and nonvolatile components can be delivered to a very sensitive detector such as a flame ionization detector.

For a recent review see Reference 25.

Liquid Chromatography

High-pressure liquid chromatography (HPLC) has recently become a useful tool in separation and quantification of organic compounds, especially the nonvolatile ones. The amino acid analyzers are the precursors to the modern HPLC units. The system consists of a high-pressure pump or pumps and columns coated with various materials and detectors. The pumps can operate at thousands of pounds per square inch of pressure and are designed to give smooth, nonpulsing flow. The column may be packed with a large variety of packing materials, and solvents of many types can be used. Detectors usually measure UV absorbance, refractive index, fluorescence electrochemical oxidation, photodiode array, and mass spectra. The design of these detectors is still in flux.

Some of the solvent characteristics of the separation system are important. For example, the higher the viscosity, the greater the pressure drop across the column; larger pressure drops cause lower efficiency in column separations. If the HPLC method of detection is selected, it is important to have a large difference in refractive index between the solvent and solute. This is especially true with gradient operation. Some solvents have high absorbance at UV wavelengths. Solvents should be checked before use to be sure that they will not mask the solute because of excess UV absorption.

Sometimes the solutes cannot be effectively separated with any particular solvent. In such a case, altering the pH of the solvent can turn an ionized compound into an un-ionized compound that is no longer hydrophilic and that can be separated. Ion pairing is another system whereby a hydrophilic molecule can be turned into a hydrophobic compound. For example:

$$\underset{\text{Hydrophilic solute}}{RCO_2^-} + \underset{\text{Ion-pair agent}}{N(C_4H_9)_4} \rightleftharpoons \underset{\text{Hydrophobic ion-paired solvent}}{RCO_2N(C_4H_9)_4}$$

This can be achieved for amines by pairing with an alkyl sulfate.

"Reverse phase" indicates that the solvent system used is usually aqueous, although other polar solvents can be substituted and the stationary phase is organic. Other less polar solvents can be added to water to alter the polarity of the solvent. Normal phase means the stationary phase is more polar than the solvent.

Some separation systems have been concisely described (26). These are summarized in Table 76. As Bakalyar noted, "the number of selective handles available to the chemists is limited only by his imagination." This is primarily because in contrast to gas chromatography, where the stationary phase is the only phase that can be altered significantly, liquid chromatography involves an almost unlimited number of solvent–stationary phase combinations.

Reverse-phase columns are probably the most popular type of column in use because of their versatility. Ion-exchange columns are coming into greater use now for polar separations, especially for amino acids and organic acids. With proper care and precautions, present-day columns have relatively long lives.

A schematic diagram of a typical HPLC setup is shown in Figure 45.

As the equipment becomes more sophisticated, more problems can arise. One of these worth considering is sample injection valves. Symptoms can be as follows: peak broadening, poor precision, leaks, and increased system pressure. Particulates, such as crystals from buffers, can scratch seals, and blockages can occur. A dirty syringe can be a problem that may be misinterpreted as a valve problem. Pumps are precision devices that can and do leak if the seals are not properly maintained or if proper rinsing of the pumps is not done. Crystal formation can cause breakage and costly repairs. Guard columns are really short columns, generally of the same material as the packaging in the main column. The purpose is to protect the main column from deterioration by adsorption of material that would normally not elute. Replacement of guard columns should be on a routine basis and should be done before retention times on the main columns start to shorten. The guard column also is involved in component separation. Some of the theoretical aspects of this are covered in Reference 27. Most units have frits installed which filter out all particulate matter. These become plugged over a period of time. There are generally several on each unit. When the pressure starts to rise, this is an indication of plugged frits and they should be replaced or cleaned if possible. One should maximize column life by selection of buffers and solvents that do not interact with the column to destroy it. The use of high-purity solvents is essential as well as use of buffers within the tolerance range of the column packing. Some simple rules are: (a) Use the best solvents possible, (b) be sure all glassware is clean, (c) de-gas the solvent, (d) measure mixture components carefully, and (e) record every action so future troubleshooting can be done effectively. Remember that mold, bacteria, or yeast can grow in some of the solutions that are used. This

Table 76. Separation Systems for High-Pressure Liquid Chromatograph

	Phase separation			Retention time control		
Type	Mobile	Stationary	Type	Solvent composition	Solvent pH	Ion-pairing agent
Reverse Phase						
Hydrophobic adsorption	Water	Hydrocarbon–silica	Reverse-phase adsorption	CH_3OH, CH_3CN, THF, etc.	Acid–base, buffers	Ionic surfactants
Liquid–liquid partition	Water	Organic coated on hydrophobic silica	Reverse-phase partition	CH_3OH, CH_3CN, THF, etc.	Acid–base, buffers	Ionic surfactants
Reverse Phase or Normal Phase						
Liquid–liquid partition	Water or organic	Functional groups bonded to silica	Polar-bonded phase	CH_3OH, CH_3CN, THF, etc. CH_2Cl_2, BuOH, ETOAc, etc.	Acid–base, buffers	Ionic surfactants
Normal Phase						
Liquid–liquid partition	Organic	Water coated on silica	Normal phase partition	CH_2CL_2, BuOH, ETOAc, etc.		Ionic surfactants
Polar adsorption	Organic	Silica	Normal phase adsorption	CH_2CL_2, BuOH, ETOAc, etc.	Weak acids and bases	Ionic surfactants
Ion exchange	Water	Ionic groups bonded to silica	Ion exchange	—	Acids–base, buffers	

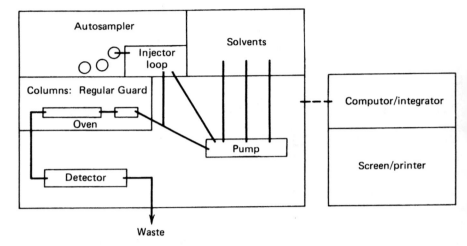

Figure 45. An HPLC setup for autosampling and for a three-solvent single-column operation.

can rapidly plug the system. Other rules to follow are: Minimize dead space and tubing lengths, and use proper detector size.

The problem may not always stem from the HPLC (or other equipment) but instead may be due to some pretreatment of the sample. For example, the proper use of Sep-pak pretreatments is crucial. If the device is overloaded or the residence time of contact to the absorbent is too short, some of the material may not be held up; thus poor or variable results will be obtained.

As with gas chromatographic columns, the smaller the column diameter the greater the enhancement of the maximum responding signal to the solute concentration. A mathematical treatment can be found (28).

A critical comparison (29) of peak height to peak area showed that for normal, good separation and peak shape, area measurement was superior. However, with poor separation or non-uniform-shaped peaks, height measurement was better.

The general use of HPLC in wine analyses has been reviewed (30).

ION CHROMATOGRAPHY

The most common ion chromatography unit is the suppressed anion–cation system. Anions are separated on a low-exchange-capacity anion resin. The eluate is a dilute base such as sodium carbonate. Following the anion column is a cation column, which is the "suppressor" column. This converts the eluted carbonate to carbonic acid and converts the sodium ion to hydrogen ion. Detection is by conductivity cell and is very sensitive. The carbonic acid has a low conductivity if the anion is ionized to any extent. The use of hollow-fiber cation-exchange resins allows for continuous regeneration of the cation exchangers. For cation separation and detection, a cation-exchange resin is used for separation, a hydroxyl-charged anion-exchange resin is used for suppression and a conductivity cell is used for detection. The solvent can be dilute nitric acid or dilute perchloric acid.

Many of the same problems that are prevalent with HPLC also apply to ion chromatography. Guard columns will avoid many contamination problems. For anion columns cyanide complexes, perchlorate ions, aromatic organics, and surfactants can contaminate; for cation columns heavy metals, surfactants, and aromatic organics may be a problem.

GEL SEPARATION

Gel filtration or exclusion chromatography is very useful in separating large molecules. Its primary use is in protein or enzyme analysis and purification. The choice of a gel depends on the solvent and the materials to be separated. Certain substances have an affinity for the gels beyond the normal separation characteristic of molecular weight.

As the name of the process implies, the large molecules are excluded from this gel and are the first to be eluted from the gel column. The distribution coefficient between the inner (interior of gel) and outer solvent areas can be termed C_d. Usually C_d is a function of molecular weight. For molecules excluded, $C_d = 0$, and for those with complete accessibility, $C_d = 1$. The elution volume V_e depends on the external volume of the gel particles V_0 and on $C_d V_i$, where V_i is the solvent internally situated in the gel:

$$V_e = V_0 + C_d V_i$$

To separate the two compounds, the separation volume V_s must be at least equal to $(C_{d_1} - C_{d_2}) V_i$ and less than V_e. This procedure has limited use in routine analysis.

SELECTIVE-ION ELECTRODES

In recent years, electrodes that selectively screen out most ions but that preferentially measure a specific ion have become popular. The advantages of these electrodes are their specificity and their ability to measure minute quantities accurately. These electrodes are in the same class as the glass electrode for measuring hydrogen ion activity (pH). The electrodes cause an electrical potential to be developed that is proportional to the concentration of the ion or gas in question. Most require a reference electrode, which provides an electrical contact to the test solution and a stable voltage for purposes of comparison. In principle, whether gas or ions are being measured, the electrodes consists of a barrier membrane, an internal filling solution, a collector contact, a proper body, and necessary electrical connections.

The Nernst equation

$$E = E_0 + 2.3 \frac{RT}{nF} \log A$$

indicates the relationship of the potential generated, E, to the concentration or, better, to that activity of the ion in question in the test solution A. For a theoretical treatment see Clerc et al. (31). The use of these electrodes can be divided into three categories: (a) direct potentiometric reading, (b) method of addition, and (c) end-point monitor for titration. Immobilized enzyme systems can be incorporated into the electrodes. Products of enzymatic reactions then can be measured. This appears to be useful in enzyme systems producing carbon dioxide where a gas-sensing electrode can be incorporated and the substrate in question measured indirectly by measuring enzyme reaction and evolution of carbon dioxide. There are other examples (32).

Gas-Sensing Electrodes

Among the gases that are of applicable interest in musts and wines are sulfur dioxide, ammonia, hydrogen sulfide, and carbon dioxide (and possibly HCN). Specific commercial usefulness remains to be proved. One great advantage is that the samples need little or no pretreatment before testing. Figure 46 shows schematically a gas-sensing electrode, a reference electrode, and a solid-state electrode.

As an example the ammonia electrode is described. The internal filling solution consists of 0.01 M ammonium chloride solution. The wine or test sample is made basic by addition of a strong base. Any ammonia present is then in the NH_3 form:

$$NH_4^+ + OH^- \rightleftharpoons NH_3 + H_2O$$

It can then go through the membrane and enter the chloride solution. The predominant species are NH_4^+ and Cl^-. The ammonia entering the solution

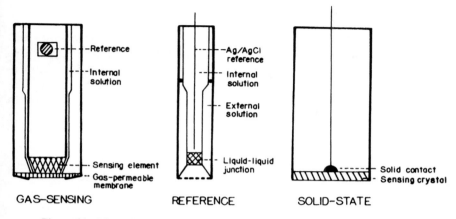

Figure 46. Schematic diagrams of gas-sensing, reference, and solid-state electrodes.

immediately undergoes the reaction

$$NH_3 + H_2O \rightleftharpoons NH_4^+ + OH^-$$

The resulting concentration of NH_4^+ is changed only slightly. The major change is a *relatively* large increase in OH^- to adjust for this.

$$H_2O \rightleftharpoons OH^- + H^+$$

The equilibrium shifts to the left and the H^+ concentration is decreased. This change in pH is sensed by the internal sensing element (or glass pH electrode). Thus the concentration of the ammonia can be measured. A steady-state equilibrium is rapidly reached between the force necessary to cause the ammonia to pass through the membrane and the difference in concentration inside and outside the membrane.

Solid-State Inorganic Crystal Electrodes

Application for the measurement of many metals with solid-state selective electrodes (Figure 46) seems to be feasible now or in the near future.

The sensing element must be a salt crystal (not soluble in the test solution) containing the anion or cation being tested for. If the crystal is sensitive, the element in question migrates through the crystal and equilibrium is reached. At equilibrium the force required to move the ions across the membrane is proportional to the concentration of the element in the test solution. The potential is measured by comparing to a silver–silver chloride electrode or some other reference potential.

In the case of fluoride measurement, the membrane is a lanthanum fluoride crystal. Another variation of this electrode is the silver halide crystal type. This can be used to measure the Ag^+ concentration but is more useful in the measurement of I^- or other halogens when the proper crystals are used. Although the crystals are Ag^+ sieves, they behave as if the halogen were passing through the crystal; based on the solubility equilibrium

$$AgI \rightleftharpoons Ag^+ + I^-$$

they can be treated as a direct measure of the I^- activity in the test solution:

$$E = \text{constant} + 2.3 \frac{RT}{nF} \log [I^-]$$

In general, solid-state electrodes cannot operate in solutions where an ion species exists that will react with the solid crystal to form a substance more insoluble than the membrane crystal (it would coat the surface and prevent access for the test ion). Adsorption of ions to container walls, interference from supporting electrolytes, and solid-state defects all contribute toward limiting the sensitivity to less than the theoretical level of the solubility of the salt pair making up the membrane (33).

Liquid Membrane Electrodes

The use of liquid membrane electrodes, also called ion-exchange electrodes, has been suggested to measure sodium, potassium, and calcium in wine. As shown in Figure 47, such electrodes consist of a hydrophobic membrane impregnated with an organic liquid ion-exchange material and an aqueous filling solution.

The organophilic membrane saturated with a water-immiscible organic solvent forms the carrier for the ion-exchanger molecules. These neutral, high-molecular-weight organic salts are free to move within the membrane. As they approach the surface, the cation portion can be exchanged with other cations and carried back and forth from the internal aqueous solution. Soon equilibrium is reached between the internal and external solutions.

The force required to maintain this equilibrium is proportional to the concentration of the specific cation in the test solution. The selectivity of electrode depends on stability of the specific salt of that cation compared to the stability of that salt with other cations. Anion-specific organic salts are also available.

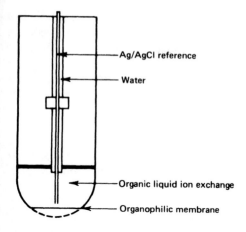

Figure 47. Schematic diagram of an ion-exchange (liquid membrane) electrode.

GLASS ELECTRODES AND pH MEASUREMENT

Glass electrodes are primarily designed to measure hydrogen ion activity. They consist of thin glass envelopes of specially formulated glass. There is a transport of potential across the glass related to the hydrogen ion activity in the test solution. The general theory is that the surface of the glass becomes hydrated and forms a gel, and the hydrogen ion transfers a charge onto this surface. The charge is passed, atom by atom, through the dry portion to the inner surface.

The most H^+-sensitive glass is that containing 1% Al_2O_3. Other formulations are more sensitive to other alkali cations.

On the inner side of the glass electrode is a buffer solution (pH 7.0). A standard calomel electrode, sealed into the buffer solution, maintains the buffer at a known potential and completes the cell. This electrode is connected to a reference electrode immersed in the test liquid (Figure 48). A readout system is furnished.

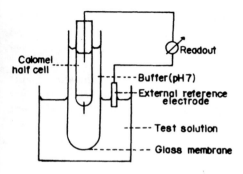

Figure 48. Schematic diagram of a glass electrode and supplementary connections.

The usual mode of operation is to place the electrodes into a solution of known pH (buffer solution) and standardize the unit. Then the electrodes are placed in the test solution. Any difference in hydrogen ion activity from the calibration pH will cause a potential difference across the glass membrane, which will unbalance the system. The amount of potential difference required to rebalance the meter can be read as pH units:

$$\text{pH} = -\log [\text{H}^+ \text{ activity}]$$

The glass electrode should be stored in slightly acid solution, and the reference electrode should be stored in 0.1 M potassium chloride solution (34). The junction must be kept wet to prevent clogging. Glass electrodes can be rejuvenated by immersing in 0.1 N hydrochloric acid or by alternatingly dipping into acid then base. Reference electrodes can be unplugged by (a) replacing filling solution, (b) soaking overnight in 0.1 M potassium chloride solution, (c) applying pressure to the filling hole or vacuum in the tip, (d) boiling the junction for 10 min in dilute potassium chloride, and (e) sanding the tip with 600 emery paper. The treatments should proceed from steps (a) to (e), stopping with treatment that unclogs the junction.

COMPLEXOMETRY

Chelation can be defined as an equilibrium process in which a metal complexes with a ligand. The ligand can consist of a relatively simple organic compound such as citric acid or even an inorganic compound such as polyphosphate. The general equation can be written as follows:

$$\text{M} + \text{L} \rightleftharpoons \text{ML}$$

For chelation to occur, the steric and electronic configurations of the metal and the ligand must match, and the surrounding physical and chemical conditions must be suitable. Under ideal chelating conditions the metal and ligand will form five- or six-member rings. Calcium, magnesium, zinc, copper, and iron have bond angles favorable for such ring formations and are relatively easy to chelate.

Ethylenediaminetetraacetic acid (EDTA) is one of the better chelating agents (or sequestrants). EDTA has six donor groups, and as few as four groups are capable of forming 1:1 metal complexes. As the pH increases, more donor groups are available for chelation. However, increasing the pH beyond the point of complete ionization of the EDTA does not increase its ability to chelate.

Figure 49 is a diagram of a typical chelate of a metal with EDTA where all

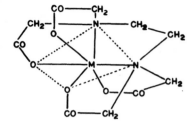

Figure 49. EDTA bonding in chelation with a metal.

six donors are used and all ring structures (solid lines) are stable five-member rings.

West (35) has reviewed the use of complexometric reactions. The remainder of this section presents some of the properties of EDTA, indicator requirements, and general information. It is essential that a complexometric reagent react stoichiometrically and "instantaneously"; in addition, "free" metal must be reduced to a low level ($>10^{-5}M$). It must not be necessary to add excess reagent, and the analyst must be able to monitor the reaction. The reaction must be highly selective, the required conditions (pH, temperature, etc.) must not be highly critical, the reaction products must be soluble and not be highly colored, and complex formation should take only one or two steps.

Complexometric indicators should be weak complexing agents (weaker than EDTA) and should change color visibly when the metal is removed, even at low concentration; they must be usable in the operating pH range and must be water soluble; and they must be specific for the metal being measured and must be found under the same conditions as the EDTA complex.

EDTA (ethylenediaminetetraacetate or Versene®) should complex with all metals containing more than unit charge, and the stoichiometric reactions should be rapid. Since pH may have an affect in the acid range it should be controlled. This material forms colorless complexes and is a strong chelating agent (CN can "mask" the heavy metals Zn^{2+}, Fe^{2+}, etc). EDTA is destroyed by moderately strong oxidizing agents, has limited solubility in water (can be precipitated by acid), can be used as a primary standard, and reacts with metals in a 1:1 ratio. There is an optimum pH at which metal reacts best with EDTA. Table 77 gives these values.

EDTA occurs in three forms. Table 78 lists the solubilities of aqueous solutions of each form and the pH of the saturated solution.

The disodium salt ($2H_2O$) may be purified and used as a primary standard for complexometric titrations. To purify, saturate a water solution with the disodium salt, add ethanol until the first sign of precipitate, then filter and add an equal volume of ethanol to the filtrate. Filter out the crystals and wash with acetone and ether. Air dry overnight, then dry at 80°C for 4 days. Purity is 99.5% and yield is 75%. The dry form is dihydrate.

Table 77. Optimum pH for Complex Formation of Certain Metals with EDTA

Metal ion	Optimum pH
Calcium	7.5
Cobalt	4
Copper (II)	3
Iron (II)	5
Iron (III)	1
Lead	3
Magnesium	10.0
Manganese	5.5
Nickel	3
Zinc	4

Table 78. Forms of EDTA and Solubility Data at 22°C

Form	Solubility (g/100 mL)	pH
EDTA, free	0.2	2.3
Disodium salt (anhydrous)	10.8	4.7
Tetrasodium salt (anhydrous)	60.0	10.6

A complete description of the theory and use of EDTA and related compounds is available (36).

ENZYMATIC ANALYSIS

A review of basic simple kinetics is given below. For more detailed information, see References 5 and 7.

$$S + Enz \underset{k_{-1}}{\overset{k_1}{\rightleftharpoons}} Enz\text{-}S \xrightarrow{k_2} Enz + P$$

where S = substrate
  Enz = enzyme
  Enz-S = enzyme–substrate complex
  P = product
  k = rates of reaction

Normally, the amount of Enz is much lower than that of S or Enz-S; thus if the first reaction rate is rapid to the right and slow to the left and Enz-S decomposes rapidly and irreversibly, the following equation is valid:

$$\frac{dx}{dt} = k_2 \times [\text{Enz}] \frac{a - x}{(a - x) + [(k_{-1} + k_2)/k_1]}$$

where a = initial [S]
x = amount of [S] transformed
[Enz] = enzyme concentration
$\frac{k_{-1} + k_2}{k_1}$ = Michaelis–Menton constant = K_m

Then

$$v = k \frac{[\text{ENZ}] \times [\text{S}]}{[\text{S}] + K_m}, \quad k = \text{overall rate constant}$$

It is generally assumed that 1°C increase causes a 10% "rate" increase.

$$\log k = a - \frac{b}{T}, \quad T = \text{absolute temperature}$$

Linear plots are obtained for log k versus 1/T as long as they are within the temperature limits of the enzyme.

pH

Certain pH ranges are optimum for enzymatic analysis and must be adhered to.

Other

Inhibitors and conditions that poison enzyme protein or cause it to become deformed must be avoided.

Sensitivity

Changes in various parameters and conditions affect the accuracy of determinations made by enzymatic analysis.

The following general questions must be answered before an enzyme procedure is chosen: Is the enzyme specific? Is the activity sufficiently high (V_{max}/K_m calculation)? Is the enzyme active, or has it deteriorated? Are the equilibria favorable? Can a useful measurement be made?

Handling and Storage of Enzymes and Other Reagents

Enzymes are stable for reasonable periods of time (up to 2 or more months) and retain a high level of activity. However, care is necessary and certain rules must be observed.

For example, if enzymes are suspended in ammonium sulfate solution they should be stored at $-4\,°C$ (do not freeze, since a significant loss of activity can occur). Lyophilized enzymes can be stored at lower conditions. Moisture must be excluded from the preparations. Each enzyme should be handled in a manner suitable to it.

Coenzymes such as NAD^+, $NADP^+$, CoA, and FAD can deteriorate quickly.

Exposures to moisture, excess heat, and light all play a role in loss of enzyme stability. Moisture is most harmful for NADH because of hydrolysis of the pyrophosphate bond (80% humidity, $22\,°C$ caused 50% deterioration in 1 week). Short periods at temperatures up to $33\,°C$ do not harm this coenzyme appreciably. NADH solutions should not be below pH 7.5. NAD^+ should be kept close to pH 7.0 because it (as well as $NADP^+$) is alkali labile.

Most coenzyme solutions, except for NADH, are best divided in small portions and frozen separately to avoid thawing and refreezing. Storage containers must be of a type that does not contaminate the contents. Pipets should not be inserted directly into the primary sample.

Water used for diluting should be deionized, then distilled before use in glass. Reagents should be made up and the solutions kept in all-glass reagent bottles or disposable plastic containers.

Coenzyme A solutions should be stored at pH 4.0, and ATP shold be stored at around pH 9.0. Cleaning of glass should be by dichromate–sulfuric acid with thorough rinsing and drying. Detergents can be used as a substitute if all is removed by hydrochloric acid rinse, followed by a water rinse.

The actual weight of protein can be calculated to get a rough amount of the enzyme present. One method involves measuring the adsorption at 260 and 280 nm and using the formula

$$E = 1.55\,(E_2 - E_1)_{280\,nm} - 0.76\,(E_2 - E_1)_{260\,nm}$$

where E_1 = extinction of cuvette with double-distilled water at 280 and 260 nm and E_2 = sample solution adsorption of the two wavelengths.

The C (mg/mL of sample in solution) = $(E \times 3.0)/0.20$. The result can vary slightly from enzyme to enzyme.

The Loury method using the Folin–Ciocalteu reagent is based on the amount of tyrosine present in protein. A standard curve is usually run with serum albumin or another protein.

The biuret test is basically the formation of a colored copper complex with peptide bonds. It is linear over a wide range of protein molecular weights. It can be affected by ammonia, glycerol, and a number of substances.

Procedure

It is recommended first, by boiling or addition of alcohol, to denature and precipitate the protein and centrifuge clear of interfering substances.

Dissolve 8 g of sodium hydroxide in carbonate-free water and make to 1-L volume (0.2 N NaOH). The biuret reagent is made by dissolving 9 g of sodium potassium tartrate in 400 mL of 0.2 N sodium hydroxide, then adding 3 g of finely powdered copper sulfate pentahydrate, followed by 5 g of potassium iodide, only after the previous salts have been dissolved completely. Make to 1 L with 0.2 N sodium hydroxide. To make 3 M trichloroacetic acid, dissolve 49.0 g in water and make to 100 mL.

Use a solution containing about 10 mg of protein per milliliter without dilution (if higher, dilute as necessary). Use a 2-cm light-path cuvette and measure absorbance at 546 nm. Pipet into test tube: 1.30 mL of water, 0.20 mL of sample, and 0.20 mL of trichloracetic acid solution.

Mix by shaking; centrifuge for 2 min, and discard the supernatant. Place the tube mouth down on a filter paper and drain. Then add 5.00 mL of water and 5.00 mL of biuret reagent. For the blank, use the same addition but no precipitated protein.

$$E = E_{sample} - E_{blank}$$

Compare results to that of a standard curve made with bovine blood albumin. Report as equivalents of bovine blood albumin (mg/L).

REFERENCES

1. W. F. Pickering, *Modern Analytical Chemistry*. Dekker, New York, 1971.
2. J. W. Robinson, *Undergraduate Instrumental Analysis*. Dekker, New York, 1970.
3. R. B. Fischer and D. G. Peters, *A Brief Introduction to Quantitative Chemical Analysis*, Saunders, Philadelphia, PA, 1969.
4. Y. Pomeranz and C. E. Meloan, *Food Analysis: Theory and Practice*. 2nd ed. Avi Publishing Co., Westport, CT, 1978.
5. H. R. Mahler and E. H. Cordes, *Biological Chemistry*. Harper & Row, New York, 1966.
6. J. Mika and T. Torok, *Analytical Emission Spectroscopy Fundamentals*. Crane, Russak, New York, 1974.
7. H. U. Bergmeyer and K. Gawehn (Eds.), *Principles of Enzymatic Analysis*. Verlag Chemie, New York, 1978.
8. W. Bertrach, W. G. Jennings, and R. E. Kaiser (Eds.), *Recent Advances in Capillary Gas Chromatography*, Vol. 1. Hüthig, Heidelberg, 1981; Vols. 2 and 3, 1982.
9. J. S. Fritz, D. T. Gjerde, and C. Pohlandt, *Ion Chromatography*. Hüthig, Heidelberg, 1982.
10. M. Sher Ali and R. W. Woods, *An Overview of High Performance Liquid Chromatogrpahy*. U.S. Dept. of Agriculture, Washington, DC, 1986.

11. A. Braithwaite and F. J. Smith, *Chromatographic Methods*. Chapman & Hall, London, 1985.
12. K. Hennig and L. Jakob, *Untersuchungsmethoden für Wein and 'ähnliche Getränke*. Ulmer, Stuttgart, 1973.
13. R. L. Schneider, *Eastman Org. Chem. Bull.* **47**(1), 1-12 (1975).
14. B. Parker, *Am. Lab.* **9**(10), 93-99 (1977).
15. R. W. Yost, *Am. Lab.* **6**(1), 63-71 (1974).
16. *Duolite Ion-Exchange Manual*, 2nd ed. Diamond Shamrock Corp., Redwood City, CA, 1969.
17. M. Roth, *Clin. Biochem., Princ. Methods* **1**, 219-230 (1974).
18. H. Brandenberger, *Clin. Biochem., Princ. Methods* **1**, 260-274 (1974).
19. G. Zweig and J. R. Whitaker, *Paper Chromatography and Electrophoresis*, Vol. 2. Academic Press, New York, 1971.
20. E. Stahl, *Thin Layer Chromatography*. Springer-Verlag, Berlin and New York, 1969.
21. E. J. Shellard, *Quantitative Paper and Thin-Layer Chromatography*. Academic Press, New York, 1968.
22. W. Jennings, *Gas Chromatography with Glass Capillary Columns*, 2nd ed. Academic Press, New York, 1980.
23. H. Maarse, *Found. Biotech. Ind. Ferment. Res.* **3**, 71-97 (1984).
24. R. L. Grob, *Modern Practice of Gas Chromatography*. Wiley, New York, 1977.
25. M. L. Lee and K. E. Markides, *Science* **235**, 1342-1347 (1987).
26. S. R. Bakalyar, *Am. Lab.* **10**(6), 43-61 (1978).
27. E. Lundanes, J. Dohl, and T. Greibrokk, *J. Chromatogr. Sci.* **21**, 235-240 (1983).
28. H. H. Lauer and G. P. Rozing, *Chromatographia* **15**, 409-413 (1982).
29. R. N. McCoy, R. L. Aiken, R. E. Pauls, E. R. Ziegel, T. Wolf, G. T. Fritz, and D. M. Marmion, *J. Chromatogr. Sci.* **22**, 425-431 (1984).
30. M. E. Evans, *J. Liq. Chromatog.* **6**, 153-178 (1983).
31. J. T. Clerc, H. J. Degenhart, and E. Pretsch, *Clin. Biochem., Princ. Methods*, **1**, 446-459 (1974).
32. G. J. Moody and J. D. R. Thomas, *Analyst (London)* **100**, 609-619 (1976).
33. J. Kentoyannakos, G. J. Moody, and J. D. R. Thomas, *Anal. Chim. Acta* **85**, 47-53 (1976).
34. C. C. Westcott, *Am. Lab.* **10**(8), 71-73 (1978).
35. T. S. West, *Complexometry with EDTA and Related Reagents*, 3rd ed. BHO Chemical, Poole, England, 1969.
36. R. Pribil, *Analytical Applications of EDTA and Related Compounds*. Pergamon Press, Oxford, 1972.

Index

Acetal, 140, 143
Acetaldehyde, 117, 140, 224, 225, 324
  amounts in wine, 142
  causes of formation, 141
  chemical procedure, 143-145
  colorimetric, 147-148
  enzymatic, 147-148
  gas chromatographic, 146
  α-hydroxysulfonate, 224, 232
  Jaulmes-Espezel reactions, 142-143
Acetic acid, 50, 51, 52, 53-57, 60, 82, 196, 266, 267, 288, 324. *See also* Volatile acidity
  conversion to tartaric acid, 60
Acetobacter, 54, 160, 161, 162
2-Acetohydroxybutyrate, 148, 151
Acetoin (3-hydroxy-2-butanone), 140, 141
  amounts in wine, 149
  methods of separation, 150
  procedure, chemical, 151
    gas chromatographic, 151-153
2-Acetolactate, 148, 151
Acetone, 324, 325
Acetovanillin, 218
Acid esters, 266, 267
Acids:
  changes during ripening, 50
  determination, gas chromatographic, 61-62
  fixed, 59-61
  interconversions, 52
  kinds, in must and wine, 50
  nonvolatile, 60-75
  sensory effects, 50
  separations, other, 75
    paper and thin-layer chromatography, 75-76
  total, 51-52

Acrolein (2-propenal), 155
Addition, method of, 271, 284, 338-339
Adonitol, 131-132
Adsorbants, common, 340
Adsorption spectrophotometry, elements, flame effect, 339-340
  equipment, 336, 337
  optimum flame and wavelengths, 336
  theory, 333, 336
Aeration-oxidation, *see* Gases; Sulfur dioxide
Aflatoxins, 255-256
Agmatine, 174
Alanine, 141, 172, 173
β-Alanine, 174
Alcoholic fermentation, 80-81
Alcohols, general importance, 80-81
  in wine, 80
Aldehyde dehydrogenase, 108
Alkalinity of ash:
  amount in wine, 267, 268
  cation-anion balances, 266-267
  procedure, 267-268
Altroheptulose, 46
Aluminum, 6, 268, 273, 279, 280, 292, 338, 344
Amines, 172, 174
  biogenic, 186-187
Amino acids:
  separation methods, 172, 173, 179-181
  amounts, 173, 182
p-Aminobenzoic acid, 174
α-Aminobutyric acid, 141, 172, 173
β-Aminobutyric acid, 174
α-Amino nitrogen, 181
  amounts in musts and wines, 184
  formol titration, 181-182
  ninhydrin method, 182-183

Ammonia, 172, 264, 266, 267, 348, 354–355
  procedures, 178–179
Amyl alcohol, active, 81, 113, 114, 118–119, 141
n-Amylamine, 174
Analysis:
  methods of choice, 2
  selection, 1–3
  sensory, 6, 11
  statistical, 6–7
  texts on, 3
Anions, 264, 288–296
  limits, 291
Anisole, 203
Anthocyanins, 65, 197, 225. *See also* Grape pigments
  acylation, 197
  distribution, Cabernet Sauvignon, 207
  extinction coefficients, 209–210
  glucosides, 197
  procedure, bisulfite bleach, 209
    malvidin diglucoside, 210–211
    pH shift, 209
  structure and nomenclature, 197
  sulfur dioxide binding, 208, 225
Anthocyanogens (leucoanthocyanins), 198, 215–216
  amounts, 215
  procedures, 215–216
Anthoxanthin pigments, *see* Flavonols
Anticeptic, 222, 235, 241, 242, 243, 244
Antimony, 6, 279, 280
Apple, detection of, 252
Arabinose, 21, 24, 47–48, 130
Arabitol, 130, 131–132, 133
Arginine, 172, 173, 175, 252
  procedure, 183, 185
Arsenic, 6, 268, 279, 280–281
Artificial wines, 252
β-Asarone, 250–251
Asbestos, 256
Ascorbic acid (Vitamin C), 73, 76, 234, 302, 303
  oxidase, 232
  procedure, 74–75
Ash, 31
  amounts in wine, 264, 265, 266
  calculated, 264
  procedure, 265–266, 268
Asparagine, 173
Aspartic acid, 141, 173

Atomic absorption spectrometry (AAS), 268, 270, 271, 272, 274, 275, 276, 277, 278, 279, 280, 281, 282, 283, 284, 285, 286, 287, 295
Azeotrope, 295, 325
Azinphos-methyl, 257

Bacteria, 223, 225, 236, 242
Bacterium mannitopoeum, 130
°Balling, *see* °Brix
Barium, 279, 281
Baumé, 22–23
Bayleton (tridinefon), 257, 258
Beer, 289, 306, 310–311
Beer-Bouger equation, 329
Beer-Lambert law, 334
Benolate, *see* Benomyl
Benomyl, 257, 258, 311
Bentonite, 268, 272
Benzaldehyde, 148, 197, 200
Benzoic acids, 200, 222, 240, 241, 242
  procedure, 242
Benzyl alcohol, 243
Benzyl bromoacetate, 243
Benzyl ether, 243
Beryllium, 281
Betaine, 250
Biotin, 174
Bisulfite complex, 223–225. *See also* Acetaldehyde; Sulfur dioxide
Blue fining, *see* Ferrocyanide and cyanide
Boron, 6, 281
Botran (dichloran), 258
Botrytis cinerea, 47, 75, 129
Brettanomyces sp., 160
Brightness, 317
°Brix, hydrometer corrections, dessert wines, 29–30
  methods of determination, 15
  nomograph, 35–36
  overcropping, effect of, 14
  procedures, 19–21
  specific gravity, conversion, 15–17
Brom cresol green, 60–61
Bromide, 6, 288, 289–290
Bromopropylate, 257
2,3-Butanediols, 81, 102, 121, 126–127
  amounts, 127
  isomers, 126–127
  procedures, 127–129

1-Butanol, 81, 102, 140, 141, 146, 324
2-Butanol, 115
2-Butanone, 102, 140
Butenal, 146
n-Butylamine, 174
Butyric acid, 54, 58
γ-Butyrolactone, 115

Cabernet sauvignon, 172
Cadaverine, 174, 186
Cadmium, 6, 268, 279, 281–282
Caffeic acid, 200, 218, 224
Caffeoyl tartaric acid (caftaric), 168, 169, 197, 202
Calamus oil, 250
Calcium, 264, 268, 270, 271, 272–275, 334, 337, 338, 356, 360
  amounts, 269, 272
  carbonate, 272
  instability, 272
  procedures, 268, 273–275
  source, other than grape, 272
  sulfate, 272
Captan, 256, 257, 258
Caramel, 155, 252
Carbamates, 258
Carbendozim, 257
Carbon dioxide, 55, 302, 304, 306–310
  bottling, 307, 310
  causes of, 306
  procedures, 306–310
  saturation, wines, 306–307
  toxicity, 308
Carbonic anhydrase, 308–309
Carbonyl compounds, 140–155
  source, 140, 141, 148, 150
Carbryl (Sevin), 257
Cash still, 55
(+) Catechin (2,3H-trans), 198
Catechins, 198, 211–215
  amounts, 212
  effect on mold, 214
  procedure, 214–215
Cation-anion balance, see Alkalinity of ash
Cations, 264, 268–288
  losses, during fermentation, 268
  ratios, authenticity, 30–31
Cellobiose, 47
Cesium, 279, 282, 337
Chemical additives, 222–259, 263

Chlordane, 256
Chloride, 271, 288, 290–292
Choline, 174
Chromatography, adsorption, 339–340
  gas-liquid, 61–62, 246, 254, 258, 343–348
  gel separation, 354
  HPLC, 46
  ion chromatography (IC), 268, 352–354
  liquid, 46, 64–65, 244, 349–352
  paper, 60–61, 340–341
  partition, 339–340
  separation, theory, 339
  silica gel, 339
  super critical, 348–349
  thin-layer, 46, 61, 244, 252, 255, 256, 258, 339–340
Chromium, 61, 280, 282, 338
Cinnamaldehyde, 148, 197, 200
Cinnamic acid, 200
Citramalic acid, 70
Citric acid and citrate, 52, 59, 60, 61–62, 130, 266, 267
  methods available, 70–71
  procedure, 70–71
Cobalamine, 174
Cobalt, 279, 282, 360
Color:
  artificial, 252
  discrimination of, 316–317
  procedure, 317–319
  terminology, 317
Colorimetric, 214, 254, 272, 274, 275, 276, 277, 278, 284, 285, 294
Complexometry, 272, 273, 274, 275, 358–360
Concentrate, 252
Conductivity, 265, 292
Coniferylaldehyde, 200
Copper, 222, 247, 268, 277–279, 280, 302, 338, 344, 360
  amounts, 269, 278
  legal limits, 5, 277
  procedure, 278–279
  source, 278–279
  stability, 278
Coulson detector, see Gas-liquid chromatography
p-Coumaric acid, 200, 218, 224
Coumarin, 148, 200, 203, 250, 251–252
p-Coumaroyl tartrate (coutaric), 168, 169, 197, 202

m-Cresol, 218
o-Cresol, 218
p-Cresol, 218
Crotonaldehyde (2-butenal), 155
Cufex, 247
Cuvette material, 330, 331
Cyanic acid, 175
Cyanide, *see* Ferrocyanide and cyanide
Cyanidin, 197
 -3,5-diglucoside, 210
 -3-monoglucoside, 207
 -3-monoglucoside acetate, 207
 -3-monoglucoside *p*-coumarate, 207
Cyanidinol, 198
Cysteine, 173, 311, 312
Cystine, 173
Cytosine, 226

Dalapon, 257
p,p'-DDT, 256, 257
Degreasing, 321
Dehydro-L-ascorbic acid, 74
Dellé procedure, 36
Delphinidin, 197
 -3-monoglucoside, 207
 -3-monoglucoside acetate, 207
 -3-monoglucoside *p*-coumarate, 207
Delphinidinol, 198
Demeton-*s*-methyl sulfone, 257
Density meter, 15. *See also* Ethanol, procedure, density meter
Diacetyl (2,3-butanedione), 148, 150
 amounts, 149
 analysis and separation methods, 150
 procedure, chemical, 150–151
  gas chromatographic, 151–153
1,3-Diaminopropane, 186
Diathianon (Delan), 256
Dibenzyl ether, 243
Dichlorofluranid (Euparen), 256
Dichromate oxidation, 99–104
Dieldrin, 256
Dielectric constants of solvents, 323, 324
Diethyl carbonate, 245
Diethyl dicarbonate, effective level, legality, 222, 244
Diethylene glycol, 246–247
Diethyl malate, 169
Diethyl pyrocarbonate, *see* Diethyl dicarbonate

Diethyl succinate, 165, 169
Diethyl tartrate, 169
3,5-Diglucosides, separation of, 211–212
2,3-Diketo-L-gluconic acid, 74
α-Diketones, and α-hydroxyketones, 150
Dimethyl dicarbonate, 244–245
 procedures, 245
Dinocap, 258
Distilland, 325
Distillate, 325
Distillation:
 boiling points, 323–324
 macrostill, 88–91
 microstill, 89–91
 still types, 323–324
Disulfide bonds, 225
"Dry" wines, 36
Duiren, 257

Ebulliometer, 85–88
EDTA, *see* Complexometry
Ellagitannin, 197
Emission cavity analysis, 234
Emission detector, *see* Gas-liquid chromatography
Emission spectroscopy, direct reading, 286–287
Endrin, 256
Enzymatic analysis, 43–45, 58, 234, 308, 360–363. *See also specific compounds*
 kinetics, 360–361
 protein analysis (Loury), 362–363
 storage conditions, 362
(−)-Epicatechin (2,3H-*cis*), 198
(−)-Epigallocatechin (2,3H-*cis*), 198
Erythitol, 130, 131–133
Esculetin, 200
Esters:
 acids, 266
 amount in wine, 162–163
 hydrolysis, 159
 procedure, gas chromatographic, 166–168
 total volatile, 165–166
 sensory, 164–165
 source, 159–160, 162
 toxicity, 165
Ethanolamine, 174
Ethanol (ethyl alcohol), 31, 80, 141, 274, 277, 278, 279, 292, 306, 324, 325
 alcoholic fermentation, 80–81

conversion, % v/v, proof, etc., 30-31, 100-101
efficiency of fermentation, 82, 84
hydrometer errors, 92, 94
hydrostatic balance, 97
legal limits and tax rates, 84-85
limit, yeast fermentation, 82
loss, due to temperature, 83
nomograph, 35-36
procedure:
    density meter, 97-98
    dichromate oxidation, 99-102
    ebulliometer, 85-88
    enzymatic, 105-108
    gas chromatographic, 104
    HPLC, 105
    hydrometer, 91-94
    pycnometer, 94-95
    rapid oxidation (Rebelein), 103-104
    refractive index, 98-99
    sensory, 84
    separation, by distillation, 88-91
2-Ethoxyhexa-3,5-diene, 236
Ethyl acetate, 117, 160-161, 162, 165, 324
    amounts in wine, 161
Ethyl acid malate, 61
Ethyl acid tartrate, 61
Ethyl alcohol, see Ethanol
Ethylamine, 174
Ethyl bromoacetate, 243
Ethyl butyrate, 163, 164, 165
Ethyl caprate, 159, 163, 164
Ethyl caproate, 159, 163, 164
Ethyl caprylate, 159, 163, 164
Ethyl carbamate (urethane), 175, 244
Ethylene diaminetetraacetate (EDTA, Versene), masking, 358
    bonding, structure, 359
    pH optimum, 360
    theory, 358-359
Ethyl guiacol, 218
Ethyl laurate, 159
2-Ethyl phenol, 218
4-Ethyl phenol, 218
Ethyl sorbate, 236
Ethyl sorbyl ether, 236
Ethyl succinate, 160
Ethyl syringinol, 218
Ethyl vanillin, 148
Eugenol, 200, 203, 218

methyl ester, 218
Euparen, 256
European Common Market, 226-229
Europium, 279, 283
Evaporation, see Distillation
Extract:
    definition, 29-30
    nomograph, 35-36
    procedure, 30-31
Extraction, 322-323
    photometric methods, 268

Falsification formula, 30-31
Fehlings solutions, 37
Ferbam, 258
Ferrocyanide and cyanide, 222, 247-250, 283, 285, 286
    legality, 247
    procedures, 247-252
    toxicity, 247
Ferulic acid, 197, 200, 218
Feruloyl tartrate, 168, 198
Fixed acidity:
    change, causes of, 59
    definition, 59
    minimum limits, 60
Flame ionization detector, see Gas-liquid chromatography
Flame photometry, 268, 270, 272, 274, 275, 279, 280, 285
    equipment and theory, 333-337
Flavan-3,4-diols, see Anthocyanogens
Flavanones, 199, 216
Flavonoid phenols, 198, 212
    formaldehyde reaction, 212
    formation, 201
    nomenclature, 198-199
    properties and structure, 198-199
    reaction, separation from non-flavonoid phenols, 212-213
Flavonols (anthoxanthin pigments), 196, 199
    amounts, 216
Flooding, 325
Fluoride, 6, 288, 292-293, 356
Fluorometry, 43, 250-251, 279, 287, 332-333
Folic acid, 174
Food and Agricultural Organization (FAO), 256
Formic acid, 53, 60, 75, 324
Fraction, 325

Fructose, 47, 129, 130
β-Fructosidase, 46
Fucose, 47
Fumaric acid, 50
Fungicides and pesticide residues, 222, 256
  procedure, 258–259
  tolerances, WHO, 257–258
  toxicity, 257–258
Fusel oil, see Higher alcohols

Galactose, 21, 24, 47–48
Galacturonic acid, 21, 24, 75, 224–225
  methyl ester, 21
Gallic acid, 200, 203, 213, 218
(+)-Gallocatechin (2,3H-trans), 198
Gallotannin, 197
Gases, 302–315
  purification of, 313
Gas-liquid chromatography (GC), 235, 240, 241, 242, 243, 244, 245, 246, 249–250, 251–252, 253, 343–348. See also Chromatography
  bleed, 342
  chromatograph, 343
  detectors, 345–348, 349
  liquid phases, 342–343
  solid supports, 341–342
Gel separation, 353
Gentiobiose, 47
Gentisic acid, 200, 218
Geraniol, 134
Gluconic acid, 50, 75, 130
Gluconic kinase, 75
Glucose, 43, 130, 224, 225, 274, 277, 278. See also Reducing sugar
Glucose-6-phosphate dehydrogenase, 43–44, 75
Glucuronic acid, 50, 75
Glutamic acid, 141, 172, 173
Glutamine, 172, 173
2-S-Glutathionyl caftaric acid, 197
Glycerol, 61, 81, 117–119, 120, 130, 131
  amounts, 121
  flurometric, 126
  kinase, 123
  procedure:
    chemical, 120–123
    enzymatic, 123–125
    gas chromatographic, 125–126
  source, 119–120

Glycine, 173
Glycol, 141
Glycolic acid, 50
Glycoproteins, 175
Glyoxal, 141
Glyoxylic acid, 224
Gold, 279
Gondola trucks, 9
Grape juice, 302, 311
  adulteration, 252
Grape pigments:
  amounts, 206–208
  bisulfite reaction, 209
  pH effect on, 209
  separation procedure, 208
GRAS, 226
Gum, 21
Gutzeit method, 280–281

Hafnium, 279, 283
Halogenated acids, 243–244
Hansenula, 160
  anomala, 217
Henderson–Hasselbalch equation, 235
Henry's law, 235, 306
HEPT, see Theoretical plates
Heptanal, 140, 141
Heptanoic acid, 59
Heptanol, 141
Hesperidin (-7-rhamnoglucoside), 199
Hesperitin, 199, 216
trans, trans, -2,4-Hexadienoic acid, see Sorbic acid or sorbates
Hexanal, 140, 141, 146
3,5-Hexadien-2-ol, 236
l-Hexanol, 81, 114, 119, 141
Hexokinase, 43–44
Hexosamines, 175
n-Hexyl acetate, 163, 164, 165
n-Hexylamine, 174
Higher alcohols (fusel oils), 112–115
  amounts, 113–115
  Ehrlich mechanism, 113
  gas chromatographic, 117–119
  legal limits, 115
  mechanism of formation, 113
  procedure, chemical, 116–117
High performance (pressure) liquid chromatography (HPLC), see Liquid chromatography

Histamine, 172, 174, 186
  amounts, 186
  procedures, 167-168
Histidine, 172, 173
Holdup, 326
Hollow cathode, 337
Hotrienol, 134
Hubach apparatus, 247, 248
Hue, 317
Hydrogen cyanide, 172, 247, 249
Hydrogen sulfide, 302, 304, 313
  methods of analysis, 311-313
  source, 311-312
  toxicity, 311
Hydrolyzable tannin phenols, 213-216
Hydrometer, alcohol, see Ethanol, procedure
Hydrometer, Brix, cylindar, 19-20. See also °Brix
  scale conversion tables, 22-23
  temperature correction, 18-19
2-Hydroxyacetylfuran, 252
2-Hydroxybenzaldehyde, 218
4-Hydroxybenzaldehyde, 148, 218
p-Hydroxybenzoic acid, 200
  esters, 222, 242-243
Hydroxybenzoic acids, see Benzoic acids
Hydroxycinnamic tartrate-glucose esters, 168
Hydroxymethylfurfural, 153, 252
  amount, 153-154
  causes of, 153
  procedures, 154-155
3-Hydroxy-2-pentanone, 148
Hydroxyphenylpyruvic acid, 141
Hydroxyproline, 173
Hydroxypyruvic acid, 141
α-Hydroxysulfonate, see Bisulfite complex

ICAP-AES, see Inductively coupled argon plasma-atomic emission spectroscopy
Inductively coupled argon plasma-atomic emission spectroscopy (ICAP-AES), 268, 277, 279, 282, 284, 286
meso-Inositol, 130, 134
Inverse voltrometry, 279
Iodide, 288, 293
Ion chromatography, see Chromatography, ion
Ion-exchange, 271, 272, 275, 290, 326-328, 329
  active component, 326
  capacity, 328
  choosing resin type, 327
  equivalent brands, 327
  structure, 326
Ion-selective electrodes (ISE), 271, 272, 279, 285, 286, 287, 289, 292, 293, 305-306, 309-310, 311, 352-358
Iron, 6, 222, 247, 265, 271, 273, 275-277, 302, 338, 358, 359, 360
  amounts, 269
  fermentation losses, 275
  procedures, 268, 276-277
  source, 275
  stability, 276
Isoamyl acetate, 160, 162, 163, 164, 165
Isoamyl alcohol, 81, 102, 113, 114, 115, 141, 252
  procedures, 116-119
Isoamylamine, 174
Isobutyl acetate, 163, 164, 165
Isobutyl alcohol, 81, 102, 112, 113, 114, 115, 117, 141
Isobutylamine, 174
Isobutyraldehyde, 141
Isobutyric acid, 160
Isoleucine, 141, 173
Isomaltose, 47
Isopropylamine, 174
Isopropyl syringinol, 218
Isotopic analysis, 252
Isovaleraldehyde, 140, 141
Isovaleric acid, 160

Juice separators, 10

Kaempferol, 196, 199
α-Ketobutyric acid, 141
α-Ketoglutaric acid, 75-76, 141, 224-225
α-Ketoisocaproic acid, 141
α-Ketoisovaleric acid, 141
a-Keto-β-methylvaleric acid, 141
Kloeckera, 222

Lactate dehydrogenase, 123
Lactic acid and lactates, 40, 52-53, 61, 266, 267, 288, 289
  bacteria, 172, 223-224, 236
  isomers, 68
  malolactic paper chromatography, 60-61
  procedure, 69-70
Lactobacillus sp., 187

Lactose, 47
Lane and Eynon, 37–39
Lanthanum, 279
Laws, German, 5
Lead, 6, 268, 279, 283–284, 338, 360
   amounts, 283–284
   arsenate, 257, 283
   methods, 283–284
   source, 283
Legal limits, 5–6
Leucine, 141, 173
Leucoanthocyanins, see Anthocyanogens
Leuconostoc gracile, 187
Leuconostoc oenos, 172
Lieb-Zacherl apparatus, 231–232
Light sources, 330
Linalool, 134
Lindane, 257
Linuron, 257
Liquid chromatography, 349–352. *See also* Chromatography
   detectors, 349
   ion-pairing, 349
   reverse phase, 350
   separation systems, 350, 351
Lithium, 6, 279, 280, 285
Loury method, *see* Enzymatic analysis, protein analysis (Loury)
Luff-School, 39. *See also* Reducing sugar
Lysine, 173, 175

"Macération carbonique," 82
Magnesium, 256, 265, 268, 271, 274, 275, 338, 358, 360
   amounts, 269, 275
   calcium ratio, 270, 275
   methods, 268, 275
   tartrate stability, effect on, 275
Malathion, 256
Malic acid and malates, 59, 60, 61, 62, 130, 266, 267, 288
   amounts, 67
   bacterial attack, 67
   procedure:
      enzymatic, 67–68
      other, 67–68
   respiration, 67
L-Malic dehydrogenase, 67
Malolactic fermentation, 50, 52, 61, 67, 223, 306

Malonyl aldehyde, 214
Maltose, 47
Malvidin, 197
   -3,5-diglucoside, 207, 210–212
   -3-monoglucoside, 207
   -3-monoglucoside acetate, 207
   -3-monoglucoside p-coumarate, 207
Malvidinol, 198
Mancozeb, 258
Maneb, 258
Manganese, 285–287
Mannitol, 81, 129–130
   amounts, 130
   procedures, 130–134
   source, 129
Mannitol-PO_4-dehydrogenase, 124
Mannitol-PO_4-phosphatase, 129
Mannoheptulose, 46
Mannose, 47
May wine, 251
Melibiose, 47
Mercaptan, 311
Mercury, 6, 286
Mesoinositol, 81, 131–132, 134
Metals, 6
Metatartaric acid, 63
Methanal, 140
Methanol, 80, 102, 108–109, 117
   amounts, 109–110
   legal limits, 109
   lethal dose, 109
   procedure:
      chemical, 109–111
      gas chromatographic, 111–112
   source, 108–109
Methionine, 173
Method of addition, 338–339
Methylamine, 174
4-Methylaminodiazole, 252
Methyl anthranilate, 162, 164
   procedure, 167–169
2-Methyl-1-butanol, *see* Amyl alcohol, active
3-Methyl-1-butanol, *see* Isoamyl alcohol
4-Methyl guiacol, 218
2-Methylpropanal, 140, 146
2-Methyl-1-propanol, *see* Isobutyl alcohol
Methyl salicylate, 203
Michaelis–Menton constant, 361
Molecular sieves (zeolites), 313
Molybdenum, 286

Monier-Williams, see Sulfur dioxide
Monoamine oxidase inhibitors, 187
Monobromacetic acid, 222, 289
  legality, 243
  procedure, 243-244, 289
Monocaffeoyl tartrate, 168, 197
Monochloracetic acid, 222, 243
  legality, 243
  procedure, 243-244
Mono-p-coumaroyl tartrate, 168
Monoferuloyl tartrate, 168
Monoiodoacetic acid, 243
Mucic acid, 50, 130
Myricitin (−3-rhamnoside), 196, 199

β-D-NAD-oxidoreductase, 46
Naringenin, 199
Naringin, 199, 216
  (-7-rhamnoside), 199
Nephelometry, 331-332
Nernst equation, 354
Nerol, 134
Neutron activation analysis, 279, 280, 282, 284, 285, 286, 287, 289
Nickel, 6, 268, 280, 286, 338, 360
Nicotinamide adenine dinucleotides, absorption spectra, 106-107
Nicotinic acid, 174
Nitrate, 172, 175, 244, 288
  amounts, 190
  procedure, 188-189
Nitrites, 175, 189
5-Nitrofurylacrylic acid (NFA), 222, 244
Nitrogen:
  amounts, 177
  total, 176-178
Nitrogen compounds, amounts, 172-176
  growth factors, as, 172
Nitrogen (gas), 302, 306, 313
Nitrosamine, 244
trans-2-Nonenal, 140
Nonflavonoid phenols:
  determination of, 216-219
  esters of, 217
  formation or source of, 199-201, 217
  nomenclature, 200
  physical properties and structure, 200
  procedures, 217, 219
  separation methods, 219
Nucleic acids, 175

Nucleotides, 174-175
Nylon, 215

Oechsle, 21-23
Omethoate, 257
Optical density, 317-319
Organic acids, 14. *See also individual acids*
Ornithine, 173
Oxalic acid, 50
Oxaloacetic acid, 141
Oxidation-reduction potentials, 302-303
Oxonium salts, 208
Oxygen, 302-303
  electrode, 304-305
  nitrogen stripping, 304-305, 313
  oxidation of juice or wine, 302-303
  pickup, 302, 303
  procedure, 304-305
  saturation concentrations, 303, 304

Paar density meter, see Ethanol, procedure, density meter
Pantothenic acid, 174
Parabens, see p-Hydroxybenzoic acid, esters
Paraquat, 257
Parathion, 254, 257
Patulin, 255
Pear, detection of, 252
Pectic acid, 24-28
Pectin, 14, 21-28
  esterase, 25, 26, 109
  formation, 24, 25
  molecular weight distribution, 26
  precipitation, alcohol, 21
  procedure, 26-28
  transeliminase, 25-27
Pediococcus cerevisiae, 172
Pelargonic acid, 59
Pentanal, 140, 146
2,3-Pentanediol, 150
2,3-Pentanedione, 148, 150, 151
2-Pentanone, 140
Peonidin, 197
  -3,5-diglucoside, 210
  -3-monoglucoside, 207
  -3-monoglucoside acetate, 207
  -3-monoglucoside p-coumarate, 207
Peptides, 175
Pesticides, see Fungicides and pesticide residues

Petunidin, 197
  -3-monoglucoside, 207
  -3-monoglucoside acetate, 207
  -3-monoglucoside p-coumarate, 207
Petunidinol, 198
pH, 50, 52-53, 224, 266, 270, 274, 295, 357-358
  factors affecting, 52-53
  measurement, 53
  ranges, 52
Phenethylacetate, 165
Phenethylaldehyde, 141
β-Phenethylamine, 174, 186
Phenol, 218, 302. See also Phenolic compounds
Phenolate ion, 204
Phenolic compounds, importance, 196, 229, 254, 302
  legal limits, 203
  lists of, 197, 198, 199, 200, 202, 207, 218
  oxidation, 204
  sensory, 196, 203
  toxicity, 203
Phenols:
  distribution, grape parts, 204
  effects on wine, 203-204
  Folin-Ciocalteu, 204-206
  Folin-Denis, 204
  hydrolyzable, 213-216
  Neubauer-Löwenthal, 204
  nonflavonoid, 197, 200-203
  procedure:
    automated, 206-207
    Folin-Ciocalteu, 205-206
    sugar correction, 206
    sulfur dioxide correction, 141, 173
    total, amounts, 204, 206
Phenylalanine, 141, 173
2-Phenyl ethanol, 81, 114, 115, 119, 141
2-Phenyl ethyl acetate, 162, 163, 164, 165
Phenylpyruvic acid, 141
Phlobaphene, 215
Phloroglucinol, 214
Phosdrin, 257
Phosphate, 208, 265, 266, 267, 274, 288, 291, 293-294
  amounts, 291, 293
Phosphoglucose isomerase, 44
Pichia sp., 162, 222
Pinot noir, 168

Piperonal, 148
Piquat, 257
Polagraphy, 277, 279, 284, 285, 286, 306
Polygalacturonase, 25-26
Polymethylgalacturonase, 25-26
Polyphenoloxidase, 303
Polysaccharides, 14, 21
Polyurinides, 21
Polyvinylchloride (PVC), 252-253
Polyvinylpolypyrrolidone (PVPP), 244, 254
Polyvinylpyrrolidone (PVP), 187, 254
Potassium, 14, 264, 265, 270-271, 279, 335, 338, 356
  acid tartrate, 270
  amounts, 269
  cation exchange, 270
  ferrocyanide, see Ferrocyanide and cyanide
  in grapes, causes, 270
  metabisulfite, 222, 223
  procedure, 270-271
  sodium ratio, 268, 271-272
  sorbate, 102
Potentiometric stripping analysis (PSA), 279, 282, 284
Pressure drop, 326
Proline, 172, 173, 252
  oxidase, 175
  permease, 175
  procedure, 185-186
  sparing of, yeast, 175
Propanal, 140, 146
n-Propanol, 81, 102, 113, 114, 115, 324. See also Higher alcohols
Propanone, 140
trans-2-Propenal, 140, 146, 155
Propionic acid, 50, 54, 58, 160
n-Propyl acetate, 163, 164
n-Propylamine, 174
Proteins, 14, 172, 173-174
  bentonite fining, 174
  heat treatment, 174
  procedure, 187-188
  tests for, 188
Protocatechuic acid, 200
Proton-induced X-ray fluorescence (PIXF), 268, 281, 284, 286
Purines, 175
Putrescine, 174, 186
Pycnometer, see Ethanol, procedure, pycnometer

Pyridoxine, 174
Pyrimidines, 175
Pyruvate carboxylase, 118
Pyruvate kinase, 123
Pyruvic acid and pyruvate, 50, 75, 76, 123, 141, 224–225

Quercitin, 196, 199, 203
Quercitrin (-3-rhamnoside), 199

Raffinose, 47
Rebelein "5 minute method," see Ethanol, procedure, rapid oxidation; Reducing sugar
Rectification, 295, 326
Reducing sugar, automated, 40–41
  configurations, 37
  HPLC, 42
  pill test, rapid, 43
  procedure, chemical, 37–40
    enzymatic, 43–47
  Rebelein "5 minute method," 39–40
Reflux, 326
Refractometer, 15
  temperature corrections, 15, 18
Rhamnose, 21, 24, 47–48
Riboflavin, 174
Ribose, 47–48
Ripper, see Sulfur dioxide
Ronilan (Vinclozolin), 257, 258
Roval, 258
Rubidium, 279, 280, 286
Rubigan (Ferarinol), 258
Rutin, 199, 203, 216
  – 3-rhamnoside, 199
R_f values, 340, 341

Saccharomyces sp., 159, 160, 217
  strain Montrachet, 83
Safrole, 250
Salicylic acid, 200, 218, 222
  procedure, 241
  source, 241
  use levels, 241
Samarium, 279
Sampling:
  grapes, 8–10, 16
  preparation, 8
  wines, 11
Sarcina hansenii, 129, 150

Saturation, 317
Scandium, 279, 286, 287
Schizosaccharomyces sp., 82
  pombe, 67
Scopoletin, 200
Sedoleptulose, 46
Selective-ion electrodes, 271, 281, 293, 353–358
  gas sensing, 354–355
  glass, 357–358
  liquid membrane, 356–357
  solid state, 355–356
  theory, 353–354
Selenium, 3, 286–287
Sensory properties, 21, 24, 36, 67, 84, 153, 159, 164, 165, 196, 203, 226, 235–236, 242, 244, 246, 247, 253, 311, 319
Serine, 141, 173
Serotonin, 174
Sevin, see Carbryl
Sherry, dry, 28
Shikimic acid, 199–201
Silicon, 286, 287
Silver, 279, 287, 356
Sinapaldehyde, 200
Slope calculation, 331
Sodium, 264, 268, 270, 271–272, 279, 334, 337, 338, 356
  amount, 269, 271
  legal limits, 271–272
  medical aspects, 270–271
  method, 268, 271–272
  sources, 271
Sodium azide, 222, 246
Sodium hydrogen sulfite, 271
Sodium metabisulfite, 222, 223, 271
Sodium sulfite, 222, 271
Soluble solids, 14–36. See also °Brix
  definition, 14
Sorbic acid or sorbates, 222, 235–241, 271
  active form, 235
  amounts required, 235
  bacterial action on, 235–236
  correction for, 58
  gas chromatographic, 240–241
    U.V., 238–240
  legal limits, 236–237
  procedure:
    automated, 238–239
    chemical, 236–238

Sorbic acid or sorbates (*Continued*)
  sensory, 235–236
  sulfur dioxide, reaction, 235
  toxicity, 236
Sorbitol, 81, 129–130, 252
  amounts, 129, 131
  methods, 131–132, 134
  procedure, gas chromatographic, 131–132
  structure, chemical, 129
Sorbyl alcohol, 236
Soxhlet's reagent, 39
Specific gravity, wine, 29–32
Spectrophotometry, 237, 240, 250, 272, 278, 279, 281, 317–319, 328–331
Spermidine, 174, 186
Spermine, 174, 183
"Steril," 243
Stills, *see* Distillation
Strontium, 257, 338
"Stuck" fermentations, 130
Styrene, 253–254
Succinic acid and succinate, 52, 60–62, 266, 267
  amount, 72
  procedure, 72–73
  source, 72
Sucrose, 36, 47, 130
  procedure, 44–47
Sugar, 36–48, 229, 279. *See also* Reducing sugar
  fermentable, 14
  levels, 36
  methods, 46–48
  sensory, 36
Sugar alcohols, *see* Mannitol; Sorbitol
Sulfate, 14, 265, 267, 272, 274, 275, 288, 295–296, 311
  amount, 291, 295
Sulfur dioxide, 222, 266–267, 286, 295, 302
  accepted daily intake, 226
  bisulfite reaction, substrates for, 224–226
  equilibrium, 223–224
  forms, solubility, 223
  levels and limits, 223, 226–228
  pK values, 223–224
  procedures:
    free, 229–232
    total, 232–235
  sensory, 226
  sources, 222–223
  toxic levels, 226
Super critical chromatography, 348–349
Syringaldehyde, 148, 200, 218
Syringic acid, 200, 218
Syrup fermentation, 83

Taberié formula, 31, 33
Tailing, 342
Tannins, *see* Phenolic compounds
Tantalum, 279, 287
Tartaric acid and tartrates, 33, 51–52, 130, 266, 270, 275, 277, 288
  amounts, 62
  double precipitation, 63
  minimum levels, 62
  procedures, 63–66
  rapid method, 64–69
  sensory, 63
Teflon, 309, 344
Terpenes, 134–135
α-Terpinol, 134
Thalium, 279
Theoretical plates, 326
Thermal conductivity detector, *see* Gas-liquid chromatography
Thermionic detectore, *see* Gas-liquid chromatography
Thiamine, 174
  pyrophosphate, 148
  sulfur dioxide, reaction, 225
Thief, wine, 11
Thiometon, 257
Thiophanate, 257
Threonine, 172, 173
Throughput, 326
Thujone, 250–251
Tin, 6, 268, 287
Titanium, 287
Titratable acidity, *see* Total acidity
Tokai, 196
Tolerances, pipets and volumetric flasks, 322
Total acidity (titratable acidity), 29
  conversion factors, 52
  errors, 52
  procedure, 51
Trace elements, *see various cations*
Transreflectometry, 318
Trehalose, 47, 130
Trifluratin, 257
Tryptamine, 174

Tryptophane, 173
Tryptophol, 115, 119, 141
Tungsten, 279
Turbidimetry, 295, 331–332
Tyndall effect, 331
Tyramine, 174, 186
  amounts, 186
Tyrosine, 141, 172, 173
Tyrosol, 115, 119, 141, 218
  acetate, 218

Umbelliferon, 200
United States Bureau of Alcohol, Tobacco and Firearms (BATF), 85
United States Food and Drug Administration (FDA), 226
Uracil, 226
Urethane (ethyl carbamate), 175

Valeraldehyde, active, 141, 146
Valeric acid, 54
Valine, 141, 173
Vamidothion, 257
Vanadium, 287
Vanillic acid, 200, 218
Vanillin, 148, 203, 218
Vermouth, 250
Vinyl chloride, 252–253
4-Vinyl guiacol, 218
4-Vinyl phenol, 218
Vitamin C, *see* Ascorbic acid

Vitamins, 172, 174
Vitis labruscana, 162
  vinifera, 159, 164, 168, 208, 210
Volatile acidity, 53, 266–267
  correction for, sorbic acid and sulfur dioxide, 54, 56–57
  errors, 54–55
  legal limits, 53
  procedure, 55–59
  sensory, 54
  source, 54

Water, 264, 324, 325
Westphal balance, *see* Ethanol, hydrostatic balance
White Riesling, 168
Wine spoilage, 129
Woodruff, 251
World Health Organization (WHO), 226, 256, 257

X-Ray, fluorescence, 268, 276, 277, 279, 285, 289
Xylitol, 130, 131–132, 134
Xylose, 47–48, 225

Yeast, 160, 172, 175, 235, 242, 253, 311

Zinc, 6, 268, 280, 288, 337, 338, 359, 360
Zineb, 258
Zygosaccharomyces sp., 222